T0407749

PHYSICS RESEARCH AND TECHNOLOGY

DECAY OF MOTION

THE ANTI-PHYSICS OF SPACE-TIME

PHYSICS RESEARCH AND TECHNOLOGY

Additional books in this series can be found on Nova's website
under the Series tab.

Additional e-books in this series can be found on Nova's website
under the e-book tab.

PHYSICS RESEARCH AND TECHNOLOGY

DECAY OF MOTION

THE ANTI-PHYSICS OF SPACE-TIME

BERND SCHMEIKAL

New York

For permission to use material from this book please contact us:
Telephone 631-231-7269; Fax 631-231-8175
Web Site: http://www.novapublishers.com

Additional color graphics may be available in the e-book version of this book.

Library of Congress Cataloging-in-Publication Data

Decay of motion : the anti-physics of space-time / [edited by] Bernd Schmeikal (Wiener Institute for Social Science, Documentation and Methodology (WISDOM), R&D, Wien, Austria).
pages cm -- (Physics research and technology)
Includes bibliographical references and index.
ISBN: 978-1-63117-809-2 (hardcover)
1. Space and time--Philosophy. 2. Motion. 3. Quantum theory. 4. Quantum entanglement. I. Schmeikal, Bernd, 1946- editor. II. Series: Physics research and technology.
QC173.59.S65D395 2014
530.11--dc23
2014013780

Published by Nova Science Publishers, Inc. † New York

CONTENTS

Foreword		**vii**
Acknowledgment		**ix**
Preface		**xi**
Prologue		**xiii**
Chapter 1	Philosophy	**1**
Chapter 2	Entry to Foundations	**5**
Chapter 3	Phenomenology of Immediacy	**45**
Chapter 4	Polarized Braids and Little Primordial Frames	**63**
Chapter 5	Emergence of Primordial Minkowski Frames	**101**
Chapter 6	Majorana Space-Time Spinors	**155**
Chapter 7	Color Braids	**171**
Chapter 8	Motion and Method	**189**
Chapter 9	Envisioned Memory	**195**
Technical Appendix: Iteration of Algebra		**235**
References		**257**
Author's Contact Information		**265**
Index		**267**

FOREWORD

Will the light ever be fully understood? Light can be seen. Some of us can even feel it. Whatever we may say about photons, bosons, and whatever we regard as being worth to be called light, is always based on what we see. Who sees the light will understand it. You may wonder when I say to you, the reader, that I regard it as very helpful if you would read a book, written not by a physicist, not by mathematicians, but by a librarian during and after World War II. I mean Erhart Kästner, he was stationed in Greece during the war. At many instances he described the 'light of nature' in the most competent and convincing manner. He spoke to humans who are able to see the innermost of the heavens, the earth and their relation to the old naked gods of Greece. Innocence and thrilling presence, there, are one. Light, we assume, gives mass to what we discern as particles; as the nature has no other way than to discern energy to us as particles. But what gives mass is a peculiar light, the light of the Higgs boson. It has coalesced with a man and a man's name like the light in the valleys of Chaironeia and Dodona in the bluegreen hills below the Parnassos where you enter the realm of Demeter, where we laid the murderous encounter of Oedipus with his father, like the honey-yellow shining archaic wall with the name Apollon. You will never forget it, Kästner assures.

That ancient light is without age, but it gives freedom of motion and orientation to those small things that jump around in any process of nature, within the primordial chaos of time. It manages to tell those small vivid creatures of energy that are not yet able to differ between left and right, between up and down, line and area, charm and volume, to make a difference, to create a difference by mere subconscious interaction. Such we may call a touch of primordial strings of energy with another. And by so interacting these beings that have the potential to interact trans-locally, create features of logic transformation. They prove their material action in space and time is a cognitive action and is not without logic, not without the orders of mathematics. Polarized strings of a certain or uncertain length interact by touch and at the same time perform the elementary mathematical operations. They create both, extension and cognition, space, time, logic and thought. But it is a ventured hypothesis that those laws of interaction created by the polarized strings bring forth brazen laws of physics sustainable for evermore. The space-time, the Minkowski space and its Clifford algebra will transmute. Matter will experience metamorphosis and perhaps in some way we will partake in this process.

For everyone there is a place, said Kästner, where he or she feels at home, rescued and utterly secure. This is for me the Volcano island of Nisyros in the Greek Aegean. It could be foreseen that a preface written in Nisyros would differ much from a one written in Vienna. Nisyros is a small Volcano where fire and water are wattled in harmony. Fire, water and light are flowing together. During the last very hot summers when on so many islands the woods caught fire, in Nisyros there was not a single outbreak and the wonderful flora survived almost untouched. Reason is that the pumice stone is so porous that it sucks a lot of water from subsurface. The Nisyrians say their island is paradise. Now, if you read the book in front of you, provided you understand a little mathematics and physics, you will come into contact with something of which hitherto you probably have only dreamt or not even that. Space, time and thought will become one, extension, matter, symbols and cognition will no longer stay to be separate divisions of existence. Clearly, the price you have to pay in order to understand as much as you can, is rather high. You must have gone through the door that separates living from dying. A proper abidance will support catharsis in this dangerous world where most of us are familiar with the unity between life and death anyway; stress and various exertions, the ubiquity of diseases and conflicts made it sure.

The only relevant actor in the universe is nature. Whitehead used to call it 'the process of nature', and unfortunately only a few could follow his philosophy. Nature has put the morphogenetic structures of orientation and thought into the cosmos, that is, into herself and into the human cognition. How could she have done differently? We are part of nature and so is human thought. In those pages you are now looking into, I have created an unspecified domain, a space for my own, and I've put the symbols into that utterly non-metric, space without attributes. The symbolic strings are relational, and it is their relation that brings about the space. They take the form of bit strings, strings with polarity $+1$ or -1. They form electrons, neutrinos, photons, fermions, bosons, atoms or what you will, but not only these. They represent the well-formed terms of interaction. So the symbolic strings of the linear writing interact like particles. They touch each other, that is, those characters which carry the polarity, experience confrontation, juxtaposition and comparison. Thus they are multiplied by logic identification. They are the carriers of those little arrows, the 'phase gates' that pile up the space-time we live in. This probably rather chaotic translocal process of interaction constitutes what we call today the Clifford algebra of Minkowski space, a holy little thing. My construction is radical as it reconstructs what nature has constructed and partly deconstructs what man has created in idiocy. This is not new. Richard Feynman was among the first who vocalized that we physicists are idiots, and I do not exclude myself from this strange community of scientists. The bits and pixels, the stars and compasses we have in our brains bring forth an order which is in good accord with the phenomena of particle physics.

ACKNOWLEDGMENT

When a branch of science separates from the main stream it does not break off its stem. But it deliberates the historic flow of the theory and mathematics. For scientists who contemplate such a course of events it means to maintain a close contact to both the main stream of knowledge and the new that begins to rise above the horizon. To accomplish the task I needed a seasonable environment and appropriate electronic transactions. Karl Müller provided me with both. I am grateful for his help and his patience. To help create a mathematical universe out of a self-referent nothing, one is badly in need of a vigilant partner who is open and generous rather than arcane and mystical. This partner was for me Louis Kauffman. I am much obliged for his guidance and his rapid responses. I also appreciated very much the subject-specific clues Bertfried Fauser, Zbigniew Oziewicz and José Vargas gave me. They often found various different viewpoints from where I could look at the matter with a fresh mind. Finally I want to give Bettina a hug for helping me lengthily with the editing.

Bernd Schmeikal, Nisyros 1st of June 2013

PREFACE

The book came into being while turning away from the Big Bang theory and our traditional concept of space-time. To tell you the truth, there was no Big Bang, and our universe never came into being, but it is just existing without asserting any claim for beginning or end. In this sense, the universe is utterly different from a book, since a book has a first and a last page. Even worse, the world is perpetually existing beyond time, thereby locally creating time. Creation, as we use to call it, is recreation, and nothing is lost. Yet, what is lost is lost entirely. What we call a particle trajectory is permanently reintegrated into the presence of the universe. This is a quite unusual view of physical affairs. While in 'space-time-physics' the Big Bang, or call it the act of creation, is remote and out of reach for us, in this new view it is part of the ongoing process of nature. It is present and therefore accessible to present action. The process of the universe is thus acting in a highly nonlinear, omnipresent manner. But this, nevertheless, does allow for those linear appearances of reality that make our living so comfortable. If you walk five miles in an hour, you walk ten in two hours, if you walk slowly below light-velocity. And adding one ton of coal to another one, we get two tons, and this doubles the time we can heat the coal oven.

The idea is not entirely unknown to cosmologists. In 2003 Arvind Borde, Alan Guth and Alexander Vilenkin published their "singularity theorem".[1] Such theorems have a prominent history. The authors took a look at 'past eternal universes' and pointed out that "*inflation alone is not enough to explain the universe, there needs to be a whole new physics to explain 'correct conditions at the boundary'*." I will give an introduction to this whole new physics which is not relying on any boundary of an inflationary space-time, for the space-time, as we used to conceive it, simply does not exist. Space-time is not a frame for fields, but quantum fields bring forth the space-time. It is a mere local illusion. Since the early writing of Stephen Hawking on "A Brief History of Time" we seem to be required, in a way like exemplary constructivists, to "*decide that there wasn't any singularity. The point is that the raw material doesn't really have to come from anywhere*" says he (Hawking 1988, p. 129). But we need not even decide. There simply wasn't any. Now there are many writings on ex nihilo creation. The Astronomical Society of the Pacific[2] put '*Nothing*' on the net, a contribution by Alexei Filippenko and Jay Pasachoff advancing science literacy through astronomy. Many well-known authors like Bernulf Kanitscheider (1990, p. 344) and Christopher Isham (1993) have written about creation of the universe, zero energy universes (Marcelo Berman 2009) and

---

[1] http://creationwiki.org/Borde-Guth-Vilenkin_singularity_theorem#cite_note-7.

[2] http://www.astrosociety.org/pubs/mercury/31_02/nothing.html.

quantum cosmology (Hawking and Penrose 1996, p. 85; Gordon McCabe 2005). But the fact is that we have no global time. The universe is re-iterating and reproducing itself in bosonic presence. It is of course changing, but that change is beyond time. The arrow of time is a mere local creation of nature and mind.

In pure mathematical terms it is indeed possible to show how a zero-energy universe comes upon. It is even quite natural to show how the Schrödinger equation and various algebraic modules of the Minkowski space-time emerge out of a mathematical void, that is, out of some self-referent domain of polarizable, iterative zero strings in no-space. But the essential feature of creation in quantum fields is not the zero energy, but the absence of time. To understand this viewpoint requires a new understanding of the philosophy of Parmenides and Melissos which got lost after Platon and Aristoteles. That is why this monography is predicating on *anti-physics*. It first skips everything that we once have believed. To be able to do so requires some familiarity with the "meaning of nothing" and the ending of time. The author is a Kundalini-Yogi, which is perhaps the reason why perception in the void and experience beyond time are quite familiar encounters for him. To fully understand the process of nature, we have to perceive beyond time and metric, oriented space. That is the source of intuition.

The most essential motivation and constituent for a new quantum cosmology, however, stems from Richard Feynman's early works on quantum electrodynamics. In this connection the author came upon a refreshing paper. Smilga (2010) has pointed out that Richard Feynman's most fundamental work on QED (1949) gave an action-at-a-distance formulation of quantum electrodynamics. This is true. In his introductory pages Feynman said: "We begin by discussing the solution in space and time of the Schrödinger equation for particles interacting instantaneously. The results are immediately generalizable to delayed interactions of relativistic electrons and we represent in that way the laws of quantum electrodynamics. We can then see how the matrix element for any process can be written down directly. In particular, the self-energy expression is written down". This seems to have fallen into oblivion, since today we are mindlessly interpreting Feynman's diagrams in the sense of a local relativistic or non-relativistic theory of action. Smilga shows how *in Feynman's action-at-a-distance formulation of quantum electrodynamics a relation between the coupling constant and the ratio of densities of states in the physical state space and in a two-particle product representation can be established. This relation is satisfied, when the physical state space corresponds to an irreducible representation of the Poincare group*. Studying motion in 'a delayed choice quantum eraser' is suggesting once more to investigate this phenomenology of action-at-a-distance in QED and QCD. We live in a universe where instant interaction brings about relativistic motion. The reality of this universe is a self-referent interactive presence of quantum phenomena. It changes and modifies itself in accordance with the forces of nature. But it has neither a past, nor a future apart from its presence. All that we denote as past or future events can only be conceived as organized in those structures of memory that are indeed part of the present. As long as we do not see this we are blinded by those memories which are part of presence and we are lured and driven by what we imagine to be the future. To understand physics, no LHC is needed. A vigilant mind is enough.

Bernd Schmeikal, Vienna, Christmas 2013

PROLOGUE

There is an old, apparently slight dichotomy in the design of reality coming up from the school of Eleatic philosophy whose members are Parmenides, Melissos of Samos and Zeno. Like Parmenides – and not to forget, like the Buddha[1] – Melissos argued that there is a reality ungenerated, indestructible, indivisible, changeless, and motionless. Melissos design differs from Parmenides, and that difference is important. It is the one that physicists were following much too long. Parmenides' view, however, is that there is only one 'moment', the present, while Melissos argues for an infinite number of moments stringed of the thread of time. We are here defending the viewpoint of Parmenides by a modern theory.

Parmenides had advised us to follow in our quest for insight that which is, rather than that which is not. In his poem »On Nature« he put this in simple ancient Greek words: »For things that are not can never be forced to be; but keep your thought from that way of inquiry, and do not let habit [derived] from much experience force you along this path, to direct your unseeing eye and ringing ear and tongue; but judge by reasoning the much-contested disproof expounded by me.« (Fragment 7, sentences 1 to 6). [2] Wishful thinking is not a good way to proper realization, since the interval between that which is and that which is not leads us astray. As others have supposed before me, Parmenides was an enlightened man (Hans von Steuben 1981, pp. 93-105). But today Pre-Socratics are rarely understood. Civilised thought left their path very early. Already Plato and Aristotle denoted philosophers from Elea as people who 'stopped the world run' (von Steuben, p. 48, Franz Schupp, p. 105) and were 'not-nature-thinkers' (*αφυσικοι*).[3] Schupp, like many others, thought that Parmenides

---

[1] Nyanatiloka Mahâthera describes to us »The Immutable« translating and explaining the words of the Buddha as follows: *Truly, there is a realm, where there is neither the solid, nor the fluid, neither heat, nor motion, neither this world, nor any other world, neither sun nor moon. This I call neither arising, nor passing away, neither standing still, nor being born, nor dying. There is neither foothold, nor development, nor any basis. This is the end of suffering. There is an Unborn, Unoriginated, Uncreated, Unformed. If there were not this Unborn, this Unoriginated, this Uncreated, this Unformed, escape from the world of the born, the originated, the created, the formed, would not be possible. But since there is an Unborn, Unoriginated, Uncreated, Unformed, therefore is escape possible from the world of the born, the originated, the created, the formed.* (Nyanatiloka Mahâthera 1952).

[2] I am using here one of the three English translations, namely the one by Arnold Hermann – »*To Think Like God: Pythagoras and Parmenides. The Origins of Philosophy*«, pp. 155-162 (amended) – to be found at the webpage http://www.parmenides.com/about_parmenides/ParmenidesPoem.html.

[3] Schupp interpreting: "Platon, der durchaus zu den Bewunderern des Parmenides zählte, nannte Melissos und seine Jünger die »Weltlauf-Anhalter«. Aristoteles bezeichneten sie als »Nicht-Natur-Denker« im Gegensatz zu den ionischen »Natur-Denkern«. Melissos hielt er außerdem für einen ganz miserablen Logiker (Physik I 3, 186a 8-10). Daß sie »Nicht-Natur-Denker« sind, hängt mit der Abwertung empirischer Erfahrung bei Parmenides

devaluated empirical knowledge. But that was not at all the intention of Parmenides. Rather he wanted to complete our insight and lead knowledge to its purest attitude which emerges in the awareness that that which is can be seen and understood in its clear appearance only then, if there is no additional thought-image, no wish, no prejudice, no vain idea that interferes with it. This perception of that which is, without any addition of something which is not, is not different from the perceived. This is what we may call the immediate appearance of the existing. Clear perception is a neurological matter. Usually it is veiled by our confusion. So Parmenides from Elea wanted us to inquire into that which is. That which is, the clearly perceived, is moving, is changing, is in permanent metamorphosis. But this motion is performed beyond time. For it is the present that is moving, and with it is moving the past. The events past are no more. They are gone. We use to say: they are dead and gone. Yet, the past, inasmuch as it is still moving, still living in our memories, is part of the present. The temporally ordered events of the past, the 'poset' of time, is part of the present. After all, the ancient Egypt is in Egypt today, is in our heads and libraries. Such a view of change beyond the linear time requires that the present moment, the active field of presence is permanently separating from moments to pass and reintegrating them into itself, so that the observer is convinced that there was something before the present and that this is over and out. But as a matter of fact, the presence just said good bye to the past, before it separated from it and ate it. By separating from the observed, the observing presence creates time and integrates the past into itself.

So this monograph has finished with a 'history of physical events' of which the present is just a very thin layer, a point-like happening within some historic universe or universal history. But the past, all the past of the universe, is just part of a universal presence, of: physical reality. Is there any physics, if the past is just part of the present? Yes, there is. But first we have to understand how we construct time and what it is that we project onto the moving presence, in order to derive event histories. After all, time is a cognitive affair, time is thought, time is a peculiar way to bring order into something which otherwise would appear as chaos. But there is no universal time-frame into which we can put all the events perceived by us. In this sense all the knowledge I pose in front of us, is deeply *αφυσικοι*. Yet it is more physical, more real than anything else we had before. If you use it, you will understand the light, the universal appearance of boson fields.

But before we go into this very deeply, we go back to the fragments of Melissos who was among the first who were struggling with that strange truth laid in front of us by Parmenides of Elea. I have shown us a reprint of a translation of Parmenides poem on 'the One' – *το εν* – in the second volume on 'primordial space'. Now we quote a translation of the fragments originating in the thought chain of Melissos handed us over by the philologist John Burnet (1920) whose editions have been considered authoritative for hundred years now. As soon as we get hold of this basic idea together with the differences that it brought on, we can proceed with quantum physics and cosmology.

und seinen Nachfolgern zusammen. Da die Bedeutung des empirischen Bereichs jedoch auf die Dauer nicht unterdrückt werden konnte, ging schon die Philosophie der Zeitgenossen des Melissos in eine ganz andere Richtung."

Chapter 1

PHILOSOPHY

THE ONE COMING UP FROM MELISSOS OVER SIMPLICIUS AND BURNET

In his Life of Perikles, Plutarch tells us, on the authority of Aristotle, that the philosopher Melissos, son of Ithagenes, was the Samian general who defeated the Athenian fleet in 441/0 B.C.; and it was no doubt for this reason that Apollodoros fixed his floruit in Ol. LXXXIV. (444-41 B.C.). Beyond this, we really know nothing about his life. He is said to have been, like Zeno, a disciple of Parmenides; but, as he was a Samian, it is possible that he was originally a member of the Ionic school, and we shall see that certain features of his doctrine tend to bear out this view. On the other hand, he was certainly convinced by the Eleatic dialectic, and renounced the Ionic doctrine in so far as it was inconsistent with that. We note here the effect of the increased facility of intercourse between East and West, which was secured by the supremacy of Athens. The fragments which we have come from Simplicius,[1] and are given, with the exception of the first, from the text of Diels. (1a) If nothing is, what can be said of it as of something real?

(1) What was ever, and ever shall be. For, if it had come into being, it needs must[2] have been nothing before it came into being. Now, if it were nothing, in no wise could anything have arisen out of nothing.

(2) Since, then, it has not come into being, and since it is, was ever, and ever shall be, it has no beginning or end, but is without limit. For, if it had come into being, it would have had a beginning (for it would have begun to come into being at some time or other) and an end (for it would have ceased to come into being at some time or other); but, if it neither began nor ended, and ever was and ever shall be, it has no beginning or end; for it is not possible for anything to be ever without all being.

---

[1] "It is, however, to Simplicius more than anyone else that we owe the preservation of the fragments. He had, of course, the library of the Academy at his disposal, at any rate up to A.D. 529". (Burnet 1920, p. 24)

[2] "Needs must" means "necessity compels". In current usage this phrase is usually used to express something that is done unwillingly but with an acceptance that it can't be avoided; for example, *I really don't want to cook tonight, but needs must, I suppose.* Origin: The phrase is old. In earlier texts it is almost always given in its fuller form - *needs must when the devil drives*. I.e. if the devil is driving you, you have no choice. This dates back to Middle English texts, for example *Assembly of Gods*, circa 1500: http://www.phrases.org.uk/meanings/needs-must.html

(3) Further, just as it ever is, so it must ever be infinite in magnitude.

(4) But nothing which has a beginning or end is either eternal or infinite.

(5) If it were not one, it would be bounded by something else.

(6) For if it is (infinite), it must be one; for if it were two, it could not be infinite; for then they would be bounded by one another.

(6a) (And, since it is one, it is alike throughout; for if it were unlike, it would be many and not one.)

(7) So then it is eternal and infinite and one and all alike. And it cannot perish nor become greater, nor does it suffer pain or grief. For, if any of these things happened to it, it would no longer be one. For if it is altered, then the real must needs not be all alike, but what was before must pass away, and what was not must come into being. Now, if it changed by so much as a single hair in ten thousand years, it would all perish in the whole of time. Further, it is not possible either that its order should be changed; for the order which it had before does not perish, nor does that which was not come into being. But, since nothing is either added to it or passes away or is altered, how can any real thing have had its order changed? For if anything became different, that would amount to a change in its order. Nor does it suffer pain; for a thing in pain could not all be. For a thing in pain could not be ever, nor has it the same power as what is whole. Nor would it be alike, if it were in pain; for it is only from the addition or subtraction of something that it could feed pain, and then it would no longer be alike. Nor could what is whole feel pain; for then what was whole and what was real would pass away, and what was not would come into being. And the same argument applies to grief as to pain. Nor is anything empty: For what is empty is nothing. What is nothing cannot be. Nor does it move; for it has nowhere to betake itself to, but is full. For if there were aught empty, it would betake itself to the empty. But, since there is naught empty, it has nowhere to betake itself to. And it cannot be dense and rare; for it is not possible for what is rare to be as full as what is dense, but what is rare is at once emptier than what is dense. This is the way in which we must distinguish between what is full and what is not full. If a thing has room for anything else, and takes it in, it is not full; but if it has no room for anything and does not take it in, it is full. Now, it must need be full if there is naught empty, and if it is full, it does not move.

(8) This argument, then, is the greatest proof that it is one alone; but the following are proofs of it also. If there were a many, these would have to be of the same kind as I say that the one is. For if there is earth and water, and air and iron, and gold and fire, and if one thing is living and another dead, and if things are black and white and all that men say they really are,--if that is so, and if we see and hear aright, each one of these must be such as we first decided, and they cannot be changed or altered, but each must be just as it is. But, as it is, we say that we see and hear and understand aright, and yet we believe that what is warm becomes cold, and what is cold warm; that what is hard turns soft, and what is soft hard; that what is living dies, and that things are born from what lives not; and that all those things are changed, and that what they were and what they are now are in no way alike. We think that iron, which is hard, is rubbed away by contact with the finger; and so with gold and stone and everything which we fancy to be strong, and that earth and stone are made out of water; so that it turns out that we neither see nor know realities. Now these things do not agree with one another. We said that there were many things that were eternal and had forms and strength of their own, and yet we fancy that they all suffer alteration, and that they change from what we see each time. It is clear, then, that we did not see a right after all, nor are we right in believing

that all these things are many. They would not change if they were real, but each thing would be just what we believed it to be; for nothing is stronger than true reality. But if it has changed, what was has passed away, and what was not is come into being. So then, if there were many things, they would have to be just of the same nature as the one.

(9) Now, if it were to exist, it must need to be one; but if it is one, it cannot have body; for, if it had body it would have parts, and would no longer be one.

(10) If what is real is divided, it moves; but if it moves, it cannot be. In Theory of Reality It has been pointed out that Melissos was not perhaps originally a member of the Eleatic school; but he certainly adopted all the views of Parmenides as to the true nature of reality with one remarkable exception. He appears to have opened his treatise with a reassertion of the Parmenidean "Nothing is not" (fr. 1a), and the arguments by which he supported this view are those with which we are already familiar (fr. 1). Reality, as with Parmenides, is eternal, a point which Melissos expressed in a way of his own. He argued that since everything that has come into being has a beginning and an end, everything that has not come into being has no beginning or end. Aristotle is very hard on him for this simple conversion of a universal affirmative proposition; but, of course, his belief was not founded on that. His whole conception of reality made it necessary for him to regard it as eternal. It would be more serious if Aristotle were right in believing, as he seems to have done, that Melissos inferred that what is must be infinite in space, because it had neither beginning nor end in time. As, however, we have the fragment which Aristotle interprets in this way (fr. 2), we are quite entitled to understand it for ourselves, and I cannot see anything to justify Aristotle's assumption that the expression "without limit" means without limit in space. Reality Spatially Infinite: Melissos did indeed differ from Parmenides in holding that reality was spatially as well as temporally infinite; but he gave an excellent reason for this belief, and had no need to support it by such an extraordinary argument. What he said was that, if it were limited, it would be limited by empty space. This we know from Aristotle himself, and it marks a real advance upon Parmenides. He had thought it possible to regard reality as a finite sphere, but it would have been difficult for him to work out this view in detail. He would have had to say there was nothing outside the sphere; but no one knew better than he that there is no such thing as nothing. Melissos saw that you cannot imagine a finite sphere without regarding it as surrounded by an infinite empty space; and as, in common with the rest of the school, he denied the void (fr. 7), he was forced to say reality was spatially infinite (fr. 3). It is possible that he was influenced in this by his association with the Ionic school. From the infinity of reality, it follows that it must be one; for, if it were not one, it would be bounded by something else (fr. 5). And, being one, it must be homogeneous throughout (fr. 6a), for that is what we mean by one. Reality, then, is a single, homogeneous, corporeal plenum, stretching out to infinity in space, and going backwards and forwards to infinity in time. Opposition to Ionians Eleaticism was always critical, and we are not without indications of the attitude taken up by Melissos towards contemporary systems. The flaw which he found in the Ionian theories was that they all assumed some want of homogeneity in the One, which was a real inconsistency. Further, they all allowed the possibility of change; but, if all things are one, change must be a form of coming into being and passing away. If you admit that a thing can change, you cannot maintain that it is eternal. Nor can the arrangement of the parts of reality alter, as Anaximander, for instance, had held; any such change necessarily involves a coming into being and passing away. The next point made by Melissos is somewhat peculiar. Reality, he says, cannot feel sorrow or pain; for that is always due to the addition or subtraction of

something, which is impossible. It is not easy to be sure what this refers to. Perhaps it is to the theory by which Anaxagoras explained perception. Motion in general and rarefaction and condensation in particular are impossible; for both imply the existence of empty space. Divisibility is excluded for the same reason. These are the same arguments as Parmenides employed.

Opposition to Pythagoreans: In nearly all accounts of the system of Melissos, we find it stated that he denied the corporeality of what is real,--an opinion which is supported by a reference to fr. 9, which is certainly quoted by Simplicius to prove this very point. If, however, our general view as to the character of early Greek philosophy is correct, the statement must seem incredible. And it will seem even more surprising when we find that in the Metaphysics Aristotle says that, while the unity of Parmenides seemed to be ideal that of Melissos was material. Now the fragment, as it stands in the MSS. of Simplicius, puts a purely hypothetical case, and would most naturally be understood as a disproof of the existence of something on the ground that, if it existed, it would have to be both corporeal and one. This cannot refer to the Eleatic One, in which Melissos himself believed; and, as the argument is almost verbally the same as one of Zeno's, it is natural to suppose that it also was directed against the Pythagorean assumption of ultimate units. The only possible objection is that Simplicius, who twice quotes the fragment, certainly took it in the sense usually given to it. But it was very natural for him to make this mistake. "The One" was an expression that had two senses in the middle of the fifth century B.C.; it meant either the whole of reality or the point as a spatial unit. To maintain it in the first sense, the Eleatics were obliged to disprove it in the second; and so it sometimes seemed that they were speaking of their own "One" when they really meant the other. We have seen that the very same difficulty was felt about Zeno's denial of the "one."

Opposition to Anaxagoras: The most remarkable fragment of Melissos is, perhaps, the last (fr. 8). It seems to be directed against Anaxagoras; at least the language seems more applicable to him than anyone else. Anaxagoras had admitted that, so far as our perceptions go, they do not agree with his theory, though he held this was due solely to their weakness. Melissos, taking advantage of this admission, urges that, if we give up the senses as the test of reality, we are not entitled to reject the Eleatic theory. With wonderful penetration he points out that if we are to say, with Anaxagoras, that things are a many, we are bound also to say that each one of them is such as the Eleatics declared the One to be. In other words, the only consistent pluralism is the atomic theory. Melissos has been unduly depreciated owing to the criticisms of Aristotle; but these, we have seen, are based mainly on a somewhat pedantic objection to the false conversion in the early part of the argument. Melissos knew nothing about the rules of conversion; and he could easily have made his reasoning formally correct without modifying his system. His greatness consisted in this, that not only was he the real systematiser of Eleaticism, but he was also able to see, before the pluralists saw it themselves, the only way in which the theory that things are a many could be consistently worked out. It is significant that Polybos, the nephew of Hippokrates, reproaches those "sophists" who taught there was only one primary substance with “putting the doctrine of Melissos on its feet." «

Chapter 2

ENTRY TO FOUNDATIONS

THE ENDING OF TIME

What is happening in the past? Nothing is happening in the past. We emphasized it in Kos, at the International Conference of Numerical Analysis and Applied Mathematics 2012. Everything that is happening, happens in the present, - even the past. Every event, wherever and whenever this could be perceived by the human mind, was perceived in the presence. The past gains its reality through the presence of perception. The past is real by the aid of our presence. What is presence? Presence is light.

If we speak of the Old Egypt, every proof of the reality of those past events that signify the ancient culture can only be found in the present, in our present considerations, our libraries, our archaeological sites, in the steles, in the pyramids. Every little thing that makes the ancient episodes real, is but in the here and now. How is the here and now turned into such a whole that can be denoted a here and now? It is turned into a whole by those phenomena of quantumelectrodynamics (qed), or better, by improved quantum chromodynamics (qcd). Most of us would believe that this is mere philosophy. But it is not, it is exact. It is a pure mathematical matter. The here and now can be seen as an ideal, a set of events synchronized by unbound light. But there is little unbound light. Bosons appear as entangled. They are polarized and entangled. Pascal Jordan in 1947 used the light ray of polarized quanta as an example to explain the decay of the historic dimension in microphysics. Probing the electron, he stated at first:

"In order to be able to observe a physical structure (shape) by means of physical instruments (our sense organs are such instruments too) some impact of this structure (shape) on the instruments is necessary; and therewith some backlashes of the instruments on the observed are inevitably concatenated. Macrophysics need not see a fundamental problem therein, because it may imagine that the measuring instrument is still more subtle than the observed object – as for example a telescope compared to a planet. But if the observed itself is only a single quantum-physical elementary particle, this resort is strictly locked up, and we must face up to consider the act of observation as an incising invasion that can place a new onset and can smear the memorization of its history of the scanned structure to a large extent." (Jordan 1947, p. 39f.) In this way, the particle is imagined to become decoupled from its history and thus loses its affiliation to the macroscopic flow of time.

Jordan then explains why it is not possible to determine the impulse of a particle by measuring two locations on its trajectory and dividing their distance by the corresponding time-interval, as the second measurement of location would partly destroy the information gained after the first. Therefore, he thought, one had to look after a possibility to gain the magnitude of the impulse by a single measurement. In fact that worked out by 'utilization of the energy-impulse-conservation-law'. When this is applied to collision-events of mass-points, we obtain the same mathematical relations as in interference-laws of plane waves. So it becomes possible to consider a mass-point, as soon as it is correlated with a determinate impulse, as a plane wave with a wavelength λ such that the equation $p\lambda = h$ is satisfied. A complete objectifying of "encompassing all observable sides and all measurable qualities of an electron at the same time, according to that, cannot exist". Localization of an electron excludes its simultaneous appearance as a vast undulation. The wave seems to 'collapse'. As another example Jordan described the action of a Nichol prism on a polarized ray of light-quanta. Each application of the Nichol signifies an intervention that makes the state of the quantum independent of its previous history. He sums up "that a microphysical shape – in contrast to a macro-physical object – does not possess a fate independent of the observation-process. Each sharp observation is an invasion that enforces a new decision and cuts off the future destiny of the shape in an amnesic manner". Restrictions of objectivity brought in by complementarity in quantum-mechanics were necessarily connected with a breaking-up of the causal texture. Such was the view then. Yet, we should be aware; Jordan wisely spoke of a 'sharp observation'.

What was the prevailing belief then? The complementary behavior of wave-like and particle-like was linked with the idea of a destruction of particle-history. It was thought that any attempt to verify the particle nature by measuring position ultimately must enforce an irreversible disturbance and loss of information about the momentum. Any such measurement of locus was supposed to effectuate a collapse of the wave function. But such 'state reduction' and 'quantum leap' can be seen as a loss of coherence caused by the entanglement of the wave function with that of the measurement apparatus. Paths that interfered before the observation thus can become distinguishable by the measurement, so that no interference appears. The interference pattern can however be regained, if an observer succeeds to erase the information that makes the difference. Going into these phenomena of entanglement and quantum erasure, the particle history is regained. The photon and electron do no longer appear as necessarily clipped off their history. [1]

With the realization of weak measurement and a lot of so called EPR-quantum steering experiments it became evident that single particles can have a history within an observer system where complementary observables are measured. John Wheeler, Yakir Aharonov, Aephraim Steinberg, Howard Wiseman and many other physicists had become attentive to the asymmetry of quantum mechanics with respect to time. "In quantum mechanics it is considered inconceivable to consider the past at all", said Steinberg. "The absence of trajectories in quantum mechanics means that one supposedly has no right to discuss where the particle 'was' prior to that measurement." (Steinberg 2003, p. 5) But as the fundamental laws of motion are as time-reversible as those of classical mechanics, one should wonder why

[1] The 'Quantum eraser' – is a delayed choice experiment proposed by Scully and Drühl in 1982. Originally the delayed choice experiment is a thought experiment proposed by John Archibald Wheeler in 1978. It was presented by Yoon-Ho Kim, R. Yu, S.P. Kulik, Y.H. Shih, Marlan O. Scully in 1999.

it would be less reasonable to use measurements in order to make statements about some particle's history than to project its motion towards the future. By pondering retrodiction together with prediction Aharonov, Vaidman and others worked out a procedure of 'weak measurement' in which our knowledge about both the initial and the final conditions of a quantum system (QS) can be used to draw conclusions about the motion in between. So Steinberg and coworkers arrived at this beautiful picture, most of us have seen sometime on some webpage.

Figure 1. 3D plot of a photon in a double-slit[2] (Kocsis, Steinberg et al, 2011).

The method of weak measurement (Aharonov and Vaidman 1990), said Steinberg, allows one to put past and future on an equal footing. In the standard approach of von Neumann (1955) measurement of an observable A_s during some finite interval of time can be effected by the interaction Hamiltonian $H = g(t)A_sP_p$ via the time-dependence $g(t)$ by moving a pointer p, or respectively acting on the canonical impulse-totality P_p of some probe. For some suitably normalized $g(t)$ the position of the pointer is supposed to change by an amount proportional to the expectation value of A_s. In a good measurement device the probe is capable to follow the different eigenstates of the observable. In this case we would say that the probe and the QS had become entangled, and irreversibility of the measurement would have been caused by decoherence of the system wave function. What about the feedback? How is the probe acting on the system? The magnitude P_p might be uncertain. The magnitude of this uncertainty determines the size of an uncertain force exerted by the interaction Hamiltonian on the QS. If the probe were in an eigenstate of momentum, the measurement would bring forth an entirely predictable time evolution of the system. That is, if the pointer momentum were determined with no uncertainty, the pointer position would be entirely uncertain. We would not obtain a measurement value. Pondering over this cybernetic affair, Aharonov argued it would be carrying on to consider an intermediate regime, where so to say two QSs would meet weakly, and at least some small amount of information could be gained during the measurement without collapsing the wave function, and even without disturbing significantly its evolution.

[2] http://physicsworld.com/cws/article/news/2011/jun/03/the-secret-lives-of-photons-revealed (1.2.2014)

"Over and over again in the past decade or two, experiments in fields such as quantum optics have revealed phenomena which surprise even those of us who ought by now to know quantum theory reasonably well." (Steinberg 2003, p. 2)

"Too many physicists have fallen prey to the reassuring but nihilistic thesis that since so many before us has failed; we would be wasting our time to seek any deeper understanding of quantum theory than is contained in our beautiful equations." (Steinberg 2003, p. 2)

"A moment's thought suffices to realize that as difficult as prediction of the future may be, prediction of the past is not necessarily any easier (even aside from the semantic issue, which leads us to adopt the term "retrodiction" for inferences about the past)." (Steinberg 2003, p. 2)

The orthodox view of quantum mechanics holds that what has been measured can be known, and what has not is "unspeakable." If a particle is prepared in a certain wave packet, that function is to be considered a complete description, and any additional questions about where the particle "is" are deemed uncouth, at least until such a measurement is made. The absence of trajectories in quantum mechanics means that one supposedly has no right to discuss where the particle "was" prior to that measurement. Yet the fundamental laws of quantum mechanics are as time-reversible as those of Newton, and one quite reasonably wonders why it is any less valid to use a measurement to draw inferences about a particle's history than to make predictions as to its future behaviour. Such considerations led Yakir Aharonov and his coworkers to a formalism of "weak measurements" which allows one to discuss the state of evolving quantum systems in a fundamentally time-symmetric way. This chapter draws heavily on their ideas, whose main elements I will introduce below. I will analyze how weak measurements can be applied to several experimentally interesting situations. (Steinberg 2003, p. 3)

MORPHEMES, IDEMPOTENCE, ITERATION

In mathematical physics we have learned how to rotate a Dreibein by some orthogonal transformation, and that has become a strong habit. We draw a little arrow that points into some direction. We give it an origin and a unit length and call it a unit vector. Next, that unit is considered as orthogonal to a second and both orthogonal to a third one and this gives us what in German language we call a Dreibein. Such a Dreibein can be denoted by a number sequence $1, 2, 3$ or a set $\{1,2,3\}$, and we know since ancient times, there exists an orthogonal rotation about some main diagonal which can turn $\{1,2,3\}$ into $\{2,3,1\}$. If we look at this procedure in informatics terms, such a turnover $\{1,2,3\} \rightarrow \{2,3,1\}$ is a mere recoding. So, what appears as a rotation in Euclidean space can be seen as a recoding or renaming in semantic space. This is a very important observation, since in the innermost layer of stable macroscopic motion which is the unstable quantum motion, the process of nature does not make a difference between those orientations that we are used to regard as different at the macro-level. We have reached an agreement to call this peculiar recoding of space an automorphism of coordinates. So it turns out that quantum motion, at a peculiar level of scale, discloses to us a decay which signifies the invariance of motion under such automorphisms and respectively what we introduced to call reorientation. We are quite familiar with the 'plane reorientation' which leads us to the symmetry and symmetry breakage of the dihedral group $D_{2d} \simeq D_4$. This represents the reorientation symmetry of the Zweibein. We have perhaps also understood the action of reorientation automorphisms in Euclidean space which gives us octahedral symmetries O and O_h the latter of which includes total space inversion. But those strange attractors of the macroscopic matter, as are brought forth by the quantum motion, both involve and are based on a much further reaching instability, namely on a decay of solid state preserving motion to the greatest possible extent. The strange attractors of solid matter feign us to what we are used to call space-time. A decay of motion to the greatest possible extent involves a total reorientation not only of the space-time, but also of its geometric algebra. That is, we are now forced to allow for transpositions and respectively recoding among base units that have different grade. In those cases quantum motion involves chaos to such a degree that the energy can perform jumps from a grade one to a grade two or even higher grade monomial of space. As we shall go on, this simple feature of motion will become more and clearer. This phenomenology is the mystery behind all our observations of something that we call the standard model of particle physics.

I beg the reader not to go into the trap and believe that we are forced to presuppose the concepts of the Euclidean or Minkowski space to describe the phenomena of disintegrated quantum motion. What I prefer to denote as 'primordial quantum motion' involves a discrete concept beyond time and without metric. We are investigating the conditions under which quantum motion brings about those metric and oriented spaces or frames which we are observing and constructing in our laboratories. To understand this, let us begin very small. Let us discover some basic features connected with plane reorientation. In his book "Theory of Consciousness" the philosopher Gerhard Frey (1980) investigated the archaic thought-form of the quartered circle, the "Gedankenkreis". He discussed this morphogenetic structure of the human mind in at least three different contexts. Namely, first, within the pre-Socratic astronomic context, second, in terms of a superposition of 'logic conceptual pairs' and, third, as a concept related to self-reference. This morpheme or 'morphogenetic structure' of cognition has also been investigated by the psychologist Jean Piaget (1982) who considered it as an important component of what he called the operational structure of thought. It is probable that this structure represents something like an interface between matter and mind. It may have developed as a cognitive structure in the Upper Paleolithic. Frey has shown how the systematic meaning of self-reference becomes evident in statements that must be called metaphysical in the proper meaning of the word. Concepts with a quite general meaning such as existence, god, substance and 'being' must be self-referent. We can find this in Parmenides statement: 'being is'. Likewise 'non-being' must be self-referent, that is, 'non-being is not'. This points at both a static and a dynamic aspect when Parmenides says that 'non-being is not' and Heidegger 'das Nichts nichtet'. The *primary movens*, god or mind, must be a moving cause for itself, substance is defined as that which has its origin, its Arché, in itself (Frey p. 158). We shall find out step by step that this has consequences for mathematics and mathematical physics.

Frey had identified the quartered circle as the archaic form of self-reference. He showed that it gives rise to a prominent difference in mathematics which we identify as the algebraic difference between idempotence and iteration. Think about the basic structure as a circular arrangement of 'being is' → 'non-being is' → 'non-being is not' → 'being is not'. In this way we begin with the cyclic thought-form of a single self-referent concept.

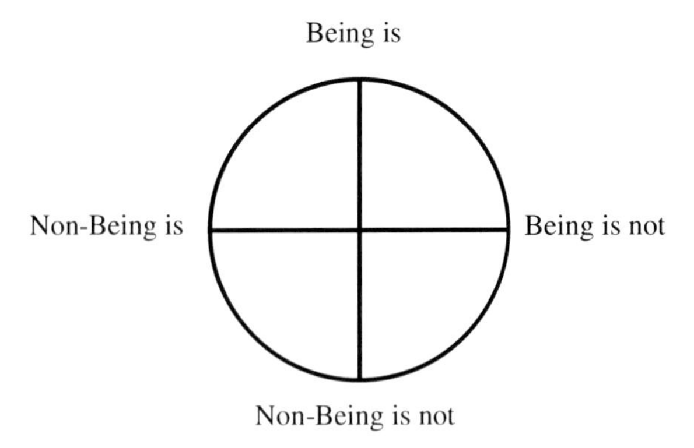

Figure 2. Thought-circle – 'Gedankenkreis'.

Now consider any concept, say, τ that we can apply to any other $\tau(\alpha)$. Now we can apply τ to itself and obtain $\tau(\tau(\alpha))$. Now we can have $\tau(\tau(\ )) = \tau(\ )$ for some or for all α in which case the concept τ acts as an idempotent.

$$\tau(\ ) = \tau(\tau(\ )) = \tau(\tau(\tau(\ )) = \ .\ .\ .\ ) \quad \text{idempotent structure, or we get} \tag{1}$$
$$\tau(\ ) \neq \tau(\tau(\ )) \neq \tau(\tau(\tau(\ )) \neq \ .\ .\ .\ ) \quad \text{iterative structure}$$

In this case we get a sequence of different expressions $\tau(\alpha), \tau(\tau(\alpha)), \tau(\tau(\tau(\alpha))\ .\ .\ .\ )$. So we can connect the archaic orientation morpheme with the conception of a temporal sequence and respectively the peculiar role of the complex plane in quantum physics. We shall realize that the dihedral structure must play a much more general role for the Minkowskian phenomena of physics, namely, in the form of a product $D_4 \times D_4$, we enter the domain of sequences capable to involve four locations in an elementary string of motion.

Hierarchy of Awareness – Perceiving and Observing Systems

Let us denote by the symbol u a »universal element«. It could mean a field, a particle, a structure, a quantum system, an element of some universal algebra such as an idempotent or a nilpotent, a, category, an allegory or anything else. It is at first only implicitly defined, but it will reveal its properties as soon as we specify its definition in mathematical terms. So we use the predicate 'universal' in a more informal, inexact way. For instance we know that in any universal geometric algebra there can exist infinitely many algebraic forms for idempotents. Consider some of the more familiar forms. For example, with the identity Id and base unit e_1, the term $a = ½\ (Id + e_1)$ is a well known idempotent which is primitive in the Pauli algebra $Cl_{3,0}$, but is not primitive in the Clifford algebra of the Minkowski space-time $Cl_{3,1}$. So we have two universal algebras $Cl_{3,0}$ and $Cl_{3,1}$ and universal elements u with given properties, e.g. 'being primitive in the algebra'. Now we may ask about the cardinality of the subset of idempotents in those algebras. Some of us know there exist orthogonal transformations of the a, carried out by some graded group-elements g, such that the element gag^{-1} is again an idempotent not primitive in $Cl_{3,1}$. The form $u := ½\, Id + 0{,}42e_1 - 0{,}17e_2 - 0{,}17e_3 - 0{,}17e_{14} + 0{,}069e_{24} + 0{,}028e_{34} + 0{,}012e_{124} - 0{,}028e_{134} - 0{,}07e_{234}$ is one out of continuously many idempotents in the Minkowski algebra. In this sense we can regard the u as a universal element in a universe of Clifford algebras. But we may just as well mean by u a 'universal element' in the entirety of algebras. However, if we speak of u as a fermion constitutive for matter, we regard it as a universal element in the material world. Still, in our theory, it is a cognitive concept. So we have a material electron which appears to be a rather simple thing, a charged point in some cavity and a mathematical electron which may appear as a rather complicated term in a still more complicated mathematical system. There are indeed many appearances, many cognitive phenotypes of 'an electron'. We hope that they all give us an electric charge minus one. Already in the first volume of 'Primordial Space' (p. 45) I said that any idempotent a can be represented by an extremely – even infinitely – complex term, it may even be as complex as the material world. Suppose the atoms of the material universe could be counted, – I am not sure if it is possible, but some give us a number, some say 10^{78}, others 10^{85} and still others 10^{89}, some factoring in the black holes, others denying countability because of transmutations and metamorphoses, dark energy and so on – in any case the number n of idempotents a is infinite and uncountable, and the number of mathematical terms that are capable to represent idempotents as well. Imagine, we have considered only the idempotent form in Clifford algebras, and that step already led us to an uncountable universe of idempotents. But there are infinitely many other entries to idempotents. For example, 'closure', as we conceive it today, is a function $cl\colon \mathcal{P}(S) \to \mathcal{P}(S)$ from the powerset $\mathcal{P}(S)$ of a set S to itself. It has, among others, the property $\forall_{X \subseteq S}\, cl(cl(X)) = cl(X)$. Thus the closure-operator cl is an idempotent function. This provides another one of infinitely many entry-points to the universe of idempotents.

Compiling the first page of this chapter, all we are saying is: abstract from all qualities, properties, structures and forms, then you might deal with a universal element. What is so special about the u is at first, that it can be an element in another universal element, and second that it may be an element in both the material world and the cognitive world, and third that it refers to itself. Inasmuch as it is both material and cognitive, it discloses some surprising features of similarity in relations. This goes back to the fact that both material systems and synchronous symbolic arrangements share certain geometric and logic qualities. As we shall go deeper into the problem of representation of physical locations and their possible entanglement and disentanglement, this similarity of relations will become conspicuous. To give you an idea of the high level of abstraction and forgetfulness of a universal element, consider a reflection made by Louis Kauffman: "The Universe is constructed in such a way that it can refer to itself. In so doing, the Universe must divide itself into a part that refers and a part to which it refers, a part that sees and a part that is seen. Let us say that R is the part that refers and U is the referent. The divided universe is RX and $RX = U$ and RX refers to U (itself). Our solution suggests that the Universe divides itself into two identical parts each of which refers to the universe as a whole. This is RR. In other words, the universe can pretend that it is two and then let itself refer to the two, and find that it has in the process referred only to the one, that is itself." In this thought the R is a 'universal element', and you can easily detect the undetermined or fuzzy degree of abstraction that leads to such an idea of a universe that pretends to be two. Nevertheless, the concept has to be taken serious, and we shall investigate this way of thought in greater depth.

Let u be a universal element (1)

Universal elements refer to themselves by a relation which preserves or reproduces themselves, but does not change them. We call this property the 'awareness of a universal element of itself'. Thus the property of awareness is defined by idempotence or 'idempotency':

$u\,u = u$ awareness $\Leftrightarrow$ idempotence of u (2)

A universal element that experiences change by referring to itself may give rise to iteration

$u\,u = v$ iterative self reference $u\,u\,u = u\,v = w\ldots$ (3)

So we make a difference between idempotence and iteration which parallels the two pre-socratic viewpoints of Parmenides and Melissos. Next and below in the hierarchy we have perception

$p\,u = u$ perception of a universal element by an observer p (4)

which does not change the observed element. There can be many universal elements preserved by the observation, as for instance in $p\,u = u, p\,v = v, p\,w = w$ a. s. o. in which case the observing element p can be thought to act like a unit element for $u, v, w,$ a.s.o. We call p the 'perceiving' element. The perceiving element is a special form of a universal element that we call an 'observer'. The observing element does not preserve the observed, but effectuates a change:

$o\,u = v$ observation of a universal element by an observer o. (5)

By observing u the observer o turns the observed u into another universal element v. We can look at the 'observer equation' (5) as a classical quantum measurement where it is thought that the state function of a mixed state 'collapses'. It represents so to say a strong measurement. Whereas the 'perception equation' (4) represents the ideal of a 'weak measurement'. Statements (1) to (5) should be regarded as fundamental statements on which we can base a theory of interaction and entanglement. This theory then provides the natural motivation for to explain the emergence of that frame into which we formerly put the (high energy physics) events. In the former quantum mechanics we took the space-time as a predetermined mathematical object, some metric space, where the process of nature goes on. Space-time, locus and impulse, time and energy were presupposed to formulate equations of motion such as the Schrödinger- and Dirac-equation. In the anti-physics approach we explain the emergence of space and time by the fundamental processes of awareness, perception, observation and the natural motion arising thereupon. This Ansatz puts our old question, how matter brings upon space-time (GR) in a slightly different form.

Equation (2) signifies the view of Parmenides, whereas (3) denotes the interpretation of Melissos. The latter gives rise to a fundamental problem of both philosophy and physics.

Suppose that u refers to itself as is given by equation (3), but does not involve any memory. Someone perceives $u\,u = v$ but v has no more link to u. The element u has vanished from the universe and left v as the only memory of that which once had been. That is, there is no memory about u. Nowhere in the universe does anyone or anything know about u. So we are left with the v. Next, v does refer to itself and gives us w, so that we have the equality $v\,v = w$. This really lets us alone with a big problem. For the w might be equal to u or it may be different. We cannot decide, because there is no memory about u.

This leads us to another important question: namely, how can a memory be organized within the present such that it can recollect the past? And can it at all be the case that it recollects all the past? After all the present transcends the past. It seems that part of the present must differ from the past. Or can it be that all of the present somehow represents all the past? Is the world a memory of itself? Is the world its own memory? If it is, it seems, the process of motion must be highly nonlinear. Every interaction, even the weakest leaves its track in the present structures such that those contain all the memory. Or is much of the memory getting lost "in the course of time"? These are difficult questions. I guess there is no immediate answer to them, but we can delve deeper into the matter. Let us start it up all anew!

How can a universe that changes in presence, but not in time, bring about time? How can such a change be evoked? How is it induced? If all that is needed, is part of the present, those 'memories' on which the process is operating, have to be utterly 'in presence', and thus they have to be synchronous structures rather than diachronic events. Are we aware of the enormous difference to the former models of physics? Formerly we had a past universe that was thought as outspread in space-time. The world went on in time, and the present could exercise no more influence on that past, because the past was over. But in the other view which goes back to Parmenides the past is accessible. What we are used to call the past acts on the present. The 'past times' are accessible through the memories they left as reflexive imprints in the present. The past is not different from the memories that are leaves in the present phenomena of nature. In such an omnipresent universe motion is beyond time, and time appears only as a derived secondary order of memorized events. Memorization thus can be understood as a peculiar process of visualization and realization. The universe moves in presence, and its motion has a driver. This driver can best be imagined as a universal self-reflexive domain. The present operates on itself and thereby brings about change. The fuel of evolution is reflexivity. However, if reflexivity plays the leading part in the process of nature, change can be supposed to be the outcome of a highly nonlinear interplay. If there emerge 'historic invariants' in such a nonlinear system, this can be due to some outstanding importance of peculiar eigenforms and properties of idempotence. We shall now have to go into this. So what is the meaning of reflexivity? Reflexivity denotes a property that refers to the presence of a relationship of some object, entity, actor, event or phenomenon to itself. In enumerating these nouns from object to phenomenon, we are signifying a certain degree of hardness. Namely the apparently inviolable object is the creation of an observer who is thought to be strictly separate from the observed. But this is not possible. The observer, in a way, is the observed, and his action is reflexively linked with society. It is also reflexively linked to the universe. The abuse of the word 'object' is usually not done with the word 'phenomenon'. Phenomena tend to escape the harshness of the objectifying look that likes to cut the observer off from the reflexive world of presence and change. But every observer has a relation to himself and is reflexively linked with the world. Objects, having the largest degree of hardness in the hierarchy of measurements, are nevertheless participating in a network of interactions and gain their stability from a cybernetic process. Any object can be taken as a sign and token of that process. Mathematically, the object appears as an eigenform, a fixed point of transformations in a reflexive domain. Thus, an object need not be sharply distinguished from the process by which it is stabilized. In a reflexive domain, every entity refers to itself and acts as a transformation on that space. Some of those entities may turn out stable and serve us as 'objects'. Any entity A may act on some other B and thereby bring on C. We write for this

$AB = C$ A acts on B and brings forth C (6)
$BA = D$ B acts on A and brings on D

Reflexive domains are open so that new elements may be created that bring forth further entities. As every element of that space is reflexive and at the same time represents a transformation the domain must have eigenforms. The reason is exact. Namely, every entity G acting on some x in the reflexive domain, must refer to the relation of x to itself, and is therefore a function of xx. That is, the G acting on x must be an F having argument xx. From this there follows a fixed point theorem in its most basic form:

Define G by $Gx = F(xx)$ then (7)
$GG = F(GG)$ so GG is an eigenform or fixed point of F.

In this way, a transformation of the entity x in a reflexive domain comes along as a function of self-reflexion of x. Therefore, every element of a reflexive domain has an eigenform. We can elaborate on this definition and refine it a little as Louis Kauffman (2009, p. 128) has done so

Theorem 1 – General Fixed point theorem: Let D be a reflexive domain with 1-1 correspondence $I: D \to [D, D]$. Then every F in $[D, D]$ has a fixed point. That is, there exists a p in D such that $F(p) = p$.

Proof: Define $G: D \to [D, D]$ by the equation $Gx = F(I(x)x)$ for each x in D. Since $I: D \to [D, D]$ is 1-1, we know that $G = I(g)$ for some g in D. Hence $Gx = I(g)x = F(I(x)x)$ for all x in D. Therefore, letting x be g, we get $I(g)g = F(I(g)g)$, and so $p = I(g)g$ is a fixed point for F.

As we go deeper into our problem of creation of space and time, we shall repeatedly fall back upon these properties of iteration, idempotence, and self-reference in reflexive domains.

UNDEFINED DOMAINS AND POLARITY STRINGS IN THE VOID

In order to reconstruct physics within anti-physics, we should not carry our minds too far away from what we have learned and what we are used to believe. We rather should work out a soft transition from what we already know to that which is impossible to be known. There are things we cannot substantiate. That will perhaps mobilize our imagination and inhibit our becoming dogmatic. Those of us who have heard Richard Feynman's UCLA lectures on

Quantum electrodynamics or who read his book "QED – The Strange Theory of Light and matter" (Feynman 1985) know that light takes every possible way. QED has therefore been denoted as *Feynman's overall space-time approach* by philosophers (Mario B. Valente 2011, p. 8). The field fills up space before it shows us where it quantizes. This discovery led to the invention of "the diagram", as Feynman called it, and to the method of the path integral. For instance, if we calculate the cross-section for the creation of an electron-positron pair, we need to estimate momenta $-i\hbar\nabla$ with the nabla operator $\nabla= e_x\, \partial/\partial x + e_y\, \partial/\partial y + e_z\, \partial/\partial z$ in the rest frame of some electron. This involves a lot of knowledge about space-time. All our theories and theorems dealing with QED or QCD presuppose the existence of space-time. This is a methodological mistake. We can never understand the phenomenology of quantum fields if we take space, time and their frames for granted. Because it is the QED phenomena which build up all our means to define and to measure space and time. Without light and matter no space-time. So we have to begin anew. We have to derive our concepts of space and time from quantum chromodynamics and quantum electrodynamics in a void that does not yet have metric and orientation, no measure, no topology! After all, the boson fields, as we know, go every possible path. They literally grope their way and establish a contact with the whole universe of possible events in space and time, while they create it. Creation of space-time, creation of Schrödinger- and Dirac equation go together with the creation of fieldquanta. Macroscopic matter is the outcome of some QCD and some QED.

Let us look at primordial space D as an undefined domain of locations. It is not a space. We know that in this domain there are locations where something is going on. But we do not know what it is that is going on and how many locations there are in this space. We simply cannot find it out. We cannot determine the cardinality of this set and we have no means to decide over any partial order on that space. So we have no functions from the power set of space to itself which could serve as closure and interior. We have no topology, no reasonable neighborhood system. Our situation is very similar to that of a speleologist who has lost his lamp and climbs into a dark cave. With him perhaps a glowworm has entered the cave. Now he sees that there is an entry to another cave. He is now able to realize the contour of the cave he is in, but cannot figure out the contour of the other. So, by the activity of the glowworm, he becomes aware of the existence of two adjoining locations one of which is slightly discernible, while the other is not. So he writes on his scratchpad $\ldots + \; - \ldots$ denoting thereby a sequence of locations which includes an illuminated cave and a dark cave. The $+$ adjoining the $-$ is indicating to him that there is a relation between the two such as a neighborhood relation. But in principle we do not restrict the connection between two locations to a relation of metric proximity. It may rather be a *translocal relation*. Two speleologists may use a rope for transport and communication. So they establish a relation between two varying locations, that is to say, between two different caves in the system. A further contribution to such a spatial relation may be brought forth by the action of cavern water. This water may provide an accidental connection between different far distanced locations which can nevertheless be used for various transactions by our researchers. In physics, a *translocal connection* between locations represents what we call an *entanglement*.

Now we may really have a void universe that is nevertheless capable of all kinds of actions. We may look at it as an entirely undefined domain. But all of a sudden that domain looks at itself and divides itself into two identical parts each of which refers to the domain as a whole. That awareness of itself is what we call energy, but we need not yet go into this. We do that later. For the time being, we just observe – and we shall later also have to speak about the observer – that this domain delivers two sequences having forms

$$\begin{array}{l} ...+-+-+-+-... \\ ...-+-+-+-+... \end{array} \quad (8)$$

and they represent, say, an electron and a positron. It seems all very crazy, since the two sequences actually represent one and the same thing. They are just shifted by one step of location. Their assumed difference becomes apparent only by direct confrontation. We shall call such confrontation a 'touch'. By mere touch it becomes apparent that we have here an electron and a positron, and the two annihilate. But that shift of location can only be realized if we put the two sequences in relation to each other. What do we have to do in order that we become aware of the annihilation? We just assume that $+$ is the same as $+1$ and $-$ is standing for -1. Then, by vertical addition we obtain a zero sequence which is just a sequence of void locations. By the way, we do not know where these locations are, how many light years away from the center of the galaxy. May be such a sequence of locations contains only eight determinate locations as in our present example (or call them digits or positions or characters), and these are somewhat irregularly distributed over our undefined domain so that they have a certain potential to misguide the observer. Yet, their translocal proximity may allow them to annihilate. So we obtain a figure of placeholders where we do not have any information about their coordinates, orientations and scales

$$... \qquad\qquad ... \quad (9)$$

within the undefined domain. For the present we claim, this signifies two bosons or else two fermions in a nonlinear void. As you can easily figure out, we have chosen here the Breit-Wheeler process as a possible entry-point to our considerations. I have seduced you to ponder this example from HEPhy with intention. As you know we are able to annihilate electron-positron pairs since long. Paul Dirac was the first who studied that event within the framework of quantum mechanics. Both theory and praxis of pair-annihilation are fully developed. Not so for the inverse process of pair creation. The possibility that an electron-positron pair might be created from vacuum by the collision of two photons was studied by Paul Dirac (1930) and by Breit and Wheeler (1934). A third process was investigated in the works of Fritz Sauter (1931) and Oscar Klein (1929) pondering the possibility to create a pair from the vacuum in a constant electromagnetic field.[3] This led us to the construction of a procedure denoted as vacuum polarization which was expected to occur in electric fields with strength above a certain critical value of

$$E_{crit} = \frac{m_e^2 c^3}{e\hbar} \simeq 1{,}3.10^{16}\, V/cm \quad (10)$$

[3] Heisenberg and Euler (1936), in their famous work on Dirac's theory of the positrons, had already found out that there is a significant difference between a magnetic and an electric background field. It is the electric field that provides a background for electron-positron pair creation. It was the analysis of this problem that was carried out by Sauter.

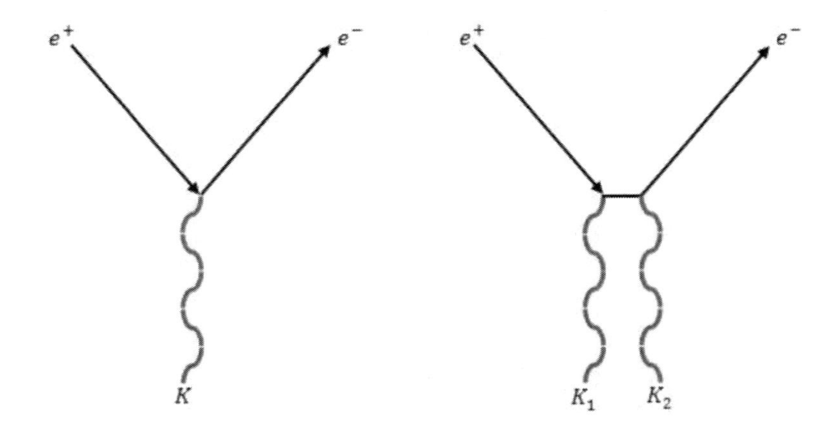

Figure 3. Laser induced and one Laser modified Breit-Wheeler process.

This large numerical value is the reason why the Breit-Wheeler pair creation has resisted empirical verification in Laser-physics until recent events. Therefore we can take the confrontation 'annihilation' versus 'creation' as a metaphoric transposition for the inversion of the epistemic process: 'quantum field theory from space-time' versus 'space-time from quantum fields'. It is this second approach that we are following here. It should be mentioned that our second approach turns out very practical due to the recent advances in Laser-technology. It has now become possible to produce routinely high energy, short duration Laser pulses. So there is springing up a renewed interest in QED. New experiments have been made at the Stanford Linear Acceleration Center (SLAC) reporting on nonlinear Compton scattering and nonlinear electron-positron pair creation. Several authors (Krajewska and Kaminski 2012)[4] have carried out calculations in the nonperturbative regime of pair creation. They found the energy-angular distributions of electron-positron pair creation in collisions of a laser beam and another nonlaser photon. Consequently, they call the two fermion lines in the diagram Figure 3 the Volkov states. For there exists a further possibility of a plane-wave background field, for which the Dirac equation had been solved by Volkov (1935). It involves zero effective action. Much of the history of this development had been comprehended in the article "The Heisenberg-Euler Effective Action: 75 years on" by Gerald Dunne (2012).

But all this is much too much for us, as it presupposes either Euclidean or Minkowski space. Yet, we must first appreciate the existence of these phenomena of pair creation. So let us recall that we began with two iterant sequences in an unspecified domain, and their reentry into the void left us with a symbolic

. . .   . . .

that could represent something like an unspecified featureless vacuum, some non-space, some figure of placeholders. Historically such a figure can take in many forms. We can become aware of the full extent of the variety of such vacua by considering publications about the vacuum energy. These speculations can be based either on the *cosmological constant* problem or on the Feynman diagrams of quantum field theory and anything in between, for example decaying vacuum energy. I am sure that most physicists have noticed what Robert Oldershaw

[4] arXiv:1209.2394 [hep-ph]

(2009) wrote in the abstract to his paper titled Vacuum Energy Density Crisis. "*The theoretical vacuum energy density estimated on the basis of the standard model of particle physics and very general quantum assumptions is 59 to 123 orders of magnitude larger than the measured vacuum energy density for the observable universe which is determined on the basis of the standard model of cosmology and empirical data.*" The HEPhy vacuum energy density ρ_{HEPhy} is usually estimated within some given unit volume. This requires, as we said, considerable knowledge about what we mean by space. Physicists consider some reasonable Compton wavelength and take the cube of this length. Vacuum fluctuations have to be bounded in order to avoid, as is said, an 'ultraviolet divergence' of energy. In HEPhy it has become a ritual to regard the Planck scale as donator of such finite cutoff. From that speculation one obtains

$$\rho_{HEPhy} = \frac{\mathcal{M}^4 c^3}{h^3} = 2{,}44\,.\,10^{91}\, g/cm^3 \text{ with the Planck mass } \mathcal{M} \text{ and Planck constant } h \quad (11)$$

In cosmology, it is said, the rigor is a little bit more straightforward. The critical density, there, is defined by a transition from open to closed solutions of the cosmological model of General Relativity. There it is proposed that we should have

$$\rho_{crit} = \frac{3H^2}{8\pi G} \text{ with the current value } H \text{ of the Hubble constant and gravitational constant } G \quad (12)$$

Assuming that the cosmological constant currently amounts to about $0{,}75\,\rho_{crit}$ cosmologists would suppose that we have

$$\rho_{cosmos} = 0{,}75\,\rho_{crit} \simeq 10^{-29}\ g/cm^3$$

Though we have only one standard model of HEPhy, the calculated estimates of ρ_{HEPhy} are also scattered over an interval of 120 orders. Recently Bernard Durney (2012) based his calculation of ρ_{HEPhy} on Feynman diagrams and respectively second order terms of the S-matrix. So he was led to a QED supported estimate at the Planck cutoff which compiled the contribution of electron-positron pair production to the vacuum energy by

$$\rho_{pair} = 1{,}2\,.\,10^{-31}\, g/cm^3 \quad (13)$$

He points out that this would be in amazingly good agreement with the observations made by Riess (1998) and Perlmutter (1999) which would give an energy density of $6{,}8\,.\,10^{-30}\, g/cm^3$. Considering the void as a sequence of placeholders in an unspecified domain, following classical QFT we had to ask how the vacuum evolves back into vacuum in the presence of background fields $\mathfrak{B}$.

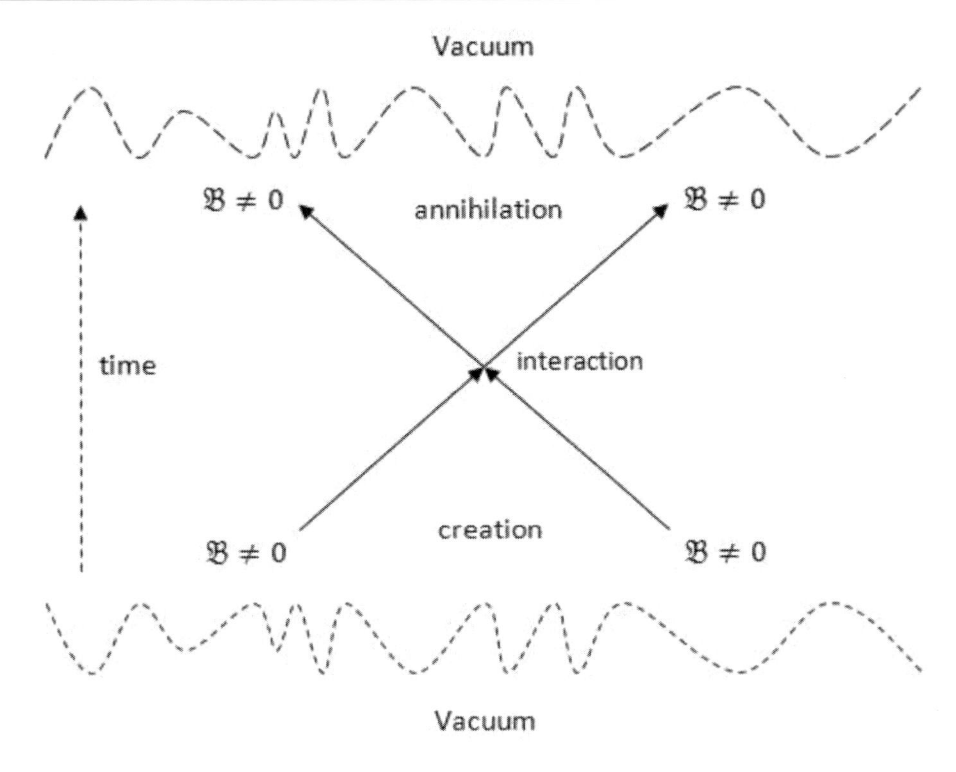

Figure 4. QED Vacuum to Vacuum transition.

Modelled after this figure, in QFT we would have defined the vacuum-vacuum amplitude in the presence of $\mathfrak{B}$ as

$$Z[\mathfrak{B}] \stackrel{\text{def}}{=} \lim_{T\to\infty}\langle 0|U_{\mathfrak{B}}(+T,-T)|0\rangle \quad \text{with time evolution operator } U_{\mathfrak{B}} \tag{14}$$

When $\mathfrak{B}$ was zero we got

$$Z[0] = \lim_{T\to\infty} e^{-iE_0(2T)} \qquad \text{with the vacuum energy } E_0 \tag{15}$$

Calculating vacuum to vacuum transition amplitudes (14) together with the corresponding cross-sections and S-matrix elements, the quantity transition amplitudes turned out to be a singular normalization factor that had the salving property to cancel when physical quantities were calculated. The 'energy' E_0 could be derived from the totality of vacuum bubbles. Formally these 'bubbles' are Feynman diagrams without external lines having self-, double, triple, fourfold a. s. o. connections. For each such connected vacuum bubble there is a certain number of vertex permutations leaving the diagram unchanged. By going over to the logarithm of the partition function $Z[0]$, we find a diagrammatic expansion for the negative free energy

$$E_0 = \frac{1}{2}\bigcirc + \left(\frac{-g}{4!}\right)3\,\bigcirc\!\bigcirc + \frac{1}{2}\left(\frac{-g}{4!}\right)^2 24\,\ominus +$$

$$+\frac{1}{2}\left(\frac{-g}{4!}\right)^2 72\,\bigcirc\!\bigcirc\!\bigcirc + \frac{1}{8}\left(\frac{-g}{4!}\right)^3 1728\,\bigcirc\!\!\!\!\triangle + \ldots \quad (16)$$

with monotonously increasing weights and integer powerfunctions of the negative coupling constant g. The problem with such a model is not to be neglected. It is one of methodology. It is said that each bubble contributes to the vacuum energy a fraction which can be identified with its factor. In finite volume, this factor could further be related to the total volume of space-time. Dividing by the volume the vacuum bubble could be interpreted as a contribution to the vacuum energy density. A second difficulty consists in the deficiency of a reasonable procedure that allows us to identify the conditions which lead to an 'ultraviolet divergence' of energy. But the third and the most serious of all difficulties in such QFT can be figured out by just looking at the arrow of time on the left hand side in figure 4. Here time is given, is predetermined. But in physical reality time is brought about by those processes of light and particle interaction which are described by the partition and correlation functions involved. Our scales for measuring spatial distances and time-intervals are given by phenomena of light propagation. They are both resulting from the quantum fields we are investigating. Think about a painter who has been conditioned to make his painting within a picture frame, and he finds out that the picture frame is part of what he wants to paint. Moreover he wants to paint a watercolor of flowing water. How will he proceed?

We shall now forget about path integrals and go slowly beyond the old epistemological error. We begin with the iterant sequences $\ldots + - + - + - + - \ldots$ and $\ldots - + - + - + - + \ldots$ and their trivial void given by a sequence of placeholders with undetermined length $\ldots \qquad \ldots$ Is it possible to define such processes like observation, creation and annihilation in such an unspecified domain D without metric and topology? The characters in the iterant sequences signify locations. They are translocal, so we have no measure for their distance. We also have no distance between the elementary sequences. But we suggest that they may or may not mate. They may arise from the same void location. They may belong together. They may be entangled. Is it possible to derive something like temporal features in such a domain? Can we ponder the emergence of undulation? Is it possible to construct a wave equation such like the Schrödinger equation? Is there a Hamiltonian? Can we construct a discrete model for primordial space generating processes? We accept that there are photons, fermions and forces, as in the QED, and now we construct step by step a method to obtain measurable space and time operators.

Having such an insecure thought in mind, being so vulnerable in the foundations of physics, we find ourselves confronted with the phenomenology of perception and with the problems brought in by the separation between observer and observed. It may turn out that quantum fields transcend that separation, so that it becomes difficult to do science as usual. Since what we scientists today are still used to, is the separation between observer and object and the apparent invariance of the objects in our perception. Most of the time, those invariant appearances of outer objects tell us less about the objects than about the internal structure of the observer's perceptional system. The observer gives stability to the outer appearance of objects. How can we put this in mathematical terms? Kauffman (2009) has proposed to use the cybernetic terminology of Heinz von Foerster (1981) saying that objects are tokens for eigenbehaviors of the observer. In the form of a simple equation we could demand that

$$O(o) = o \tag{17}$$

articulating therewith that the object o is a fixed point of the observers Observation O. The object appears as an invariant of observation. If O is denoting a process, than the o is an invariant of the process. It does not change under observation. By the way, this is the critical point for science, where science itself undergoes, so to say, a phase transition, as we all know that quantum phenomena change under observation. Most or at least many quantum events cannot be conceived as invariants of observation. Yet, we keep the equation of eigenform (17) and we say that 'the object o is an eigenform of the process O'. As we want to go on, the writer has to confess that following the advice of Parmenides he finds it difficult to introduce to the reader the iterant sequences $\ldots + - + - + - + - \ldots$ and $\ldots - + - + - + - + \ldots$ it does not even turn out trivial to show to us what is the meaning of -1 in that context. No particle has ever vanished without leaving a track. After all we have been scanning the tracks until certain photons escaped into the void, but as we said and were convinced of, the energy was always conserved. May be, however, it was not conserved locally. But then it was just shifted around. So how can we talk about -1 in terms of a process? Let us think about that by considering some historical instances. Negative numbers have been known in pre-Christian times in China. There were also some rules, how to carry out calculations with negative numbers, that came up from Diophantine. In some of the writings of Leonardo of Pisa negative numbers appear as debts. But Michael Stifel (1487-1567) still denoted them as *numeri absurdi* and *numeri fictiv infra nihil* (cited after Gericke 2004, third part, p. 248). From the ninth book of Euclid's Elements we can, however, take the most important algorithms concerning the field $\mathbb{Z}$ of integer numbers. But we can be sure that even in the late middle ages the negative integers were perceived just as 'absurd' and 'fictiv infra nihil' as the imaginary and complex numbers are seen by some of us today. As we want to represent all the 'absurd numbers' as eigenforms of processes, even the 'hypercomplex' Clifford numbers, we want to begin with the negative numbers. It goes like this: before the meaning of negative numbers appeared, in modern times, as a secure achievement of the human mind, there was a knowledge about the existence of even and odd numbers which seemed to be a bit more secure. This security can be traced back to Euclid, Aristotle and the Pythagorean 'number theory' (van der Waerden 1979, p. 396). From these sources we have inherited the rules

A sum of an odd number of odd numbers is odd.
Even minus even gives even.
Even minus odd gives odd.
Odd minus odd gives even.
Odd minus even gives odd.
Odd times even gives even.
Odd times odd gives odd.
An odd number that divides an even number, also divides half this number,
and so on.

In German language the integer numbers are called 'ganze Zahlen', and the word 'ganz' can be translated as 'integral', 'unbroken', 'undivided' and 'whole'. Interestingly this petty little matter of semantics has the potential to signify a transition from pre-Socratic to Euclidean and Aristotelian thought. In pre-Socratic Parmenides' thought the 'One' does not have parts, and is no whole either.[5] This does not come close to 'ganze Zahlen' as 'integer numbers', but rather to prime numbers. No other numbers than p and 1 divide p. The only pure whole, integer number seems to be unity. This idea has its own beauty. Every mathematician and every physicist can decide for himself, whether or not the 'One' of Parmenides is without frills. If it is not only a crazy idea, it may be a reality behind what today we still call a quantum field. How can we deal with wholeness, 'Ganzheit', in this connection? Does wholeness relate to eigenform? Here Kauffman brought in the twosome of concepts made by Henri Bortoft (1971). There is made a difference between being within a space and being inseparable in an unbroken wholeness of perception. Anything can participate in the process of nature as a thing in space, or it may partake in the unbroken wholeness of observer and observed. The latter seems to be the idea of Louis Kauffman (2001, p. 5). He writes: "*Compresence connotes the coexistence of separate entities together in one including space. Coalesence connotes the one space holding, in perception, the observer and the observed, inseparable in an unbroken wholeness. Coalesence is the constant condition of our awareness. Coalescence is the world taken in simplicity. Compresence is the world taken in apparent multiplicity.*" Sure, we can question, under which conditions observer and observed are perceived as inseparable. The perception in such unbroken wholeness turns out simple for some and difficult for others. But let us assume that we can take the world in such simplicity. Then we could believe that "*eigenform is a first step towards a mathematical description of coalescence*". This could be substantiated by the observation that observer and observed appear as united in one process that gives stability to both. Let us find out to where this thought leads us.

Process and Iterant

Think about a cybernetic process that gives stability to the observed object and at the same time stabilizes the eigenbehavior of the observer. We design, at first, a model of a recursive process in which the object is disclosed as token for eigenbehavior. We are using the spherical form rather than a square, because the square already suggests a definite type of dihedral orientation in space, a feature we want to avoid. It will turn out later that orientation is an eigenform of both inner and outer process. This figure we are using is just suggesting a relation between inner and outer. So we establish a slight difference between the presentation of Kauffman and the present author. Consider

[5] Parmenides proceeded: If One is, he said, the One cannot be many?/Impossible. / Then the One cannot have parts, and cannot be a whole?/Why not?/Because every part is part of a whole; is it not? /Yes. /And what is a whole? Would not that of which no part is wanting be a whole? / Certainly. /Then, in either case, the One would be made up of parts; both as being a whole, and also as having parts? /To be sure./And in either case, the One would be many, and not one?/True. /But, surely, it ought to be one and not many?/ It ought./ Then, if the one is to remain one, it will not be a whole, and will not have parts?/ No. [translation taken from 'Parmenides'. (2011). Web edition. eBooks@Adelaide; The University of Adelaide Library, University of Adelaide, South Australia 5005; http://ebooks.adelaide.edu.au/p/ plato/p71pa/index.html]

$F(X) =$ (X)

The process iterates an enclosure of form. We can illustrate the first steps when the process is applied to an empty circle, and is proceeding by a successive nest of spheres.

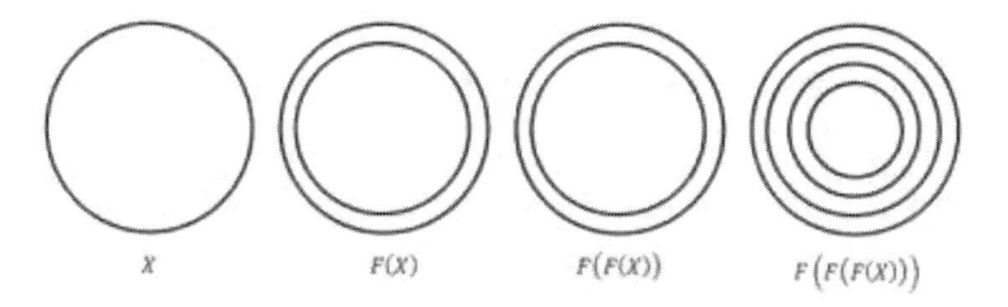

An infinite nest of spheres is invariant under any further enclosure. In the limit of infinitely many circles, we can define

$X \stackrel{\text{def}}{=} F\left(F(F(\ldots))\right) =$ . . .

$F(X) =$ . . . $= X$

In this way the infinite nest of circles X turns out invariant under the iterated addition of one more surrounding circle. It is an eigenform of the recursion. Considering thus the resulting equation $F(X) = X$, we can substitute $F(X)$ for X, and iterating the substitution we obtain

$$X = F(X) = \ldots = F\left(F(F(X))\right) = \ldots = F\left(F\left(F(F(F \ldots (X) \ldots))\right)\right) = \ldots \quad (18)$$

In this way it becomes visible how the process arises from the mathematical term of its eigenform. Kauffman states "in this view *the eigenform is an implicate order for the process that generates it.*"

We can now repeat our question: how can we talk about -1 in terms of a process? Consider the field $G_2 := \{1, -1\}$ as a domain for the process

$$\mathcal{R}(x) = \frac{1}{x} \tag{19}$$

We regard this as a recursive process. Beginning with some integer number $x_0 \in G_2$ we iterate the reentry of the resulting number

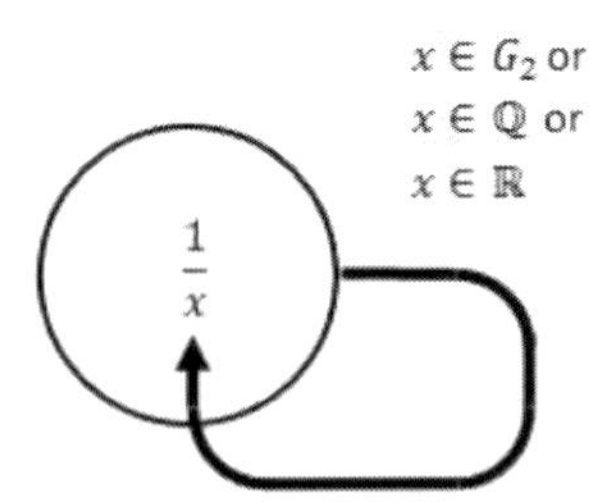

We can even insert a rational or real number as a start value. Suppose we begin with $x_0 = 2$. Then we would get the resulting series $2, \frac{1}{2}, 2, \frac{1}{2}, 2, \frac{1}{2}, \ . \ . \ .$ We realize that the process $\mathcal{R}(x)$ has two fixed points, as we have

$\mathcal{R}(x) = x \Longrightarrow x = \frac{1}{x} \Longrightarrow x^2 = 1$ from which we get two roots

$$x_1 = +1 \text{ and } x_2 = -1 \tag{20}$$

Iterating the reentry of $+1$ we obtain the series $+1, +1, +1, \ . \ . \ .$ iterating the reentry of -1 we obtain the series $-1, -1, -1, \ . \ . \ .$ Thus the iteration of $\mathcal{R}$ on $+1$ yields

$$+1, \mathcal{R}(+1) = +1, \mathcal{R}(\mathcal{R}(+1)) = +1, \ldots \tag{21}$$

And the iteration on -1 gives stationary -1

$$-1, \mathcal{R}(-1) = -1, \mathcal{R}(\mathcal{R}(-1)) = -1, \ldots \tag{22}$$

Next we affix a slight change on the process. We just change sign and define

$$\Re(x) = -\frac{1}{x} \tag{23}$$

Again we imagine that this is a recursive process operating in these three number fields. But now it turns out that our slight change has serious consequences. Beginning to iterate on the number 2 as before, we now obtain the series $2, -\frac{1}{2}, 2, -\frac{1}{2}, 2, -\frac{1}{2}, \ . \ . \ .$ and iteration of $\Re$ on $+1$ gives the alternating code

$+1, -1, +1, -1, +1, -1, \ldots$ while iteration of $\Re$ on -1 yields (24)
$-1, +1, -1, +1, -1, +1, \ldots$

So the new process $\mathfrak{R}$ gives us a discrete alternation with indefinite length. It represents the archetype of discrete oscillations. To our surprise the process has no fixed point in the original domains $G_2, \mathbb{Q}, \mathbb{R}$. But we verify that the imaginary unit satisfies the eigenform equation for $\mathfrak{R}(x)$. We have

$$\mathfrak{R}(i) = -\frac{1}{i} = i \tag{25}$$

As multiplying by i we get the identity $i\,i = -1$ which is the defining equation for the unit imaginary. It is because of the appearance of the alternating series (24) and the eigenform- or fixed point equation (25) why we have to realize that there must be a linkage between this ideal number i which we regard as the square root of minus one and the process of a discrete unitary oscillation. Note that the process has no fixed points in the original domains $G_2, \mathbb{Q}$ or $\mathbb{R}$. But if we extend our domain to the complex numbers $\mathbb{C}$, the imaginary unit $i = \sqrt{-1}$ turns out to be a fixed point and thus an eigenform of the oscillatory process $\mathfrak{R}(i) = -\frac{1}{i}$. This is an extraordinary alternative option for us to define the i, since usually we conform to the geometric definition in a Graßmann- or Clifford algebra which says that the imaginary unit i is equivalent to a unit bivector, say, $e_{jk} \stackrel{\text{def}}{=} e_j \wedge e_k$ which is the exterior product of two base units e_j, e_k having positive signature. Some of us regard this as the only appropriate introduction of the i. Thus the i takes the form of a geometric unit, namely an oriented unit area. In what follows we are not using the unit area but the temporal form of the iterant algebra in order to represent the imaginary unit. The idea goes back to Sir William Rowan Hamilton. It has been developed further by Louis Kauffman in a series of papers the most important of which are perhaps his lecture on the occasion of the 100st birthday of Heinz von Foerster (Kauffman 2011) and the IEEE-paper 'Space and Time in Computation, Topology and discrete Physics' [6] (Kauffman 1994) which could be found on his Form pages in January 2013. These can be regarded as indispensable tools for anti-physics. An early paper involving the idea of iterant views is already from the late 1970ies (Kauffman 1978).

Letting the oscillatory process $\mathfrak{R}(i)$ operate on $+1$ iteratively, we generate a 'polarity string' $\ldots +1, -1, +1, -1, +1, -1, \ldots$ applying it recursively on the other non-zero element $-1 \in G_2, \mathbb{Q}, \mathbb{R}$, we get $\ldots -1, +1, -1, +1, -1, +1, \ldots$ Now consider elementary repetitions of such a type open to the left and right. Is there any way to differ between those two? Obviously they represent two different modes of looking at one and the same form. A difference becomes visible only, if we make two copies and juxtapose them in such a way that a difference becomes visible. But then that difference is not only a property of the 'iterant', but it comes upon by a backshift and juxtaposition. It is brought about by relating the polarity string with itself. So the only way to distinguish between them is given by direct confrontation which I call 'touch' and logic comparison

$$\begin{matrix} \ldots +1, -1, +1, -1, +1, -1, \ldots \\ \ldots -1, +1, -1, +1, -1, +1, \ldots \end{matrix} \tag{26}$$

[6] TimeSpace. pdf on the page http://homepages.math.uic.edu/~kauffman/Form.html

Would we apply to (26) a backshift-operator B by half a period, we would obtain equal sequences

$$...+1,-1,+1,-1,+1,-1, \; ... \tag{27}$$

$$...+1,-1,+1,-1,+1,-1, \; ...$$

In accord with Louis Kauffman's procedere, we let $\mathfrak{J}\{+1,-1\}$ stand for an *undisclosed alternation* or ambiguity between $+1$ and -1. The $\mathfrak{J}\{+1,-1\}$ shall be called an 'iterant'. The undecided iterant possesses two iterant views: $[\![+1,-1]\!]$ and $[\![-1,+1]\!]$. It is obvious that the oscillatory process $\mathfrak{R}(i)$ can bring forth such iterant views when applied to real numbers. It can produce oscillations of real numbers, but it will not give us the fixed point as an eigenform of this process. Since this fixed point is in the complex domain. We verify

$$\mathfrak{R}(i) = -1/i = i \tag{28}$$

as the i satisfies $i^2 = -1$.

The iteration over the real numbers must be directly related to the idealized imaginary number i. We therefore introduce an algebraic temporal shift operator η having the properties

$$[\![a,b]\!]\eta = \eta[\![b,a]\!] \text{ and } \eta\,\eta = Id \tag{29}$$

so that concatenated observations are correlated with a time step, here one half period of the process. Now consider the (back)shift-operator B which brings forth the transform

$$B[\![a,b]\!] = [\![b,a]\!] \tag{30}$$

and we take the additive inverse $-[\![a,b]\!] = [\![-a,-b]\!]$, so we get

$$-B[\![+1,-1]\!] = [\![+1,-1]\!] \tag{31}$$
$$-B[\![-1,+1]\!] = [\![-1,+1]\!]$$

that is, the shift-operator has two eigenforms, namely the iterant views $[\![+1,-1]\!]$ and $[\![-1,+1]\!]$. The $-B$ can be interpreted as a combination of reversion and shift or of negation and shift. In his 2011 lecture Kauffman took a shorthand for the shift-operator B by the identity

$$B[\![a,b]\!] = [\![a,b]\!]' = [\![b,a]\!] \tag{32}$$

Just as in a number field, addition and multiplication of iterant views can be defined by

$$k[\![a,b]\!] = [\![ka,kb]\!] \text{ with a number } k \tag{33}$$
$$[\![a,b]\!] + [\![c,d]\!] = [\![a+c,b+d]\!]$$
$$[\![a,b]\!][\![c,d]\!] = [\![ac,bd]\!]$$

A unit can be defined by $Id \stackrel{\text{def}}{=} [\![+1,+1]\!]$. But notice, the eigenforms of $-B$ as are given by equations (31) do not yet give us the square root of -1 since, because of (33), we have $[\![+1,-1]\!][\![+1,-1]\!] = [\![+1,+1]\!] = Id$ and likewise $[\![-1,+1]\!][\![-1,+1]\!] = Id$. However, as soon as we define the imaginary unit as

$$i \stackrel{\text{def}}{=} [\![+1,-1]\!]\eta \tag{34}$$

and its conjugate $-i = [\![-1,+1]\!]\eta$, we have indeed

$$i\,i = [\![+1,-1]\!]\eta[\![+1,-1]\!]\eta = [\![+1,-1]\!][\![-1,+1]\!]\eta\,\eta = [\![-1,-1]\!] = -Id \tag{35}$$

So the square roots of minus one are interpreted as iterant views connected with temporal shift operators.

We say that observation brings upon time. More direct: observation is time. Suppose that observing an iterant $\mathfrak{J}\{a,b\}$ requires one step in time. That being the case will bring on a shift $[\![a,b]\!] \longmapsto [\![b,a]\!]$. Any two iterants X,Y multiplied giving XY are interpreted as "first observation of X, then observation of Y". We postpone the question "who or what is the observer in a primordial quantum field?" But we ascertain that

$$X\eta Y\eta = XY'\eta\eta = XY' \tag{36}$$

Using a shorthand for units $e \stackrel{\text{def}}{=} [\![+1,-1]\!]$ then $e' = [\![-1,+1]\!]$ and $e\,e = Id$, $e'e' = Id$, and $e\,e' = -Id$. Finally we have defined

$$i = e\,\eta \tag{37}$$

which will ultimately allow us to think of complex numbers as observations in discrete processes. This formalism represents a central tool for any further investigation.

The historic construction of the square root of minus one as an algebraic pair in conjugate functions goes back to Sir William Rowan Hamilton's "Essay on Algebra as the Science of Pure Time" (1837). This paper written sometime in 1834 had 3 main parts, the first is an introduction to philosophic aspects and theoretical approaches to algebra, the second is about "Algebra as the Science of Pure Time," The importance of the real numbers is central. Hamilton explains the existence of square roots of positive numbers, nth roots, exponentials, and logarithms. In the third part titled "The Theory of Conjugate Functions, or Algebraic Couples," Hamilton defines complex numbers by ordered pairs of real numbers. In this way the properties of addition, subtraction, multiplication, and division are preserved. Considering the above approach, take the two waves

$...\,a\,p\,.\,.\,a\,p\,.\,.\,a\,p\,...$ and its visual and audible conjugate
$...\,p\,a\,.\,.\,p\,a\,.\,.\,p\,a\,...$ and calculate with the synchronous freeze brackets

$$i[\![a,p]\!] = e'B[\![a,p]\!] = [\![-1,1]\!][\![p,a]\!] = [\![-p,a]\!] \tag{38}$$

from which we can infer that we have

$i\,i\,[\![a,p]\!] = [\![-a,-p]\!] = [\![-1,-1]\!][\![a,p]\!] = (-Id)[\![a,p]\!]$ and so
$$i\,i = -Id \tag{39}$$

On page 107 Hamilton introduced the ordered algebraic pair with the allying sentences: "In the theory of single numbers, the symbol $\sqrt{-1}$ is absurd, and denotes an impossible extraction, or a merely imaginary number; but in the theory of couples, the same symbol $\sqrt{-1}$ is significant, and denotes a possible extraction, or a real couple, namely (as we have just now seen) the principal square-root of the couple $(-1,0)$. In the latter theory, therefore, though not in the former, this sign $\sqrt{-1}$ may properly be employed; and we may write, if we choose, for any couple (a_1, a_2) whatever, $(a_1, a_2) = a_1 + a_2\sqrt{-1}$‹interpreting the symbols a_1 and a_2, in the expression $a_1 + a_2\sqrt{-1}$, as denoting the pure primary couples $(a_1, 0)$ $(a_2, 0)$, according to the law of mixture".

How can we get a wave equation from such consideration? Following the idea we introduce the eigenforms in quantum mechanics as mathematical functions that are invariant under various operators such as differentiation $D = d/dx$. The nexus of primordial time is brought into the discourse as soon as we define the process $\Re(x)$. It follows that i is a fixed point of $\Re(x)$, that is, $\Re(i) = i$ since $i^2 = -1$ is equivalent to demanding that $i = -1/i$. Using the infinite recursion we can write $i = [-1/*]$ pointing out that the imaginary unit is an infinite reentry form for the operator $\Re(x)$. In this light the wave function of quantum mechanics $\psi = e^{[-1/*](kx-\omega t)}$ provides an eigenform for three operations: time-shift and derivation with respect to space and time as impulse and energy.

Let us be aware, the iterant algebra is associative but not commutative. It contains not just the complex, but also hypercomplex numbers. First we show why concatenations generally do not commute. Let $x = [\![a,b]\!]$ and $y = [\![c,d]\!]$. Then calculate the products $(x\eta)(y\eta)$ and $(y\eta)(x\eta)$.

$$(x\eta)(y\eta) = [\![a,b]\!]\eta[\![c,d]\!]\eta = [\![a,b]\!][\![d,c]\!] = [\![ad,bc]\!] \tag{40}$$

$$(y\eta)(x\eta) = [\![c,d]\!]\eta[\![a,b]\!]\eta = [\![c,d]\!][\![b,a]\!] = [\![cb,da]\!] \tag{41}$$

Thus the commutator comes up as

$$[\![x\eta, y\eta]\!] = (x\eta)(y\eta) - (y\eta)(x\eta) = \tag{42}$$

$$[\![ad-bc, -ad+bc]\!] =$$
$$(ad-bc)[\![+1,-1]\!]$$

The time-shift is responsible for the noncommutativity of the algebra of iterative observations:

$$[\![x\eta, y\eta]\!] = (ad-bc)\ i\eta \tag{43}$$

The algebra of iterant views can be generated by elements of the form $[\![a,b]\!] + [\![c,d]\!]\eta$. It is isomorphic with the 2×2 matrix algebra

$$[\![a,b]\!] + [\![c,d]\!]\eta \Leftrightarrow \begin{bmatrix} a & c \\ d & b \end{bmatrix} \tag{44}$$

Two numbers having such a form can be multiplied. With the iterant elements $A = [\![a,b]\!]$, $A' = [\![b,a]\!]$ we get a hypercomplex number system

$$(A + B\eta)(C + D\eta) = (AC + BD') + (AD + BC')\eta \tag{45}$$

having a representation in $Mat(2,\mathbb{R})$. The iterant elements having two components $A + B\eta$ can be regarded as superpositions of two types of observations where only one, namely the $B\eta$ involves the temporal shift. The 'operator-view' moves the iterant view $[\![c,d]\!]$ to the viewpoint $[\![d,c]\!]$ while $[\![a,b]\!]$ is not affected by the shift. We can visualize this by taking horizontal and vertical snapshots of the process and draw a graphic

...$a\ a\ a\ a\ a\ a\ a\ a\ a\ a\ a\ a\ a\ a\ a\ a\ a\ a\ a\ a\ a$...
...$c\ d\ c\ d\ c\ d\ c\ d\ c\ d\ c\ d\ c\ d\ c\ d\ c\ d\ c\ d$...
...$b\ b\ b\ b\ b\ b\ b\ b\ b\ b\ b\ b\ b\ b\ b\ b\ b\ b\ b\ b\ b$...

We could imagine the horizontal oscillation as making a boundary in some synchronous freeze pattern while the vertical iterant parts a and b mark the two sides of that boundary. The wave function $\psi(x,t) = e^{i(kx-\omega t)}$ contains a micro-oscillatory system which is synchronized by the discrete oscillation $i = [\![1,-1]\!]\eta$. It is due to the existence of these synchronizations that the multiple eigenform of the exponential function $\psi(x,t)$ worked so well in reproducing correct eigenvalues for observable quantities, coordinate functions for the matter wave that approximates continuous behavior. According to Moivre's formula the exponential

$$e^{ni\alpha} = (\cos\alpha + i\sin\alpha\,)^n = cos(n\alpha) + i\sin(n\alpha) \tag{46}$$

can be expanded into a power series involving also powers of i. The exponential is synchronized via the imaginary unit to separate out its trigonometric parts. These parts can be written in the iterant form

$$\cos\alpha + i\sin\alpha := [\![\cos\alpha, \cos\alpha]\!] + [\![\sin\alpha, -\sin\alpha]\!]\eta \tag{47}$$

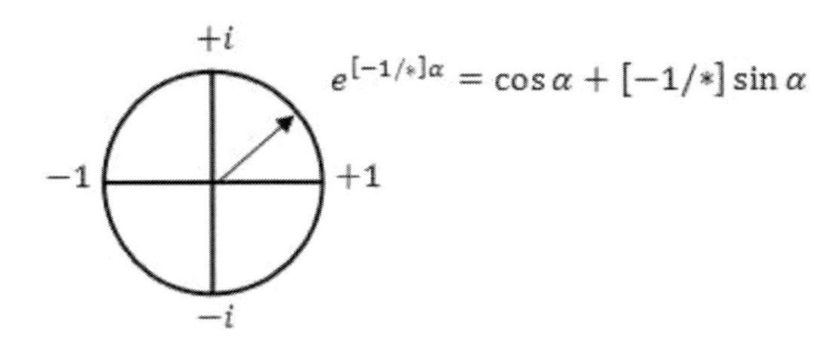

Figure 5. Complex plane as reentry form.

This is a superposition of a constant cosine iterant and the oscillating sine iterant. In this way Euler's formula turns out as the result of a synchronization of iterant processes. We can now look at the familiar picture of a unit vector orbiting in the complex plane in a new way.

We have now built a time-shift by the element $i = [\![+1, -1]\!]\eta$ and we see the wave function oscillating between $\cos(kx - \omega t) + \sin(kx - \omega t)$ and $\cos(kx - \omega t) - \sin(kx - \omega t)$.

THE SCHRÖDINGER EQUATION

Indeed, the function $\mathcal{E}(x) = e^x$ is a prominent eigenform for the operation of differentiation with respect to space-variable x. A plane wave, as we constructed, is a multiple eigenform for several operations. We can extract the energy by temporal differentiation. Reproducing the energy, we should have

$$i\hbar\partial\psi/\partial t = E\psi \text{ with } \mathrm{E} = \hbar\omega \tag{48}$$

The energy of a 'de Broglie wave. Spatial differentiation would return the impulse in terms of the wave number

$$-i\hbar\partial\psi/\partial x = p\psi \text{ with } p = \hbar k \tag{49}$$

Recall '*Licht und Materie*' (de Broglie 1939). This can help to understand the constitution of the Schrödinger equation, which is based on some further functional relationships. Schrödinger identified the measurable dynamic variables E and p with the differential operators $i\hbar\partial/\partial t$ and respectively $-i\hbar\partial/\partial x$. The position operator could be represented by just multiplying wave function with x. All the other physical quantities could be expressed in terms of the basic ones by reproducing their original classical functional relationships. The energy was decomposed into kinetic and potential parts:

$$\mathrm{E} = mv^2/2 + V = p^2/2m + V \tag{50}$$

So Schrödinger identified E with the operator

$$E \overset{\text{def}}{=} -(\hbar^2/2m)\ \partial^2/\partial \mathrm{x}^2 + V \tag{51}$$

With Schrödinger's definition it was possible to construct the wave equation for de Broglie's wave:

$$\frac{\partial}{\partial t}\psi = i\left(\frac{\hbar}{2m}\right)\frac{\partial^2}{\partial x^2}\psi - \left(\frac{i}{\hbar}\right)V(x,t)\,\psi \tag{52}$$

In principle, all these steps could also be performed on stage with the iterant algebra. For suppose we had a discrete temporal grid with mesh size Δt and we could define the velocity (which we cannot when space is not yet created) by the discrete derivative

$$v(t) \stackrel{\text{def}}{=} D\,x(t) = \frac{x(t+\Delta t)-x(t)}{\Delta t} \tag{53}$$

In this new context the time shift could be represented explicitly by the formula

$$x(t)\,J = J\,x(t+\Delta t) \tag{54}$$

Which led to a re-definition of the derivative as

$$\delta\,x(t) = J\big(x(t+\Delta t) - x(t)\big)/\Delta t \tag{55}$$

It is because of this inclusion of the time shift that the observation of position does not commute with the observation of velocity. Since the term $x(\delta x)$ differs from $(\delta x)x$. The rigor goes like this

$$\begin{aligned} & x(\delta x) = x(t)\big(J\big(x(t+\Delta t) - x(t)\big)/\Delta t\big) = \\ & = (J/\Delta t)\big(x(t+\Delta t)\big)\big(x(t+\Delta t) - x(t)\big) = \\ & = (J/\Delta t)\big(x(t+\Delta t)^2 - x(t+\Delta t)x(t)\big) \text{ and likewise} \\ & (\delta x)x = (J/\Delta t)(x(t+\Delta t)x(t) - x(t)^2)\,. \end{aligned} \tag{56}$$

So we obtain for the commutator

$$[x, \delta x] \stackrel{\text{def}}{=} x(\delta x) - (\delta x)x = (J/\Delta t)\big(x(t+\Delta t) - x(t)\big)^2 = J(\Delta x)^2/\Delta t \tag{57}$$

We know that in quantum mechanics this commutator is regarded as constant. But this is the case if the fraction $(\Delta x)^2/\Delta t$ is constant. If the time-step were a definite measurable quantity, we could infer that $(\Delta x)^2 = k\Delta t$ so that the spatial step should satisfy

$$\Delta x = {}_{\pm}\sqrt{k\Delta t} \xrightarrow{yields} x(t+\Delta t) = x(t) \pm \sqrt{k\Delta t} \tag{58}$$

which represents a Brownian process. What we believe to be a temporal wave is constituted by a stochastic process. Positions and moments are coming upon in random events of interaction. Yet, the central problem would still be there, namely that space would be anticipated. Beginning with equations such as (51) or (53) means that we anticipate the existence of a spatial frame which is, so to say, decoupled from the process of creation. The linearity of the approach raises of course some additional questions. The waveform of energy, in material observation, has to be inferred from the stepwise accumulation of this form in a synchronous freeze pattern that seems to be outcome of a nonlinear stochastic motion rather than a linear equation. Space, as we pretend to know, however, is resulting from quantum motion. Is that motion under all circumstances constituting continuous metric space? Or is it arising from some, perhaps more unusual form of continuum? Whatever may be, the iterant approach we are using does not rely on continuous, metric space. So we need not begin with any wave equation. But we can first go one step deeper to where the process of observation begins.

Let us be sure there are some unspecified locations somehow identifiable by polarity strings that pop off before our almost featureless, void background. These are thought to come about in a process of self-referent observation of polarity strings. Namely, we pose the question who or what observes any entity in a primordial domain? Is it possible that the iterants are both observer and observed? You can regard the iterant $e = [\![+1,-1]\!]$ as representing a local sequence of isospins, as a spin-chain ... ↑↓↑↓↑↓ ⋯ or anything similar. Who or what observes? To answer this, let us realize that observation, the cognitive action of comparison and creation of time are one. We articulate this by creating the time shift operator η. Observation introduces a time-shift, a separation between observer and observed, a comparison between two elements. We can now translate the following statement into a simple mathematical formula: »The iterant $e = [\![+1,-1]\!]$ in observation separates from the other iterant $e' = [\![-1,+1]\!]$ thereby reproducing unity.«

$$[\![+1,-1]\!]\eta[\![-1,+1]\!]\eta = [\![+1,-1]\!][\![+1,-1]\!]\eta\eta = Id \tag{59}$$

A polarity string $[\![+1,-1]\!]$ observing its dual iterant view $[\![-1,+1]\!]$ sees unity. There is some mystery of self-reference hidden in this algebraic conception of observation. To illustrate this let us consider a free Schrödinger field $S := \cos\alpha + i\sin\alpha = [\cos\alpha, \cos\alpha] + [\sin\alpha, -\sin\alpha]\eta$ defined as in equation (47). We can investigate what it means, if the free field S becomes engaged in a relation of self-reference and observes itself as observer on the one hand and as observed on the other. We just follow the recipe "first observation of X, then observation of Y" as introduced in the equation (36). Let us see S observing itself as an iterant view. As an iterant S has the abstract form

$$S = [\![a,a]\!] + [\![b,-b]\!]\eta = A + B\eta \text{ with } a = \cos\alpha,\, b = \sin\alpha \tag{60}$$

'First observe S then observe S yields

$$\begin{aligned} &S\eta S\eta = (A + B\eta)\eta(A + B\eta)\eta = (A + B\eta)(A'\eta + B')\eta = \\ &(AA'\eta + AB' + BA + BB\eta)\eta = (AA + BB) + (AB + AB')\eta = \text{ as } AA' = AA \\ &= [\![a^2 + b^2, a^2 + b^2]\!] + ([\![ab,-ab]\!] + [\![-ab,ab]\!])\eta = \\ &[\![a^2 + b^2, a^2 + b^2]\!] = [\![\sin^2\alpha + \cos^2\alpha, \sin^2\alpha + \cos^2\alpha]\!] = [\![+1,+1]\!] = Id \end{aligned} \tag{61}$$

So we find out, the free field observing itself as observer and observed creates unity while the observer η that brings forth time, is diminished. Observer and observed are one. The field reproduces unity. Interestingly the free Schrödinger field in self referent observation does the same as the iterant view $[\![+1,-1]\!]$ in observing its dual. It annihilates time-shift and recreates unity.

Step by step we shall go further and ponder over four dynamic elements, quad locations. These are somewhat correlated with the intuitive concept of quaternions as '4-locations' or 4-character alternating series. We shall let $\mathfrak{J}\{a,b,c,d\}$ stand for an undisclosed alternation or ambiguity between a, b, c and d. The $\mathfrak{J}\{a,b,c,d\}$ then shall be called a '4-fold iterant'. The undecided iterant possesses 24 iterant views: $[\![a,b,c,d]\!]$, $[\![b,a,c,d]\!]$ and so forth, in accordance with the symmetric group S_4. These fourfold polarity strings are supposed to make up locations. It is unimportant where in space, in which frame with what orientation these locations pop off. They are not points. That should be clear. But what are they? In equation (58) we came upon the spatial step $x(t+\Delta t) = x(t) \pm \sqrt{k\Delta t}$, an interesting quantity which nevertheless anticipates the existence of a measurable space with some definite orientation. Can we do otherwise? Can we begin with a set of locations? For nature does not know about space. It just brings it about. Nature has no concept. It creates the concept.

Considering any location in a set of locations somewhere beyond metric space. How can a string of alternating polarity come upon? We can think of the exponential map as a creation operator. With the reentry form $i = [-1/*]$ together with Euler's formula, the map

$e^{[\![-1/*]\!]\varphi}$ gives a constant iteration of unity $+1,+1,+1,\ldots$ at the phase $\varphi = 0$, but brings forth the oscillation $[\![-1/*]\!] \Rightarrow -1,+1,-1,+1,-1,+1,\ldots$ for $\varphi = \pi/2$

Events of the kind $-1,+1,-1,+1,-1,+1,\ldots$ can be imagined to emerge anywhere. Wherever and whenever such alternating iterants occur, they contribute to the constitution of locations. Locations are interacting with each other. There is a global exchange that contributes to the emergence of locations. In the sequel, sets of locations, as appear in our protocols, are denoted by A, B, C and so on. Let A, B be sets. We conceive X as a set of sets of locations and a σ-algebra over the set X as a nonempty collection Σ of subsets of X, that is $\Sigma \subseteq \wp(X)$. Σ is closed under the complements and countable unions of members and contains X. We consider the triple (X, Σ, μ) as a 'field of location-sets' and as a measureable space with measure μ. We want to introduce a metric distance on Σ by the aid of the symmetric difference of sets.

The set

$$A \vartriangle B \overset{\text{def}}{=} (A \setminus B) \cup (B \setminus A) \tag{62}$$

Is called the 'symmetric difference' between A and B. It is equal to the union of the sets A, B less their intersection

$$A \vartriangle B \overset{\text{def}}{=} (A \cup B) \setminus (B \cap A) \tag{63}$$

Such sets of locations satisfy the following equations

$$A \vartriangle A = 0 \tag{64}$$

$$A \vartriangle \emptyset = A$$

$$A \vartriangle B = B \vartriangle A \qquad \text{commutativity}$$

$$A \vartriangle (B \vartriangle C) = (A \vartriangle B) \vartriangle C \qquad \text{associativity}$$

$$A \vartriangle C = (A \vartriangle B) \vartriangle (B \vartriangle C)$$

$$A \vartriangle C \subset (A \vartriangle B) \cup (B \vartriangle C)$$

Proofs: Identities (i) to (iv) are following from the definition (62), (v) is proven

$$(A \vartriangle B) \cup (B \vartriangle C) \overset{(iv)}{\equiv} \big((A \vartriangle B) \vartriangle B\big) \vartriangle C \overset{(iv)}{\equiv} \big(A \vartriangle (B \vartriangle B)\big) \vartriangle C \overset{(i)}{\equiv}$$
$$(A \vartriangle \emptyset) \vartriangle C \overset{(ii)}{\equiv} A \vartriangle C$$

Relation (vi) follows directly from (v) by application of the equivalent definition (63) of the symmetric difference. For we have $(A \triangle B) \cup (B \triangle C) \supset \big((A \triangle B) \cup (B \triangle C) \big) \setminus \big((A \triangle B) \cap (B \triangle C) \big)$. We shall have to rely on these precious formulas, as we do not want to presuppose space-time, but we rather want to show how it is constituted so to say as a property of the free iterant field. We shall establish the problem together with the main ideas that lead to solutions by the aid of the free Schrödinger iterant field as discussed in the above approach to discrete quantum mechanics. We show how the self-referent field creates an orientation for two sets of iterants which surround a location where we conjecture creation of quanta. We denote this as a location to which we can ascribe, after some considerable investigation, a measure that leads to a metric in a most simple way.

Creation of Space in the Schrödinger Field

We begin with the free Schrödinger iterant field and the related stochastic process of Brownian motion. Let us once more clarify some fundamental geometric matters that concern orientation only. In 'Space and Time in Computation and Discrete Physics' there is made use of the graphic representation of iterants as a 'two dimensional waveform' (Kauffman 1994)

$$\begin{array}{l} \ldots a\,b\,a\,b\,a\,b \ldots \\ \qquad \ldots c\,d\,c\,d\,c\,d \ldots \\ \qquad \ldots a\,b\,a\,b\,a\,b \ldots \\ \qquad \ldots c\,d\,c\,d\,c\,d \ldots \\ \qquad \ldots a\,b\,a\,b\,a\,b \ldots \\ \qquad \ldots c\,d\,c\,d\,c\,d \ldots \end{array} \tag{65}$$

Here it is occasionally spoken of a 'freeze pattern' (2011a, p. 39)[7]. In some writings where the basic symmetries of the representation matrices of such waveforms is discovered, Kauffman (2012, section 4, p. 17 and 1998, p. 26) alludes that a given matrix 'freezes out' a way to view the infinite waveform [8]. This refers to the synchronous nature of a score that portrays the living, diachronic process. In this latter paper having the title 'Non-Commutative Worlds and Classical Constraints' the four fundamental symmetries that we can associate with the synchronous wave pattern (65) are pinned down:

$$1)\ \begin{pmatrix} a & b \\ c & d \end{pmatrix} \quad 2)\ \begin{pmatrix} b & a \\ d & c \end{pmatrix} \quad 3)\ \begin{pmatrix} c & d \\ a & b \end{pmatrix} \quad 4)\ \begin{pmatrix} d & c \\ b & a \end{pmatrix}$$

If we apply a spatial shift operator to the second row and respectively all even rows of the pattern having form $\ldots c\,d\,c\,d\,c\,d \ldots$ we obtain a slightly transformed synchronous pattern

[7] The handout "Eigenforms and Eigenvalues – Cybernetics and Physics" had 53 pages and contained a comprehensive chapter on "Iterants, Complex Numbers and Quantum Mechanics" where the term 'freeze pattern' was used. In a later much shorter publication titled Eigenforms and Quantum Physics (*Cybernetics and Human Knowing*, 2011) this chapter was omitted. But the 'freezing' survived in some more publications two of which are cited here.

[8] arXiv:1109.1085 [math-ph], viewed on 26th January 2013.

$$\begin{array}{l} \ldots a\,b\,a\,b\,a\,b \ldots \\ \ldots d\,c\,d\,c\,d\,c \ldots \\ \ldots a\,b\,a\,b\,a\,b \ldots \\ \ldots d\,c\,d\,c\,d\,c \ldots \\ \ldots a\,b\,a\,b\,a\,b \ldots \\ \ldots d\,c\,d\,c\,d\,c \ldots \end{array} \tag{66}$$

This gives us four further matrices

5) $\begin{pmatrix} a & b \\ d & c \end{pmatrix}$ 6) $\begin{pmatrix} b & a \\ c & d \end{pmatrix}$ 7) $\begin{pmatrix} d & c \\ a & b \end{pmatrix}$ 8) $\begin{pmatrix} c & d \\ b & a \end{pmatrix}$

These eight elementary arrangements of quarters represent the symmetries of the dihedral group D_{2d} for two diagonals, that is, $d = 2$. The 8 permutations of order 4 correspond with the spatial congruence group of the square D_4. This small group plays a very important role where the discrete structure of Clifford algebras is concerned (Shaw 1995). Namely, if we consider for a moment the Clifford algebra $Cl_{1,1}$ generated by the Minkowski space $\mathbb{R}^{1,1}$ with neutral signature $(+,-)$, the base units e_1, e_2 generate the dihedral group having the 8-elements $\{\pm Id, \pm e_1, \pm e_2, \pm e_{12}\}$. In our planar pattern we can imagine four locations, or quadrants, and all their possible permutations. Those are the 24 symmetries of the symmetric group S_4. If we consider only those symmetries which preserve the neighborhood relations of the four locations, that is to say, connectivity of quarters, we obtain D_4 with only one third (=24/3) of the elements. The S_4 is isomorphic with the point group O of the octahedral group, without total space involution. The S_4 can also be viewed as the rotational symmetry or spatial congruence of a cube in Euclidean 3-space, or as the automorphism group of a 'Dreibein'. What is important is that such a cube or Dreibein admits 3 planes each of which has a dihedral symmetry D_4. In group theory we say that D_4 is a subgroup of the symmetric group S_4 with index 3. Therefore our 'iterant pattern' has 8 constitutive iterant elements. The whole affair involving 4 quadrants can be seen in two different ways. This because we have two isomorphic algebras of order 2, namely $Cl_{2,0} \simeq Cl_{1,1}$. The first way leads us to the interpretation of the imaginary unit i in Clifford algebra as an exterior product of two spatial unit vectors. Such a product is called a bivector and has form $e_{ij} = e_i \wedge e_j$ with the base units satisfying $e_i^2 = Id, e_j^2 = Id$ and the anti-commutation relations $\{e_i, e_j\} = e_i e_j + e_j e_i = 0$. The bivector e_{ij} is representing a hypercomplex number satisfying the equation $e_{ij}^2 = e_i e_j e_i e_j = -e_j e_i e_i e_j = -e_j Id\, e_j = -Id\, e_j e_j = -Id$. Therefore the bivector $e_{ij} \stackrel{\text{def}}{=} e_i e_j$ has the potential to represent the imaginary unit. However, geometrically, in this picture, it represents an oriented unit area. We may ask which of the eight matrices, 1) to 8), could stand for the i? To see this, we just have to verify that the i is a period-4 element, that is, $i^4 = Id$. Identify the first matrix with the identity permutation $\mathbb{I} = (1)(2)(3)(4)$ in cycle notation. Then the 7th matrix is represented by the 4-cycle $(1\,4\,2\,3)$ and its inverse is the 8th matrix which represents the 4-cycle $(1\,3\,2\,4)$. The $(1\,4\,2\,3)$ is corresponding with i or respectively some bivector e_{ij}. This is in $Cl_{2,0}$ in positive signature $\{+,+\}$. It would be the second unit in neutral signature $\{+,-\}$, that is, $Cl_{1,1}$. In that case the second unit vector takes in the role of the imaginary and the bivector now had positive signature, that is $e_{12}^2 = Id$ in $Cl_{1,1}$. The 8th matrix is corresponding with its inverse $-i$ or respectively e_{ji} or $-e_2$ in our case.[9] These facts signify the basic kinship between the iterant algebra and Clifford algebra. As the Clifford algebras $Cl_{3,1} \simeq Cl_{2,2}$ have multivector-groups (Shaw 1995) isomorphic with a product $D_4 \circ D_4$, we could ascertain from the begin that the method of the iterant algebra can also be applied to the Clifford algebra of the Minkowski space and respectively to the matrix algebra $Mat(4, \mathbb{R})$. But let us begin humbly and investigate orientation.

The small group D_4 is the *reorientation group* of the plane. When nature is insecure whether the oriented unit e_1 is indeed the e_1 and not perhaps the e_2, if she has not decided upon the definite orientation of the oriented area e_{12}, the action of the reorientation-group becomes important. Now, let us forget about Clifford algebra for a while, as the iterant algebra provides us a bit higher complexity and richer phenomenology. Let us consider the general iterant relevant for this most basic investigation. It has the form $[\![a, b]\!] + [\![c, d]\!]\eta$ in correspondence with the synchronous pattern (65). The iterant view $[\![c, d]\!]\eta$ is combined with the shift. It drives the process. We shall call it a *propagating* iterant. The other, the $[\![a, b]\!]$, does not contain this shift. It will be denoted 'silent' or 'synchronous', it represents the 'silent part' of the iterant. In the sequel we need both the silent and the propagating. Recall what we said about the universe. The Universe is constructed in such a way that it can refer to itself. In so doing, the Universe must divide itself into a part that refers and a part to which it refers, a part that sees and a part that is seen. Let us say that R is the part that refers and U is the referent. The divided universe is RX and $RX = U$ and RX refers to U (itself). Our solution suggests that the Universe divides itself into two identical parts each of which refers to the universe as a whole. This is RR. In other words, the universe can pretend that it is two and then let itself refer to the two, and find that it has in the process referred only to the one, that is itself. The most elementary self-reference leads to a division into two identical parts that nevertheless appear as different in their relation, namely

$$\ldots + - + - + - \ldots \text{ and}$$
$$\ldots - + - + - + \ldots$$

The universe may refer to itself in a location. We do not always know, if this self-reference is only of a local nature or if it has a global appearance. Now consider a light-source, a laser emitting linearly polarized coherent light and a screen. We are used to locate these objects in our laboratory, that is, within a given frame. Now, forget about the laboratory. Let there be just two locations in nature. The laboratory assistants have fallen senseless. They lost their minds. We are all on vacation. There is just nature with a source and a receiver, just two locations. Does nature know anything about those locations? Let me guess that the following might be the case:

[9] This may be a little bit confusing at first, but we just have to realize that the Clifford algebra of the real plane $Cl_{2,0}$ is isomorphic with the Clifford algebra $Cl_{1,1}$ in neutral signature, both isomorphic with matrix algebra $Mat(2, \mathbb{R})$.

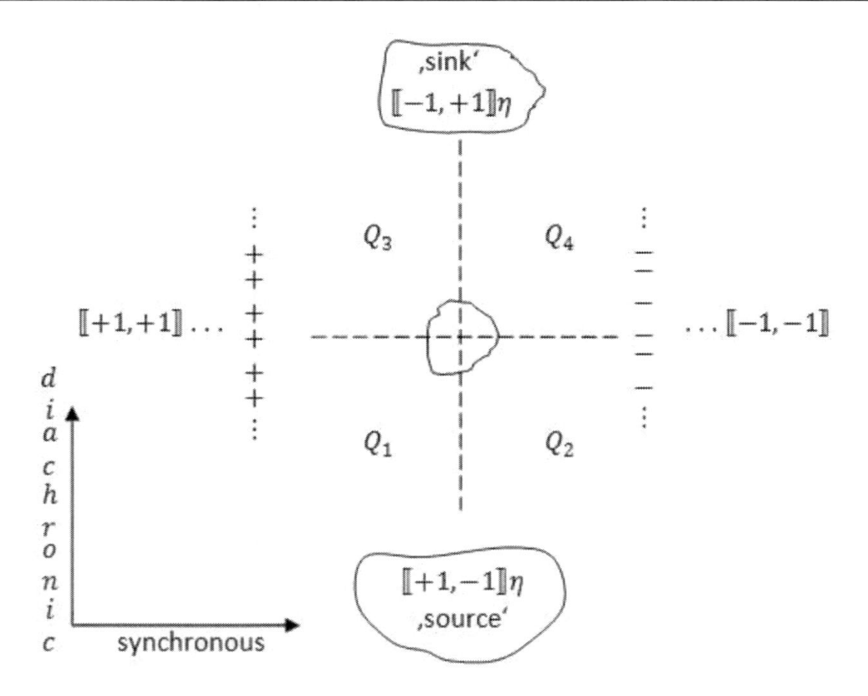

Figure 6. Iterant plane orientation of the Schrödinger equation.

There is an awareness in nature that something emits photons from a source. *We* intend to or *believe to* emit quanta which something or someone is supposed to receive at the location called 'sink'. I accentuate "believe to", because he or she believes that the photon is emitted at the source. This 'fact' is not sure! The field may come from elsewhere. - Now the two locations 'sink' and 'source' begin to speak with each other. There is a readiness to make a step and to transgress the horizontal line in the middle that indicates that sink and source are separate locations. The field transcends this separation by propagating from the unit iterant $[\![+1,-1]\!]\eta$ to the remote sink iterant $[\![-1,+1]\!]\eta$. How is that path made possible at all? It is not self evident. But the translocal communication between source and sink challenges a silent process of self-reference that gives a seeming orientation to the field. The silent iterant views actually represent the same thing, a non-alternating constant code. Seen in their relation they represent the ambiguity between left constant $[\![+1,+1]\!]$ and right constant $[\![-1,-1]\!]$ code. They can be seen as observation of some translocal undisclosed alternation $\{+1,-1\}$. The silent iterant views on the left and on the right are separated from each other by a spatial shift and they limit the space of events. Actually they are like the polarities in an iterant $[\![+1,-1]\!]$ taken apart and separated from each other by 'space'. Let us say that the 'silent' constant iterant views $[\![+1,+1]\!]$ and $[\![-1,-1]\!]$ span the synchronous dimension, whereas $[\![+1,-1]\!]\eta$ and $[\![-1,+1]\!]\eta$ are said to span the diachronic dimension. So the object of our investigation is a plane area with translocal, oriented left-right- and source-sink relations in a free Schrödinger field. We can now think about four further locations, four quarters signified by four iterants that are brought forth by the elementary iterants of the figure 6, factoring in some normalization constant $1/\sqrt{2}$, namely

(i) $Q_1 = \frac{1}{\sqrt{2}}[\![+1,+1]\!] + \frac{1}{\sqrt{2}}[\![+1,-1]\!]\eta$ (67)

(ii) $Q_2 = \frac{1}{\sqrt{2}}[\![-1,-1]\!] + \frac{1}{\sqrt{2}}[\![+1,-1]\!]\eta$

(iii) $Q_3 = \frac{1}{\sqrt{2}}[\![+1,+1]\!] + \frac{1}{\sqrt{2}}[\![-1,+1]\!]\eta$

(iv) $Q_4 = \frac{1}{\sqrt{2}}[\![-1,-1]\!] + \frac{1}{\sqrt{2}}[\![-1,+1]\!]\eta$

These correspond with synchronous freeze-patterns

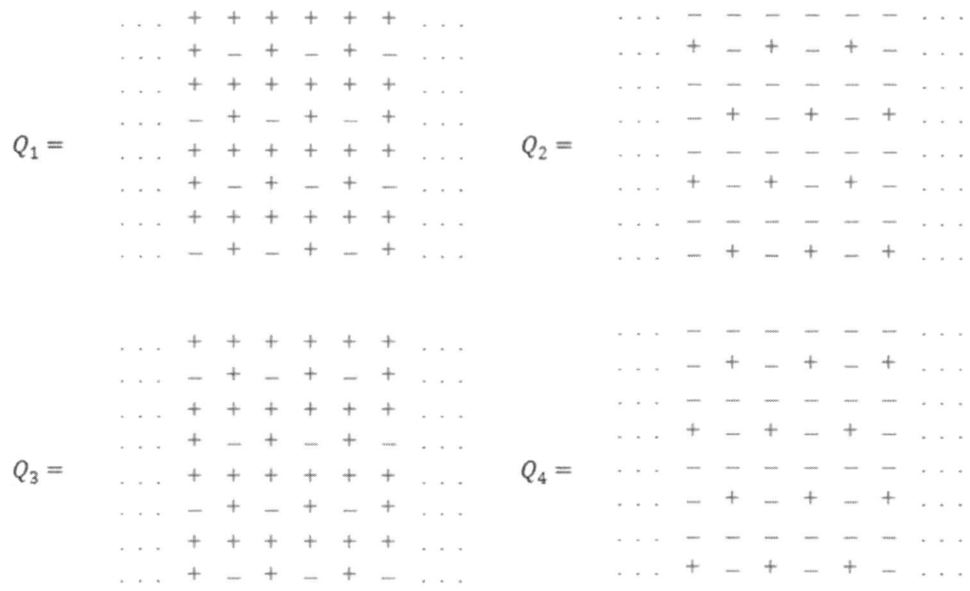

$$Q_1 = \begin{matrix} \cdots & + & + & + & + & + & + & \cdots \\ \cdots & + & - & + & - & + & - & \cdots \\ \cdots & + & + & + & + & + & + & \cdots \\ \cdots & - & + & - & + & - & + & \cdots \\ \cdots & + & + & + & + & + & + & \cdots \\ \cdots & + & - & + & - & + & - & \cdots \\ \cdots & + & + & + & + & + & + & \cdots \\ \cdots & - & + & - & + & - & + & \cdots \end{matrix} \qquad Q_2 = \begin{matrix} \cdots & - & - & - & - & - & - & \cdots \\ \cdots & + & - & + & - & + & - & \cdots \\ \cdots & - & - & - & - & - & - & \cdots \\ \cdots & - & + & - & + & - & + & \cdots \\ \cdots & - & - & - & - & - & - & \cdots \\ \cdots & + & - & + & - & + & - & \cdots \\ \cdots & - & - & - & - & - & - & \cdots \\ \cdots & - & + & - & + & - & + & \cdots \end{matrix}$$

$$Q_3 = \begin{matrix} \cdots & + & + & + & + & + & + & \cdots \\ \cdots & - & + & - & + & - & + & \cdots \\ \cdots & + & + & + & + & + & + & \cdots \\ \cdots & + & - & + & - & + & - & \cdots \\ \cdots & + & + & + & + & + & + & \cdots \\ \cdots & - & + & - & + & - & + & \cdots \\ \cdots & + & + & + & + & + & + & \cdots \\ \cdots & + & - & + & - & + & - & \cdots \end{matrix} \qquad Q_4 = \begin{matrix} \cdots & - & - & - & - & - & - & \cdots \\ \cdots & - & + & - & + & - & + & \cdots \\ \cdots & - & - & - & - & - & - & \cdots \\ \cdots & + & - & + & - & + & - & \cdots \\ \cdots & - & - & - & - & - & - & \cdots \\ \cdots & - & + & - & + & - & + & \cdots \\ \cdots & - & - & - & - & - & - & \cdots \\ \cdots & + & - & + & - & + & - & \cdots \end{matrix}$$

Figure 7. Fourfold freeze pattern of a Schrödinger wave.

and yield the equations

$$Q_1 + Q_2 + Q_3 + Q_4 = 0, \tag{68}$$
$$Q_1Q_3 = Q_2Q_4 = +Id \quad Q_1Q_2 = Q_3Q_4 = -Id$$

Characteristic for such fourfold locations: the sum of iterants is zero. That is also the case for the generating iterants in figure 6. Iterant Q_3 is the conjugate of Q_1 and Q_4 is conjugate to Q_2.

Performing in greater detail, we are aware of Schrödinger plane waves as iterants

$$S_1(\varphi) = [\![\cos(\varphi), \cos(\varphi)]\!] + [\![\sin\varphi, -\sin\varphi]\!]\eta \qquad \text{for phases } \varphi = kx - \omega t \tag{69}$$
$$S_2(\varphi) = [\![-\cos(\varphi), -\cos(\varphi)]\!] + [\![\sin\varphi, -\sin\varphi]\!]\eta$$
$$S_3(\varphi) = [\![\cos(\varphi), \cos(\varphi)]\!] + [\![-\sin\varphi, \sin\varphi]\!]\eta$$
$$S_4(\varphi) = [\![-\cos(\varphi), -\cos(\varphi)]\!] + [\![-\sin\varphi, \sin\varphi]\!]\eta$$

These are iterants in a complex plane and so their relation to the geometry of locations is not immediately visible. As we start the process in some event space of locations with no metric and orientation, we must first ask, how nature is constructing orientation. Because such orientation is basic for a Brownian process of motion as given by equation (58)

$$x(t+\Delta t) = x(t) \pm \sqrt{k\Delta t}$$

We have to think about a figure like that

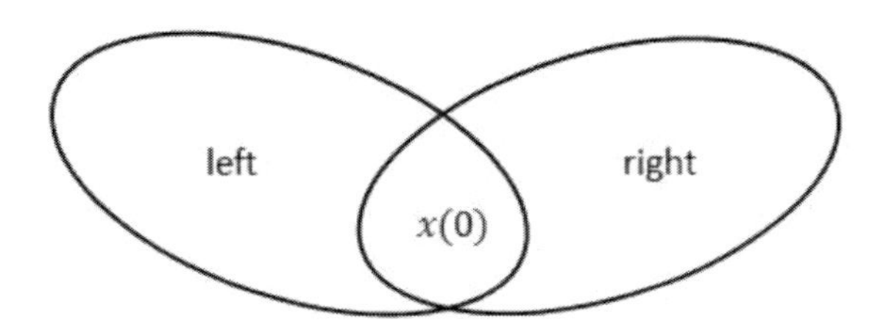

Figure 8. Orientation in a Wiener process.

If the process which is responsible for the appearance of quanta has an internal intelligence capable to differ between right hand and left hand side, there could be some further possibility to measure the symmetric difference between the two sets, one of which counting quanta coming in on the left and the other counting quanta on the right hand side. How shall we proceed? We shall let kx turn over into $-kx$. This makes $\cos kx$ go over into $\cos(-kx) = \cos kx$ and turns $\sin kx$ into $\sin(-kx) = -\sin kx$. In the sequel we also use the well-known trigonometric formulas

$$\sin(\alpha \pm \beta) = \sin\alpha\cos\beta \pm \cos\alpha\sin\beta \qquad (70)$$
$$\cos(\alpha \pm \beta) = \cos\alpha\cos\beta \mp \sin\alpha\sin\beta$$

So we are able to investigate a parity flip of the iterant $S_1(\varphi)$. We obtain a transition as follows

$$kx \to -kx \implies \qquad (71)$$
$$S_1(\varphi) \to [\![\cos(kx+\omega t), \cos(kx+\omega t)]\!] + [\![-\sin(kx+\omega t), \sin(kx+\omega t)]\!]\eta$$

Perhaps it is possible to show the relationship between the geometry of a Schrödinger wave and the iterant picture more clearly. That would be purposive if we wish to relate parity- and time inversion. Apart from the absolute value of the amplitude, the Schrödinger wave has the general iterant form

$$\psi(x,t) \simeq ([\![a,a]\!] + [\![b,-b]\!]\eta)([\![c,c]\!] + [\![-d,d]\!]\eta) = \qquad (72)$$
$$[\![ac,ac]\!] + [\![-ad,ad]\!]\eta + [\![bc,-bc]\!]\eta + [\![b,-b]\!][\![d,-d]\!]\eta\eta =$$
$$[\![ac+bd, ac+bd]\!] + [\![bc-ad, ad-bc]\!]\eta$$

with $a := \cos kx$, $b := \sin kx$, $c := \cos \omega t$, $d := \sin \omega t$. Formula (72) is an iterant decomposition of the free Schrödinger wave with respect to space and time. Now we can easily find out about the geometric meaning of the four quadrants, the 'quad-locations' of a free Schrödinger field. We get

$$\begin{aligned}
&\text{(i) } x = 0, t = 0 \quad \Rightarrow \psi(x,t) \simeq [\![+1,+1]\!] \\
&\text{(ii) } x = \tfrac{\pi}{2k}, t = -\tfrac{T}{4} \quad \Rightarrow \psi(x,t) \simeq [\![-1,-1]\!] \quad \text{with cycle duration } T = 2\pi/\omega \\
&\text{(iii) } x = \tfrac{\pi}{2k}, t = 0 \quad \Rightarrow \psi(x,t) \simeq [\![+1,-1]\!]\eta \quad \text{with time shift } \eta \\
&\text{(iv) } x = 0, t = \tfrac{T}{4} \quad \Rightarrow \psi(x,t) \simeq [\![-1,+1]\!]\eta
\end{aligned} \tag{73}$$

These fix the directions and the 'corner stones' for figure 6.

Let us investigate a parity flip. It turns b into $-b$ and thus

$$\text{Parity reversion } \mathfrak{P}: \quad \psi(x,t) \to [\![ac - bd, ac - bd]\!] + [\![-ad - bc, ad + bc]\!]\eta$$

which is the same as the $S_1(\varphi)$ in formula (71) as it should be in accordance with the trigonometric identities (70). What is the result of a time reversal? In that case we transform $d \to -d$ and so

$$\begin{aligned}
&\text{Time reversal } \mathfrak{T}: \quad \psi(x,t) \to [\![ac - bd, ac - bd]\!] + [\![bc + ad, -ad - bc]\!]\eta \\
&\text{A } \mathfrak{PT} \text{ reversion}: \quad \psi(x,t) \to [\![ac + bd, ac + bd]\!] + [\![ad - bc, -ad + bc]\!]\eta
\end{aligned}$$

which yields in explicit term

$$\psi(-x,-t) \simeq [\![\cos(kx - \omega t), \cos(kx - \omega t)]\!] + [\![-\sin(kx - \omega t), \sin(kx - \omega t)]\!]\eta$$

If we establish a correspondence between the free Schrödinger wave $S_1 = \psi(x,t)$ and the first quadrant and unit iterant $Q_1 = \frac{1}{\sqrt{2}}[\![+1,+1]\!] + \frac{1}{\sqrt{2}}[\![+1,-1]\!]\eta$ then $\psi(-x,-t)$ must be correlated with the third, that is, $Q_3 = \frac{1}{\sqrt{2}}[\![+1,+1]\!] + \frac{1}{\sqrt{2}}[\![-1,+1]\!]\eta$.

To complete the picture, let us now raise the amplitude of the Schrödinger wave to its maximum value by letting t increase to half the cycle duration

$$\begin{aligned}
&t = 0 \longrightarrow t = T/2 \quad \text{with} \quad T = 2\pi/\omega \quad \text{which carries} \\
&\cos kx \ \text{ to } \cos(kx + \pi) = -\cos kx \ \text{ and } \sin kx \ \text{ to } -\sin kx
\end{aligned} \tag{74}$$

This denotes a turnover from iterant $S_1 = [\![\cos kx, \cos kx]\!] + [\![\sin kx, -\sin kx]\!]\eta$ at time $t = 0$ to $S_4 = [\![-\cos kx, -\cos kx]\!] + [\![-\sin kx, +\sin kx]\!]\eta$ which is characteristic for a turnover from quadrant

$$Q_1 = \frac{1}{\sqrt{2}}[\![+1,+1]\!] + \frac{1}{\sqrt{2}}[\![+1,-1]\!]\eta \quad \text{to} \quad Q_4 = \frac{1}{\sqrt{2}}[\![-1,-1]\!] + \frac{1}{\sqrt{2}}[\![-1,+1]\!]\eta \tag{75}$$

Now we can ask how nature differs between left and right, between past and future, if she does at all. Is there any simple procedure that makes this partition evident? We should suppose that the reverse of the process, that makes the difference, should diminish it. Is it possible to restore the unity from the difference between left and right? When self reference leads to the iterants $e = [\![1,-1]\!]$ and $e' = [\![-1,1]\!]$ the observation

$$e\eta e'\eta = [\![1,-1]\!]e\eta\eta = [\![1,-1]\!][\![1,-1]\!] = Id \tag{76}$$

restores unity. First observe $e = [\![1,-1]\!]$ then observe the back-shifted iterant $e' = [\![-1,1]\!]$ and put them together. This gives unity. The same can be done by unifying the iterant Q_1 with its parity inverted Q_3. We obtain

$$Q_1\mathfrak{P}Q_1 = Q_1Q_3 = \left(\tfrac{1}{\sqrt{2}}[\![+1,+1]\!] + \tfrac{1}{\sqrt{2}}[\![+1,-1]\!]\eta\right)\left(\tfrac{1}{\sqrt{2}}[\![+1,+1]\!] + \tfrac{1}{\sqrt{2}}[\![-1,+1]\!]\eta\right) = \tag{77}$$
$$\tfrac{1}{2}([\![+1,+1]\!] + [\![-1,+1]\!]\eta + [\![+1,-1]\!]\eta + [\![+1,+1]\!]) = Id$$

Does the process of nature have a means to test that fundamental difference and then restore the unity? Can it count quanta that come in on the right and such that come in on the left hand side? How does it restore the memory of these events? We think about the symmetric difference in terms of a Brownian process. But nature just realizes the measure, - perhaps even without any counting.

Let us go back to Brownian motion. There is a mutual awareness of quanta that pop off on the left and on the right hand side in figure 8. In accordance with measure theory we denote the (non-empty) set of locations by Ω, and $\mathcal{P}(\Omega) = \{A \subset \Omega\}$ is the power set of Ω, the set of all subsets. Note that we do not yet have a measure for locations such as X. But we can bring in a definition of σ-algebra and a probability measure. We define as usual:

A system of sets $\mathcal{A} \subset \mathcal{P}(\Omega)$ is a σ-algebra in Ω if

(i) $\Omega \in \mathcal{A}$ (78)
(ii)
(iii) $A \in \mathcal{A} \Rightarrow A^c \in \mathcal{A}$ with complement $A^c = \Omega \backslash A$
(iv) $A_1, A_2, \ldots \in \mathcal{A} \Rightarrow \bigcup_{n=1}^{\infty} A_n \in \mathcal{A}$

Consider $\mathcal{G} \subset \mathcal{P}(\Omega)$, then we call

$$\sigma(\mathcal{G}) := \bigcap_{\mathcal{G} \subset \mathcal{A}} \mathcal{A} \tag{79}$$

the σ-algebra generated by $\mathcal{G}$.

Now that we have a σ-algebra of locations we define a set function

(i) $\mu\colon \mathcal{A} \longrightarrow [0,\infty]$ (80)
(ii) $\mu(\emptyset) = 0$
(iii) μ is σ-additive

that is, for every series $(A_i)_{i\in\mathbb{N}}$ of pairwise disjoint sets we must have
$\mu(\bigcup_{i\in\mathbb{N}} A_i) = \sum_{i\in\mathbb{N}} \mu(A_i)$. (81)

The measure μ is a probability measure if it satisfies $\mu(\Omega) = 1$. We denote probability measures by $\mathbb{P}$. The pair $(\Omega, \mathcal{A})$ is now a 'measurable space'. The triple $(\Omega, \mathcal{A}, \mu)$ is called a measure space. If the μ is a probability measure $\mathbb{P}$, we call Ω a sample space of locations and our subsets of locations turn into 'events'. The advantage of this sight is that we can realize space as a dynamic *system of events* rather than as a fixed *frame for events*.

A system of sets $\mathcal{A} \subset \mathcal{P}(\Omega)$ is a semi-ring over Ω if (82)
(i) $\emptyset \in \mathcal{A}$
(ii) $\cap$-stability
meaning $A, B \in \mathcal{A} \rightarrow A \cap B \in \mathcal{A}$, that is, stability with respect to intersection
(iii) $A, B \in \mathcal{A} \rightarrow$ there are pairwise disjoint sets $C_1, C_2, \dots, C_n \in \mathcal{A}$ such that
$A \setminus B = C_1 \cup C_2 \cup \dots \cup C_n$

Let $\mathcal{A}, \mathcal{B}$ semi-rings $\mathcal{A} \subset \mathcal{B}$ and μ: $\mathcal{B} \longrightarrow [0, \infty]$ a set function on $\mathcal{B}$ which is supposed finite on $\mathcal{A}$; $\mu(A) < \infty$ for every $A \in \mathcal{A}$. Consider the symmetric difference as defined in (62) or (63). Then we can rely on the approximation theorem for measures:

Be $\mathcal{A} \in \mathcal{P}(\Omega)$ a semi-ring and μ a finite measure on $\sigma(\mathcal{A})$. Then for $L \in \sigma(\mathcal{A})$ with $\mu(L) < \infty$ and $\varepsilon > 0$ there is some $n \in \mathbb{N}$ with pairwise disjoint sets $A_1, \dots, A_n \in \mathcal{A}$ such that the measure satisfies

$$\mu(L \vartriangle \cup_{k=1}^{n} A_k) < \varepsilon \tag{83}$$

What have we done? We described, in a classical way, not using categories or more complicated things, how we can obtain a metric measure for a location L from sequences of finite sets. This is actually what we observe when we watch the slow constitution of a location within an interference pattern of some single or double-slit experiment (Figure 9). Let us recall the formula (58):

$$x(t + \Delta t) = x(t) \pm \sqrt{k\Delta t}$$

It tells us, the longer we wait, the larger the deviation $\pm \sqrt{k\Delta t}$ from the $x(t)$. But we need not have a Euclidean measure for $x(t)$ in assumed space. We just have to be aware of the light flashes detected to the left and to the right. The lightning appears both on the left and right hand side 'on the screen'. We count the flashes of light in the sets, call them 'left', 'zero', 'right', or L_-, L_ε, L_+. Proceeding thus, we have avoided to presume Euclidean or Minkowski space. We can consider L_ε as a center of investigation, our central location, a 'blob'. Taking μ to be the counting measure, we find out: the longer we count, the larger the distance between L_+ and L_- as more quanta appear outside L_ε after some time interval $\Delta t \geq \varepsilon^2/k$. The question remains, if and how nature makes the bookkeeping. The distance may saturate or it may not. If it equilibrates at some stable measure, it gives us a ruler. We get a video clip like that:

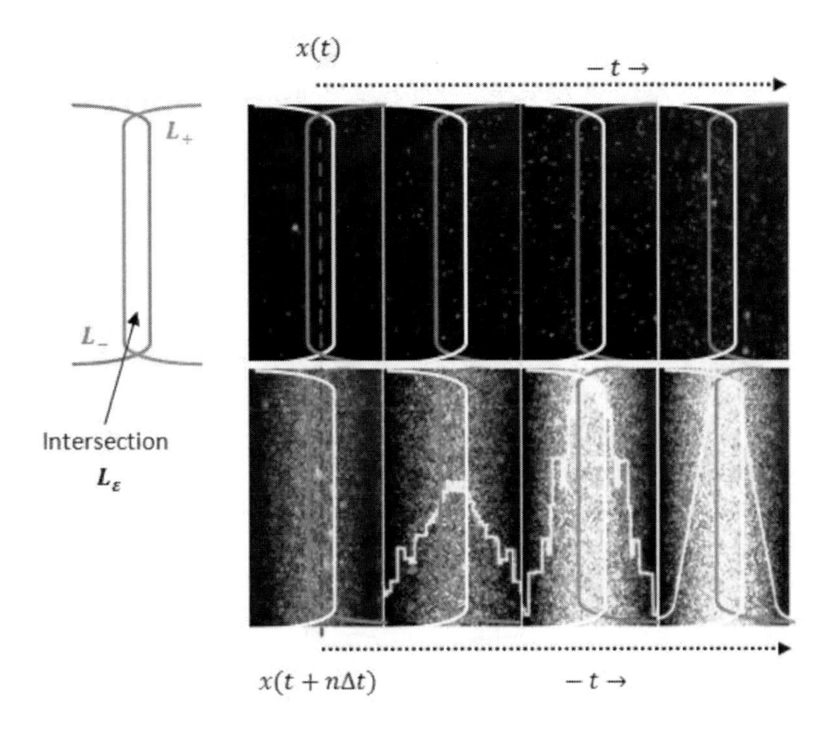

Figure 9. Constitution of a location by sets in Young's double slit experiment.

In quantum reality the appearance of light flashes depends on many events. At first it seems evident that it must depend on the number of particles emitted at the source. This is partly true. On the other hand it can be thought as being influenced by the counting measures of those local sets, that is, the energy density. A certain amount of energy must accumulate before a quantum pops off. This is something rarely mentioned. We must also consider more global interactions that can lead to a lightning. And there is something that should be questioned at any case: the assertion that the particle which left a laser or a cathode, can be identified with the particle that hits the screen. Though there is a significant correlation between the events of emitted particle and received particle, the claim of identity is a misstatement. There is no trajectory between source and sink, unless we make one.

Assuming that the process of nature has the capability to constitute the measure of $x(t+\Delta t)$ all by itself we should believe that it has a sensor. Its structure should provide some kind of sensing device for the counting measure μ. What could that be? A well known traditional sensing device is involved wherever and whenever an event happens in a reaction to pressure. If we consider some limited volume between source and screen where electrons, atoms or molecules are located as quanta, the density of these quanta gives the pressure. The pressure is somehow determined by the number of collisions of particles and thus depends on a measure of velocity and therefore temperature. The 'sensor' is a matter of thermodynamics. We have certain imaginations such as that photons do not interact, but electrons do, and so atoms and molecules. These images are partly legitimate, yet somewhat vague. Considering fields, where quanta cannot yet be counted, on the one hand, particles as countable sets on the other hand, and relations of various degrees between those sets, we know that the action of forces we are observing depends on relations rather than sets. Specific types of relations are responsible for the appearance of peculiar forces. Some of the abstract features of these important relations have been described and formalized in a chapter on the 'Dark Van der Waals Gas' in the first volume of 'Primordial Space' (2010). The van der Waals gas is one special element within a topological equivalence class. The universal mathematical features of this element have been investigated by Robert Kiehn (2004, part II, chapter 2). They are deformation invariants for all physical systems that admit a 4D representation as a space-time variety. But a drawback in these contemplations, again, is in the presupposition of a 4D. So in some way there must be found non metric, discrete counterparts that can be correlated with such fundamental space-time varieties. Before we are at all able to make first moves in the right direction, we should investigate the most fundamental phenomena and processes, namely the propagation and electrodynamic interactions of photons which, as we were convinced, do not interact with each other. Their field involves two rather unsettling phenomena of (a) delayed choice, so called quantum erasure, and (b) a strong causality disconfirmation which I use as a word instead of 'violation'. It is the first phenomenon that is already experimentally settled, - while the second is still waiting for verification, - and about this first we want to report first. Understanding 'delayed choice' can promote a deeper intuitive understanding of the fact why Feynman's very first attempts towards path integration made ends with an action-at-a-distance formulation of quantum electrodynamics. The assumption of a Schrödinger equation of motion was wholly sufficient.

Chapter 3

PHENOMENOLOGY OF IMMEDIACY

DETECTING PHOTONS

The photons are wrapped in fine woolen environments of infinite path integrals, always well hidden from our eyes. They disclose their secrets in indirect way, never disentangle from their field in any obvious manner, always showing us one of two sides. When we say that this or that photon is emitted at a titanium-doped sapphire, that it goes through a slit, passes some optical crystal and hits a screen, then this is a nice fantasy. It is not real. The photon does its best to avoid being observed and counted, and when it first comes in and hits the beta-barium borate non-linear crystal, it is destroyed. There may appear a pair of entangled photons after frequency doubling. At least we hope so, since we even baptize these new created particles. We call them 'signal' and 'idler'. But never does the photon pay us its respect and appear directly in front of our eyes. But it prefers to emerge within some active area of a typical photosensitive environment, some so called large area of InGaAs APD (avalanche photodiode) which has enough photo sensitivity so that it can set free an avalanche of charge carriers. The only practical way to detect single photons is facilitated by cooled APDs. These are operated in Geiger mode. The reverse voltage on APD is kept just below the breakdown voltage level, mostly 20-40 V, most of the time. When we anticipate arrival of a photon, for a short period, the voltage is risen above breakdown threshold. It is guaranteed that under normal conditions, no avalanche occurs, but if a photon comes in and is absorbed in the junction area of the APD, it generates an electron-hole pair. This can cause an avalanche. The electric current through the APD quickly rises to macroscopic values of, say, 100 to 1000 μA, and we can observe a pulse at the output of the amplifier. Then it is necessary to lower the voltage on the APD below breakdown, to stop the avalanche. Then the output pulse ends. This is what happens in practice. Apparently the physical process, - the real life of the photon, - is a most complicated narrative of ourselves which contains insecure events and shakiness. The InGaAs is expected to have a low dark current and high-speed response, at a cut-off frequency of 1 GHz. We hardly ever measure the photon, but rather the peculiar response of the specific electronics to some photon field.

Delayed Choice and Orientation

We want to understand how physical events take form in the presence of observation. Therefore we shall describe the measurement stations of some typical delayed choice (dc) experiments. We try to disentangle, so to say, their photonic and electronic environment, and concentrate on the photonic results which we believe to be reliable. The design of any dc-experiment is in accord with some basic assumptions of mathematical physics, either explicitly or implicitly, and it modifies these. The most fundamental idea of QED can of course be identified with its peculiar phenomenology and with the correlated mathematical method of path integrals (Feynman and Hibbs 1965, 2005). But the semantic reality of our prevailing belief-system is rather contradictory. We speak of 'which way'- and 'which path'-information, but in reality there seems to be a field, partly known, partly unknown, onto which we project our idea of an infinity of possible pathways. This tags the origin of Feynman's theory. In the old view of quantum mechanics the observation of a particle excludes the observation of a wave. The wave was thought to collapse when the quantum was created. To obtain a bright spot on a photographic plate or a blob on the screen, needs a lot of collapse. - In our new plans we are now confronted with a variety of dissenting notions.

The experimental method of spontaneous parametric down conversion (SPDC) allows us to produce entangled photons by annihilating a single one with double frequency. Observation of the entangled photons has made it possible to perceive both particle-like and wave-like behavior. The 'which way' and respectively 'both way'-information of our quantum can be marked or erased by the aid of the entangled twins even after the detection. Seen within the laboratory frame this advises the observer to be aware of even more than just some virtual causality violation. The event signifies the *presence of a time reversion*, a causality violation in the laboratory as we see it in the presence of a photon field. Photons need not go from 'here' to 'there', they just vanish 'here' and appear 'there'. They just appear. They do not re-appear. There is no need for a particle trajectory, and the 'space' between two appearances is not necessarily metric, not Poincaré. That is the essential meaning of annihilation and creation of quanta. Yet, it is obvious from the lab perspective that photons do not just appear anywhere. But they appear in some locations where their creation is made probable by, say, the preferred orientation of a laser beam and an optical lens or prism.

Our habitual knowledge dictates that the Heisenberg uncertainty relation would make it impossible to determine the pathway – which slit the photon went through – without destroying the interference pattern. However, new procedures of *weak measurement* and *delayed choice* have proven by the use of new technology that there are physical environments in which the assertion of complementarity may not be true. In the last century many physicists sensed the uncertainty relation as an obstacle and searched for experiments to circumvent it. Following a thought experiment of Wheeler, the idea of a quantum eraser had been proposed by Scully and Drühl in 1982. Later corresponding assemblies were put into realization by various researchers. Most of these consisted of some optoelectronic measurement-device with three peculiar subsystems. The first can simply be called the 'source', the second a 'quantum information processing system' – where a pair of entangled photons is manipulated by optical components – and the third a registry in the form of an electronic scanner, a detector device and a counter. Meanwhile there have been built thousands of such instruments. The first ones were rather unpretentious. The source consisted of an argon pump laser, a double slit screen with a beta-barium borate crystal (BBO), and two Glan-Thompson prisms that were somewhat misused. That is, they were used less for polarization control, but more for deflection in order to get two regions A, B for a pair of $702{,}2\ nm$ orthogonally polarized signal-idler photons created either from region A or B. The reason for this slight alienation is due to the fact that the photon pairs created by BBO leave the crystal with an opening angle of only 6 degrees. This source assembly is somewhat redundant, as the job of splitting the field is done twice: first by the double slit, second by the crystal. Thereby the emissions from regions A, B are randomized, and then we can muck around with the photons.

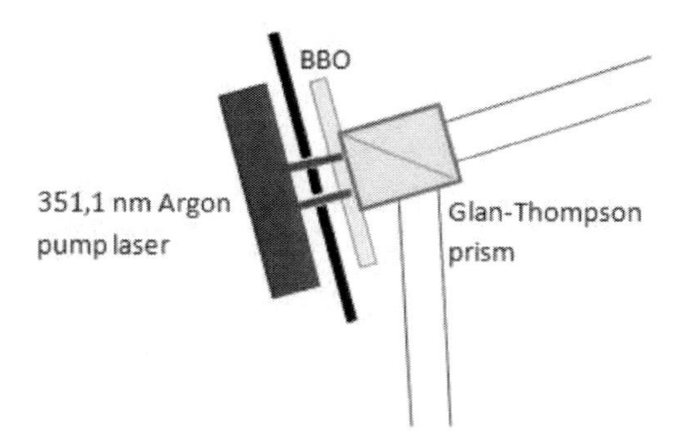

Figure 10. Photon source in a delayed choice quantum eraser.

By the time these sources of photon-twins were improved in various respects. In some later experiments made by Hannes René Böhm (2003) at the Universities of Vienna, an elaborate assembly capable of polarization control was developed. Böhm's light conducting components are endemic right at the interface between quantum physics and computer science. In his small SPDC sources, operated at $810{,}4\ nm$, before photons are coupled into a single-mode optical fiber (SMF) by the coupling lens, they pass an optional polarizer and an orange glass filter which blocks scattered UV light. To compensate arbitrary polarization rotation within the fiber, a polarization control module is connected to the end of the SMF. There are used so-called *fiber loop compensators* also known as "Bat-Ears". They consist of approximately one meter bare fiber which is wound to form three loops. By rotating the loops relative to each other, one can control the type of rotation caused by the fiber. So this tiny equipment allows for rather accurate measurements. However, in the 80s our experimental assemblies were still a little massy and unstable. Yet, the apparent paradox of delayed choice could already be seen, as the coincidence counter was already quick enough to compensate the small 8 ns reverted time-interval.

Quantum Eraser

In the *delayed choice set-up* reported by Kim, Yu, Kulik, Shih and Scully (1999) special importance was attached to the space-time orientation of the experiment such that it allowed for observation of a communication between signal- and idler-photons over intervals which according to Special Relativity are purely space-like and therefore causally separated. There exists a central counter D_0 for the signal field which provides an automatic monitoring of the whole process. This maneuverable counter is operating in a far field condition relative to the two emitting locations in the BBO. That is, there is a lens producing a collimated light beam which lets photons from the two slits appear as if they had travelled a long distance. Since there is no other device involved and no information about which way any photon might have travelled from locations A, B to detector D_0 we expect that a scanning of D_0 along the axis x will bring forth an interference-pattern. Such is indeed the case. Technical details are as usual in such experiments. The scanning is accomplished by a computer controlled step-motor. Impulses are counted at a fixed location within a given time interval of about 100 to 1000 seconds, before the counter is moved ahead by $\Delta x = 0{,}1$ mm or more. Thus it is assured that, as we use to believe, any 'single photon interacts with itself' before another one enters the device. The detector is essentially given by an avalanche photodiode and some electronics.

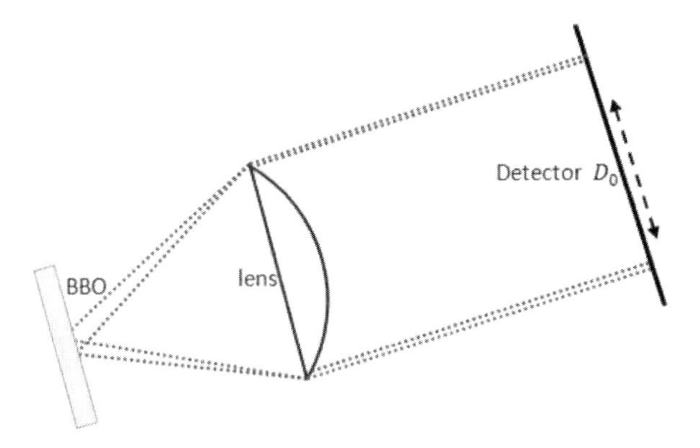

Figure 11. Signal scanning detector.

Scanning x in direction, say, e_1 of the apparent macro-space-time algebra gives us the information about position. Counting photons at that position provides a measure for the energy in accordance with the rules of first quantization with its number operator. In that region of the detector device nothing, - neither the observer, nor the lens, nor the photodiode, - will distinguish between a photon from region A and a photon from region B. Therefore we use to say that there is no 'which way' information. I am here conforming to this denotation, although it should be doubted that any photon travels on any path at all. There is rather an instant self-organization of energy and creation of quanta depending on primordial actualities of locations and environments. Seen from the macroscopic viewpoint of the observer this givenness, those primordial actualities of locations, represent what we subsume under the word geometry. So the geometry of detector D_0 opens the door to observe interference and to supervise what is processed in the other part of the device, which is perceived as a disjoint part, that is, in the interferometer. This contains four further detectors as shown in figure 12.
Idler photons from slits A, B are sent to a prism which ensures that fields from A and B disembark towards detectors D_3, D_4 at beamsplitters BS_A and BS_B, while with equal probability, we say, the photons will pervade the splitters and continue traveling to mirrors M_A and M_B. If a photon is detected at D_3, we know that it comes from location A. If it is

detected by D_4 it is from B. It is due to our observation that the prism secures, pathways A and B can be distinguished and identified, that we make a quantum-informatic statement: The detection at components D_3, D_4 provide 'which path' information, that is, identification of path A or path B with reference to the idler photon. It should be noted that the different denotation of photons as signal- and idler-photons is not essential. It is just a convention which makes a difference between the spatial directions in which they are sent off and between the ways we are processing them. Registration of coincidence between the controlling detector D_0 and D_3 discloses the joint detection rate R_{03} having the shape of a bell and being suspected to indicate the particle-like nature of the photons. We may be aware that the form of this bell-shaped distribution of quanta popping off is indicating *no less a wave* than the curve recovered by the interference pattern.

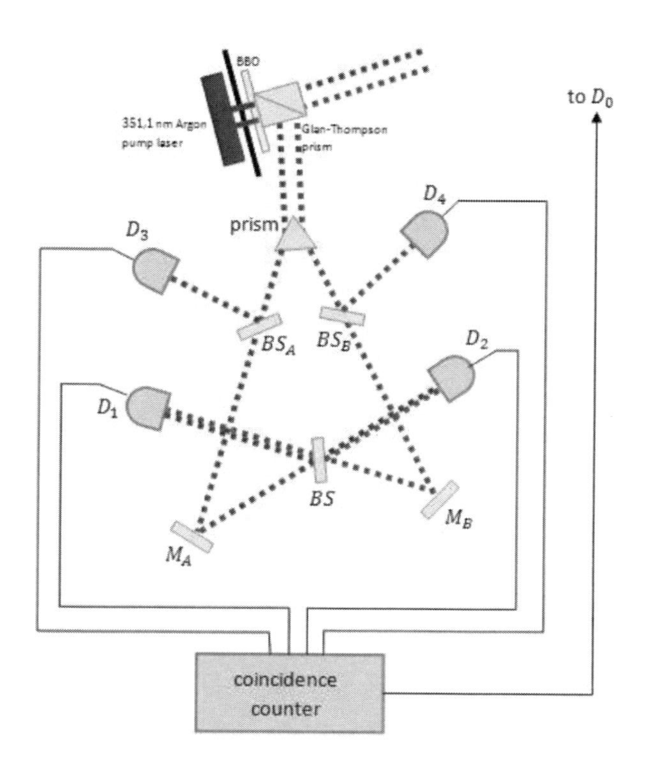

Figure 12. A Delayed Choice Quantum Eraser.

Which path information is correlated with reflection either at BS_A or BS_B. If the idler field is not reflected it will stride ahead to the mirrors M_A, M_B and be reflected at a 50:50 beamsplitter BS which erases the which path information (figure 14). The component BS represents the quantum eraser in the proper sense.

Every author of such descriptions and explanations and every interpreter of such wonderful experiments is following his or her own imaginations and thus changes words according to his or her own viewpoint. For example when I have written "that the form of this bell-shaped distribution of quanta popping off, as in figure 13, is indicating *no less a wave* than the curve recovered by the interference pattern", I have in mind the image of a *wave-particle duality* which became some kind of mental fixation for physicists of my generation.

It has been coined by Niels Bohr in 1927 (Bohr 1928) who described and legitimated the quantum-mechanic fact of 'complementarity' by the particular features of 'wave-like' and 'particle-like' events. So, lately, in correlation with a few colleagues I have developed negative attitudes towards this idea and some criticism. We always register the quanta, whether they belong to a 'no which way info' interference pattern or to a 'which way info' bell. Both *bell-like* and *interference like* pictures can be interpreted as material imprints of waves. What matters is the way we analyze the difference between wave and bell. As I am convinced of the cognitive potential and value of each scientists own way of thinking, I do not want to mystify the original thoughts of the discoverers of the delayed choice quantum eraser. I strongly advise readers to study this original paper (Kim, Yu, Kulik, Shih 2000, pp. 1,2)[1] and cite here how the authors see things themselves:

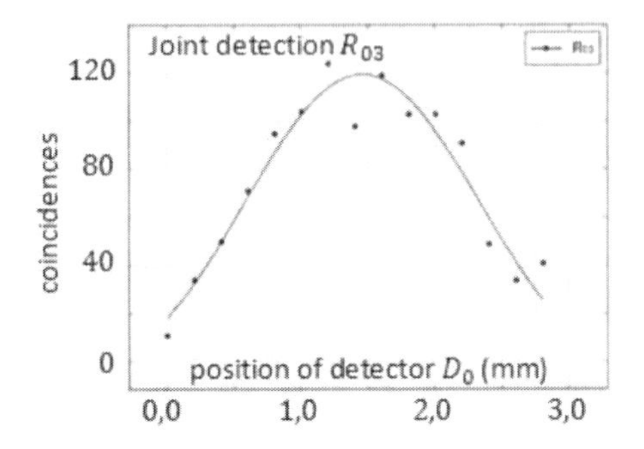

Figure 13. Blob building up at detector D_0 in coincidence with D_3[2].

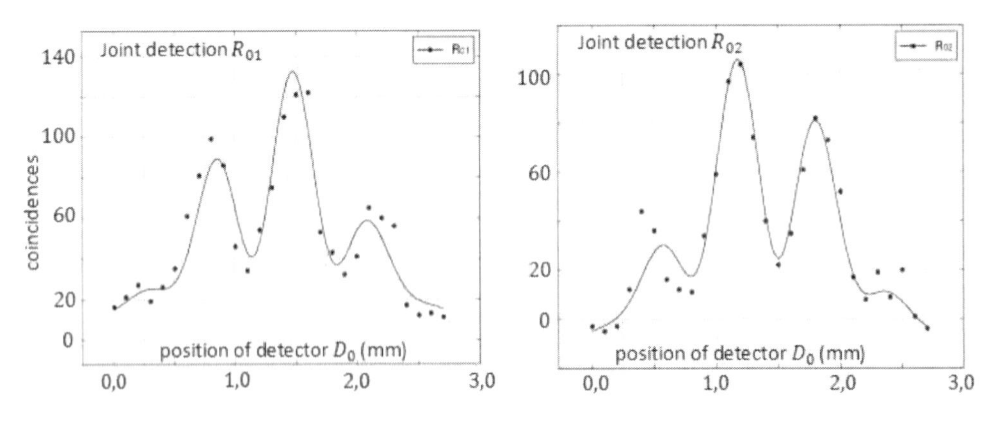

Figure 14. Interference pattern building up at detector D_0 in coincidence with D_1, D_2 after erasure.

"If the pair is generated in atom A, photon 2 will follow the A path meeting BS_A with 50% chance of being reflected or transmitted. If the pair is generated in atom B, photon 2 will follow the B path meeting BS_B with 50% chance of being reflected or transmitted. Under the

[1] Phys Rev Lett 84 1-5 (2000), and http://xxx.lanl.gov/pdf/quant-ph/9903047. (Figures 3-5; Copyright (2000) by the American Physical Society)

[2] The plots of count rates simultaneous at detectors D_0 and D_j are taken from Excerpts with »commentary by Ross Rhodes from 'A Delayed Choice Quantum Eraser'« at the page http://www.bottomlayer.com/ bottom/kim-scully/kim-scully-web.htm.

50% chance of being transmitted by either BS_Aor BS_B, photon 2 is detected by either detector D_3 or D_4. The registration of D_3 or D_4 provides which-path information (path A or path B) of photon 2 and in turn provides which-path information of photon 1 because of the entanglement nature of the two-photon state of atomic cascade decay. Given a reflection at either BS_Aor BS_Bphoton 2 will continue to follow its A path or B path to meet another 50-50 beamsplitter $BS$and then be detected by either detector D_1 or D_2, which are placed at the output ports of the beamsplitter BS. The triggering of detectors D_1 or D_2 erases the which-path information. […] It is easy to see these "joint detection" events must have resulted from the same photon pair. It was predicted that the "joint detection" counting rate R_{01} (joint detection rate between D_0 and D_1) and R_{02} will show interference pattern when detector D_0 is scanned along its x-axis. This reflects the wave property (both-path) of photon 1. However, no interference will be observed in the "joint detection" counting rate R_{03} and R_{04} when detector D_0 is scanned along its x-axis. This is clearly expected because we now have indicated the particle property (which-path) of photon 1. It is important to emphasize that all four "joint detection" rates R_{01}, R_{02}, R_{03}, and R_{04} are recorded at the same time during one scanning of D_0 along its y-axis. That is, in the present experiment we 'see' both wave (interference) and which-path (particle-like) with the same apparatus. […] The electronic output pulses of detectors D_1, D_2, D_3, and D_4 are sent to coincidence circuits with the output pulse of detector D_0, respectively, for the counting of "joint detection" rates R_{01}, R_{02}, R_{03}, and R_{04}. In this experiment the optical delay $(L_i - L_0)$ is chosen to be $\simeq$ 2.5 m, where L_0 is the optical distance between the output surface of BBO and detector D_0, and $L_i$is the optical distance between the output surface of the BBO and detectors D_1, D_2, D_3, and D_4, respectively. This means that any information one can learn from photon 2 must be at least 8 ns later than what one has learned from the registration of photon 1. Compared to the 1 ns response time of the detectors, 2.5 $m$delay is good enough for a 'delayed erasure'."

Parity and Spatial Symmetry Breakage

There are many lessons that can be learned from the quantum eraser. But we have to be aware of our habits and see the meaning of words we are using, if we want to look beyond the edge of the plate. Let us be humble and say what can be said and find out what can be decided and what not. That is, we shall try to proceed in the manner of Kurt Gödel or Alan Turing or Heinz von Foerster: logically and constructively. We use to say that a photon passing the double slit interacts with itself and constitutes an interference pattern on some screen or alongside the path of some detector (D_0) and that thereby it shows us its wave-like nature. Actually we perceive or record or make visible by some device the accidental popping off of some particles. We register a one- or two-dimensional random walk of events. That walk, a Brownian motion, is at present defended by some of us, as purely accidental, a matter of hazard, a stochastic process, in any case, and at all events, and for some others that view may have represented the worst case. Please, beware that the writer is not making you any suggestion! We just go together and ask, what is the matter. Anton Zeilinger conversing with the Dalai Lama explained that this view of the existence of purely accidental events might be an alternative to the Buddhist principle of *pratityasamutpada* according to which every result or outcome is the result of many causes and conditions.[3] Let us first point out that Einstein's statement "God doesn't throw dice" ("Der Alte würfelt nicht") has lost its constraining meaning since we know that *deterministic* equations can generate *stochastic* processes. There seems to be no contradiction between determinedness and chance. Now, what is it that we see on the screen behind a double slit? We see the image of a stochastic process. While this process of particle creation goes on, it seems to constitute the shape of a wave. But that wave does not proceed in a temporally ordered manner, but it emerges in a chaotic way. Therefore some are convinced that there is a chaotic process with an orderly mathematical shape behind these events. There are people who have gone deeply into nonlinear field theory. Are we really ready to just neglect what they say? At least we confess that the photon interacts with itself. But that is a nonlinear process, and nobody has ever seen that photon because it is a field in that state of self-interaction. In the chapter "*Wave vs. Particle Properties of Matter*" in »*Quantum Theory*« David Bohm (1989) has also considered "the idea that an electron is neither a particle, nor a wave, but is instead a third kind of object which has some, but not all, of the properties of both particles and waves". (p. 117f.) In this chapter, Bohm has still described the "*Impossibility of Simultaneous Observation of Wave and Particle Properties of Matter*". But he has also pointed at the '*Importance of Phase relations*' when he analyzed the changing phase relations among the various components of a wave packet. "Note that the center of the packet occurs at the point where a wide range of $\phi(k)$ tend to interfere constructively, whereas some distance away they tend to cancel because of destructive interference. Since each $\phi(k)$ oscillates as $\exp -i\hbar k^2 l/2m$, the resulting changes of phase of the $\phi(k)$ with time change the position of constructive and destructive interference and, therefore, govern the motion of the wave packet. The classical equations of motion are thus contained in the phase relations among the different $\phi(k)$." David Bohm came very close to a view which has been revived by the discovery of entanglement, when he stated: "It seems necessary, therefore, to give up the idea that the world can correctly be analyzed into distinct parts, and to replace it with the assumption that the entire universe is basically a single, indivisible unit." When we are today confronted with a title like »Everything is entangled« (Buniy and Hsu 2012) saying in the abstract "we show that big bang cosmology implies a high degree of entanglement of particles in the universe. [...] As a consequence, small subsystems are mostly entangled with particles far beyond the horizon ..." then we encounter this old imagination of a universal whole.- So, let us say what can be said and find out what can be decided and what not.

What can we learn from the quantum eraser? What can be decided? We can learn that under given conditions simultaneous observation of wave and particle properties of matter is possible. However we must distinguish between signal-photon and idler at runtime, whilst the physics tells us they are the same. We differ between the interference pattern indicating a wave-like constitution and a bell-distribution indicating a particle-like constitution. Does this indicate that the equation of motion is linear? No, it does not. Does it exclude that there is a chaotic process of particle creation? No, it does not. Does it violate special relativity? It seems that it does so, but it does not. If the interference indicates a wave, why should the bell indicate no wave? We say that in the first case, without which path information, the photon

[3] The 'principle of dependent origination' in Buddhism and the 'Four Noble Truths' have in common the law of cause and effect, of action and consequence. This could be thought in physics to be violated on a reduced basic level of the natural processes.

interacts with itself and thereby builds up the interference pattern. Does that include that in the other case, having which path information, the photon does not interact with itself? No, it does not. Why should the photon in one case interact with itself and in the other not interact? Just because of some feature of wave-theory? What else can we learn? What about symmetry? If there is no path info, is there orientation?

The rigor leading to the coincidence counting rates is based on the traditional Hilbert-space image, and is meanwhile well known. We assume that in the quantum eraser there are four wavefunctions $\Psi(t_0, t_j)$, corresponding to the joint-detection-measurements at detectors 1 to 4, having the following different forms (Kim, Yu, Kulik, Shih and Scully 1999, eq. 4, 5) of linear superpositions: Yu, Kulik, Shih

$$\text{(i)}\ \Psi(t_0, \text{t}1) = A(t_0, t_1^A) + A(t_0, t_1^B), \tag{84}$$
$$\text{(ii)}\ \Psi(t_0, t2) = A(t_0, t_2^A) - A(t_0, t_2^B),$$
$$\text{(iii)}\ \Psi(t_0, t3) = A(t_0, t_3^A),$$
$$\text{(iv)}\ \Psi(t_0, t4) = A(t_0, t_4^B)$$

It is derived that the two amplitudes in $\Psi(t_0, t_1)$ and $\Psi(t_0, t_2)$ are indistinguishable because of the overlap in both $t_0 - t_j$ and respectively $t_0 + t_j$, so that interference is expected in both the coincidence counting rates, R_{01} and R_{02}; however, with aphase shift of π due to the different sign in 84 (i) and (ii),

$$\text{(i)}\ R_{01} \propto \cos^2(x\pi d/\lambda f)\text{‹and} \tag{85}$$
$$\text{(ii)}\ R_{02} \propto \sin^2(x\pi d/\lambda f)\triangleright$$

Assuming slits A and B both are not infinitely narrow, but have finite width, carrying out an integral to sum all possible amplitudes along slit A and slit B, gives a standard interference-diffraction pattern for R_{01} and R_{02},

$$\text{(i)}\ R_{01} \propto \text{sinc}^2(x\pi a/\lambda f)\cos^2(x\pi d/\lambda f), \tag{86}$$
$$\text{(ii)}\ R_{02} \propto \text{sinc}^2(x\pi a/\lambda f)\sin^2(x\pi d/\lambda f),$$

where a is the width of slits A and B, d is the distance between the center of slit A and B, $\lambda = \lambda_s = \lambda_i$ is the wavelength of both the signal and idler, and f is the focal length of lens LS.

In one case where the photons have *no which way information*, the field, nevertheless, knows orientation, as it makes a difference between left and right: the $\cos^2(x\pi d/\lambda f)$ turns over into $\sin^2(x\pi d/\lambda f)$. We can interpret this as a reaction of the field to the blotting out of the which path information. When we annihilate the which way information, we have the space at the light source symmetric, two open slits, and the field reacts by making it asymmetric:

$$x \longrightarrow -x \Longrightarrow R_{01} \neq R_{02} \text{ preserved} \tag{87}$$

as we have $\sin(-\varphi) = -\sin\varphi$ & $\cos(-\varphi) = \cos\varphi$ and so $\sin^2(-\varphi) = \sin^2\varphi$, $\cos^2(-\varphi) = \cos^2\varphi$. This is not a minor negligible fact, but a significant property inherent in the process of nature, that by its very action it establishes a difference between left and right introducing the π-phase shift.

What happens in the other case when the photons do *have which way information*? The width of the emitting region on the BBO is about $0.3\ mm$and the distance between the center of A and B is about $0.7\ mm$. We observe the constitution of two blobs mapping the distance of $0.7\ mm$ between the slits. Figure 13 reports a R_{03}-joint detection counting rate between D_0 and determined path to D_3, against the $x$coordinates of detector D_0. Thus an absence of interference is indicated. The authors are reporting no significant differences between the distributions of R_{03} and R_{04} except the small shift of the center representing the distance between the center of A and B. Recall that the triggering of detectors D_3 and D_4 provide which-path information of both the idler photon 2 and the signal photon 1. Therefore, providing which way info we introduce a symmetry breakage at the light source. We have the space at the light source asymmetric, and the field reacts by making it symmetric by producing two equivalent bell-distributions (blobs) for both slits. The field has a tendency to balance spatial symmetry by reacting to asymmetry and reverse.

Apparent Time Reversal and the Big Moment Path

Fields and quanta do not have a fixed identity as either wave or particle. But they can disclose a history of transformations. Waves do not just collapse. It is interesting that this is not a new finding. After having analyzed phase relations, Bohm, in a section titled 'Quantum Properties of Matter as Potentialities' reinvents the electron: "The same electron, however, is potentially capable of developing into something more like a particle when it interacts with a position measuring device, at which time its wave-like aspects become correspondingly less important. But even while it is acting more like a particle, the electron is potentially capable of again developing its wave-like aspects at the expense of its particle-like aspects, if it is allowed to interact with a momentum measuring device. Thus, the electron is capable of undergoing continual transformation from wave-like to particle-like aspect, and vice versa. At any particular stage of its development it may further transform, while keeping its same general aspect; or it may emphasize the opposite aspect instead. The kind of apparatus with which the electron interacts determines which of these potential aspects prevails" (Bohm 1989, p. 132). Now it could appear that 'entanglement' in nowadays lucidness had not been known to the old physicists. So I would like to make us aware of a message published in the thesis of Louis V. de Broglie (1925) in section 7.3. 'The photon gas':

"*If two or more photons have phase waves that exactly coincide, then since they are carried by the same wave their motion cannot be considered independent and these photons must be treated as identical when calculating probabilities. Motion of these photons "as a wave" exhibits a sort of coherence of inexplicable origin, but which probably is such that out-of-phase motion is rendered unstable. This coherence hypothesis allows to reproduce in its entirety a demonstration of MAXWELL's Law. In so far as we can no longer take each photon as an independent "object" of the theory, it is the elementary stationary phase waves that play this role.*"[4]

[4] From a translation of : RECHERCHES SUR LA THÉORIE DES QUANTA, (*Ann. de Phys.*, 10e série, t. III; Janvier-Février 1925), *by*: A. F. Kracklauer, © AFK, 2004; page 64.

Before we go into the aspect of phase-shift desynchronization and decoherence, we should mention a few more experiments which report the surprising emergence of apparently causality violating transactions within such laser fields *exhibiting a sort of coherence of inexplicable origin*. There are some small, but beautiful experiments (Walborn, Terra Cunha, Pádua, and Monken 2002)[5] which ultimately led to versions of home-made 'Do It Yourself Quantum Erasers' (Hillmer and Paul Kwiat 2007)[6], and there are also the experiments involving large distances,- a distance of 400 meters in Vienna (Zeilinger 1999) [7] and space-like separations (144 km across open air from La Palma to Tenerife) so that communication between choice,- whether or not which path information is erased,- and the interferometer seems to be impossible. Using 'ultrafast switching' and precisely timed random setting choices conclusively ensured the 'space-like separation of all relevant events' in the experiments (Ma, Kofler, et al. 2012).

It is a good exercise to study the laboratory details, because, as a theoretician, one can see what makes the difference among all the differences, and what seems to be common in those fundamental structures. Often, after the more differentiating experiments are carried out, it turns out that even the small table experiments gave correct results. For instance Zeilinger's laser beam across open air from La Palma to Tenerife gave us results that legitimated the smaller experiments by his Viennese research group and the groups Kim, Yu, Kulik, Shih, Scully as well as Walborn, Terra Cunha, Pádua, Monken, and many others. Also the almost prophetic insight by Louis de Broglie, the famous thought experiment by John Archibald Wheeler and last, but not least, the very fine teaching aid that was so nicely prepared by Rachel Hillmer and Paul Kwiat and presented on Youtube, the *Do It Yourself Quantum Eraser* lectured by Dr. Robert Nemiroff. [8] Let us take a look at the design used by the Walborn group (figure 15).

In their version of quantum erasure, polarization is used to mark 'which way information' for the photons. QWP_l and QWP_r are quarter-wave plates aligned in front of the double slit. These convert a diagonally polarized wave to circulary polarized waves with opposed circulation. A further component that can be put into the p- and s-path is a linear polarizer which allows light with a definite polarization to pass. The 'p' giving away that the polarized photon travelling in direction p shall provide the erasing field. When the circulary polarized field passes through, say, a horizontal polarizer, the vertical component is annihilated and only the horizontal can pass. The intensity is reduced to ½ of the incoming. Now suppose we repeat Young's experiment with a horizontally polarized field travelling in signal direction s and passing through the quarter-wave plates QWP_l and QWP_r. Without them we had no 'which way information' and there appeared interference fringes at the location of detector D_1. But now, as we insert the quarter-wave plates, one turns the horizontally polarized field

[5] http://grad.physics.sunysb.edu/~amarch/ "A double-slit quantum eraser experiment" The authors "report a quantum eraser experiment which actually uses a Young double slit to create interference. The experiment can be considered an optical analogy of an experiment proposed by Scully, Englert, and Walther in: Nature (London) 351, 111 (1991)

[6] A Do-It-Yourself Quantum Eraser, Scientific America, May 2007; http://www.youtube.com/watch?v=bnxHc6OqB7U (This video is a part of the "Physics-X" lecture series, taught at Michigan Technological University by Dr. Robert Nemiroff.)

[7] "Long-distance Bell-inequality experiment with independent observers. The two entangled photons are individually launched into optical fibers and sent to the measurement stations of the experimenters Alice and Bob which are separated from each other by a distance of 400 m."

[8] http://www.youtube.com/watch?v=bnxHc6OqB7U.

into a left-handed circular polarized wave and the other into a right-handed circular polarized wave. So there is a difference correlated with 'which way information'. The interference fringes vanish and there appears a freckle, a single swatch of light most bright in the middle. It is a local cluster of quanta with a bell-shaped distribution around that center of maximum intensity. In order to demonstrate quantum erasure the 'which way information' must be erased. This can be done by inserting a horizontal polarizer between the quarter-wave plates and D_1. If we repeat the experiment thus equipped, the speck disappears and interference pattern recurs. The horizontal polarizer transforms both left circular and right circular field into a definite horizontally polarized. So there is no way to tell which slit the light had passed through. With the particle-like information removed the photons can start their travel as waves again and so bring forth an interference pattern. We can just as well use a vertical polarizer. However, now there appears a fringe pattern that is exactly out of phase compared with the one we saw with the horizontal polarizer. This is correlated with the formulas (85), and the fringes corresponding to the phase shifted pattern are called anti-fringes. Walborn et al. point out that the Heisenberg principle does not say anything about this experiment, as polarization and location are not complementary variables. They ask what was enforcing complementarity in that case? Their answer is that polarization became entangled with path. "When a photon passes through the double-slit apparatus, it enters a superposition of position states: slit 1 + slit 2. The quarter-wave plates perform an additional conditional logic operation. If the photon passes through slit 1, it emerges with right-circular polarization and if it passes through slit 2, it emerges with left-circular polarization. The photons state can be described as a new and more complicated superposition: (slit 1 AND right-polarized) + (slit 2 AND left-polarized). Because the two observables are now entangled, manipulating the information about either one automatically changes information about the other. It is completely equivalent to describe the photon's state as: (fringes AND horizontal polarized) + (anti-fringes and vertical polarized)."

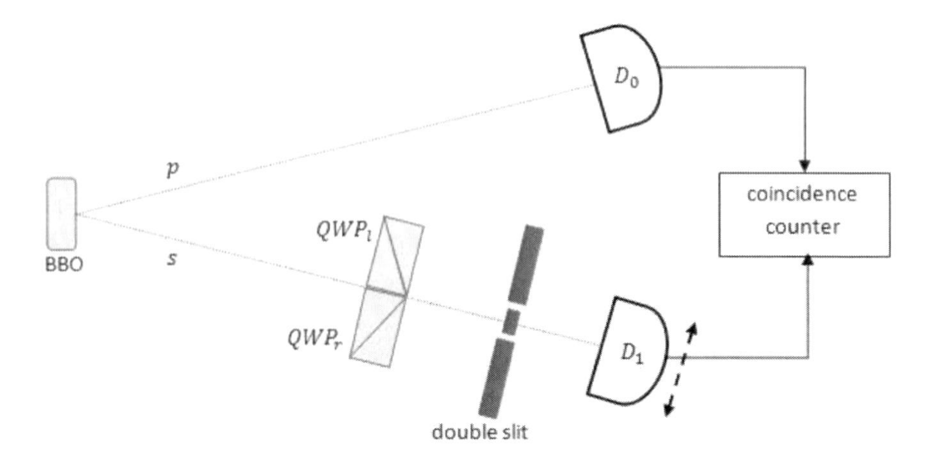

Figure 15. Original setup for the Bell-state quantum eraser by Walborn group (2002).

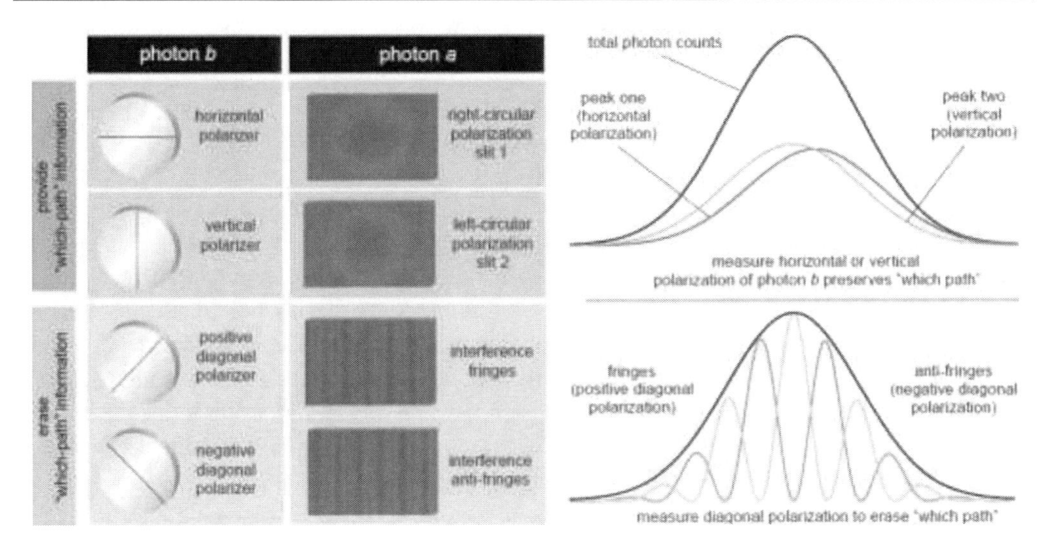

Figure 16. "Delayed-choice" paradox amounts to a change in bookkeeping, not a change in history (Walborn, et.al. 2003, figure 7).

In a delayed choice erasure, the machinery is working by inserting a horizontal polarizer not in the signal path, but in the p-path that leads to detector D_0. This detector can now be shifted away from the source as far as possible, so that we can produce a space-like separation between emission and delayed choice. In this way we decide upon the appearance of fringes or blobs some dozens of nanoseconds in the future.

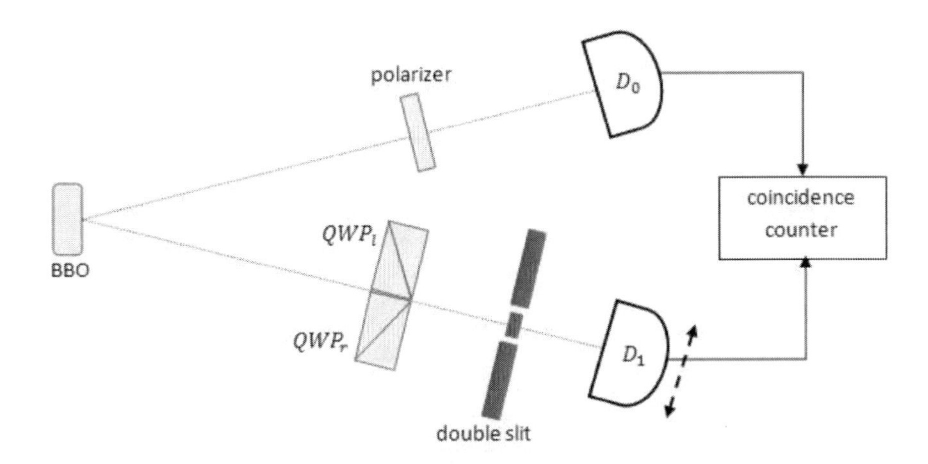

Figure 17. Delayed choice experiment with path-polarization entangled photons.

Walborn, Terra Cunha, Pádua, and Monken, Pádua, emphasize that the experimentor does not change history, but only the bookkeeping. This is correct, but nevertheless the entangled – telepathic, as was put metaphorically by them – events can be observed in conditions of perfect space-like separation, that is, the relevant events are 'causally disconnected'. It is, at the present state of empirical research, rather a linguistic matter, if we regard those events of bookkeeping as causally disconnected, and therefore as disconnected from relativity theory, or if we regard them as indicators of some kind of causality violating mechanics. After all, it seems that such 'telepathically connected coins' as Walborn and others denote them, remind us very much of a rigid body with some kind of sovereign consistency, or unsurpassable fastness. It may seem that we have to divide the universe into dynamic partitions of entangled and disentangled locations. At any case, Einstein understandably did not feel comfortable with those spooky long-distance effects. But it seems that entanglement is a relevant phenomenon operating right at the interface between matter and mind, or stated otherwise, between space-time and information. This philosophic theme also signifies the interval between Spinoza's godlike properties of extension and cognition a theme Albert Einstein was very much involved with. Here lies the origin of a dichotomy that I would denote by the pair 'orientation/disorientation'. That is to say, it is the phenomenon of entanglement that allows stable oriented matter as a peculiar form of stable oriented space-time to emerge within chaotic fields. At any case the experiments portrayed provided increasing evidence against realistic local theories. There seemed to have remained some important subclass of such theories that could still be tested with the concept of quantum steering. However, it appeared, there were some so called major loopholes to be closed, a task that challenged several experimental improvements as depicted in the setup below (Wittmann, et.al., 2012, figure 2).

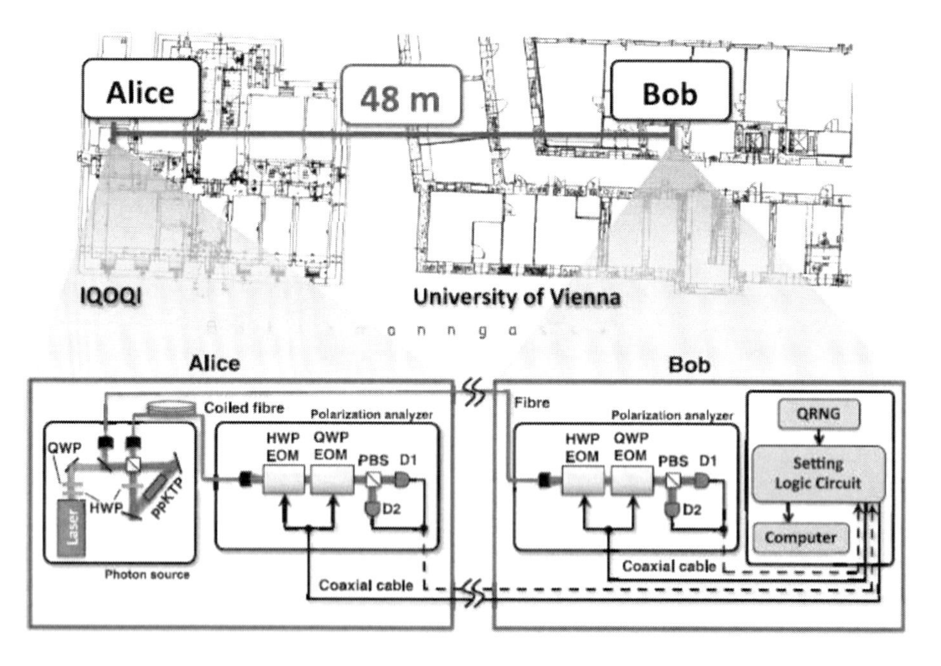

Figure 18. Experimental setup. The loophole-free steering experiment was carried out between two buildings: the Austrian Academy of Sciences (IQOQI) and the University of Vienna. A polarization-entangled photon pair is generated by Alice using an entangled photon source. For each entangled pair, one photon is kept in an 80 m long, coiled optical single-mode fibre (red line) on Alice's side, located next to the source. The other photon is sent to Bob via another optical fibre. On Bob's side, one of three measurement settings is chosen by his fast home-made quantum random number generators (QRNGs) and sent to Alice via a low-dispersion coaxial link and subsequently applied to both photons (solid black lines). Alice's polarization analyzer implements the different settings with two electro-optical modulators (EOMs) realizing ultra-fast switchable half and quarter-wave plates (HWP, QWP), as well as polarizing beam splitter (PBS) and two home-made photon detector modules based on silicon avalanche photo-diodes (D1, D2). On Bob's side there is an equivalent polarization analyzer. The results are then collected (dashed lines) in Bob's laboratory and compared in a logic circuit.

The *quantum steering*, a term coined by Schrödinger in 1935, should allow the experimenter to steer the state of a causally disconnected system, as in the Einstein-Podolsky-Rosen experiment (EPR). Again, in the improved experiments, there should have been used polarization/path-entangled photons, now between two far-distanced laboratories. And further, the three 'major loopholes' were expected to be closed. This experimental procedure was called 'loophole-free quantum steering'. By a large separation of the actors involved and some fast number generators the locality loophole and some form of the 'freedom of choice' loophole were closed. The time window of the setting information independence as determined by the distance between the laboratories, amounts to 360 nanoseconds. Hidden communication between the distanced observers is what defines the 'locality loophole' (Aspect, et al. 1982, Weihs et al. 1998, Bell 2004). Influences from and on the choice of measurement (+1 or −1) denote 'freedom of choice' (Scheidl et al. 2010, Bell et al. 1985), and unfair sampling of the measured observables defines the 'detection loophole' (Rowe et al. 2001, Tittel et al., Pearle 1970, Clauser, Horne 1974). In the experiment carried out by Wittmann et al. (2011) these loopholes were closed for the first time. In this peculiar class of steering experiments distribution of events in the presence of an 'untrusted party' was also allowed for. The experimental design had become amazingly sociological, in a way, almost game theoretic.

The freedom-of-choice loophole would allow the entangled photon pair to influence the random number generator that chooses one of the three measuring setups. "What we cannot exclude, as with any experiment, is the possibility that an earlier common cause in the overlap of the backward light cones of the two events (emission and choice of setting) influences the two events in a correlated manner. We believe, however, that such a hypothesis is outside the scope of what can in principle be tested experimentally (quoted Scheidl et al. 2010, Kwiat et al. 1994)." (Wittmann et al. 2011, p. 7)

There is no reason to assume, in this clever, almost game theoretic space-time arrangement in that experiment why there should act in the background such a disturbing influence. So we could concur entirely with the authors' conclusion that their results "show most rigorously that if one demanded that an isolated system is defined by a local quantum state, this would imply the existence of 'spooky action at a distance'. Would there not be the mere fact that there exists a most natural design of space-time where the geometric distance between the two lab locations is finite while the geometric distance between phenomena, being fields, is zero. We saw that the Dirac equation can be understood much better, if we perform the transition from Minkowski space to Minkowski algebra. Likewise the constitution of Minkowksi space with all its quantum phenomena can be much better be comprehended as soon as we ponder over its constitution by those fields in the Clifford algebra of that space. Unfortunately, mathematically no reason can be seen why translocal events from all possible locations in the universe do not influence the outcome of an observation. Rather the contrary is the case. Mathematical reasoning proposes that such strange impacts do exist without violating the observed phenomena of Special Relativity in the least. The reason for this surprising behavior of the process of nature becomes visible, if we understand that events take place in the Minkowski algebra rather than in the Minkowski space.

To complete the story about the experimental status quo, we must of course mention the work of a team of European researchers around Zeilinger who sent a quantum key based on entangled photons over a 144 km free-space link. This distance was still some orders of magnitude larger than in many other similar such experiments. The outcome represents a crucial step towards future satellite-based quantum communication.

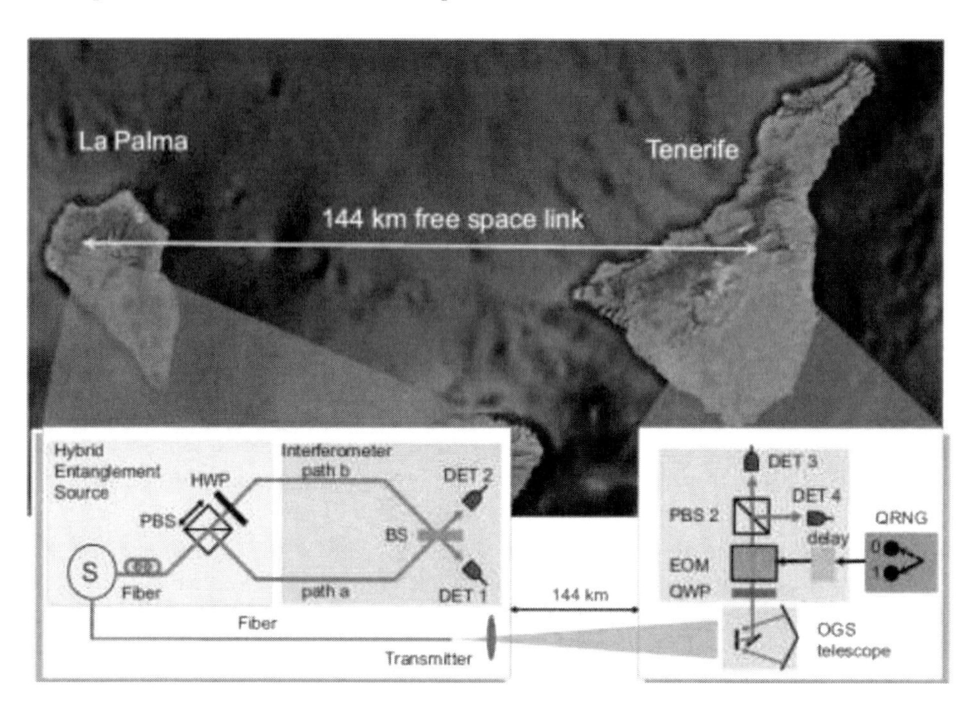

Figure 19. Satellite image of the Canary Islands of Tenerife and La Palma and overview of the experimental setup (Google Earth). The two laboratories are spatially separated by about 144 km. In La Palma, the source (S) emits polarization entangled photon pairs, which subsequently are converted to a hybrid entangled state with a PBS (PBS 1) and a half-wave plate oriented at 45°. The interferometric measurement of the system photon is done with a free-space BS, where the relative phase between path a and path b is adjusted by moving PBS 1's position with a piezo-nanopositioner. The total path length of this interferometer is about 0.5 m. The projection setup consists of a quarter-wave plate (QWP), an EOM, and a PBS (PBS 2), which together project the environment photon into either the H/V or +/− basis (with $|+\rangle = (|H\rangle + |V\rangle)/\sqrt{2}$ and $|-\rangle = (|H\rangle - |V\rangle)/\sqrt{2}$). Both the system photon and the environment photon are detected by silicon avalanche photodiodes (DET 1-4) A QRNG defines the choice for the experimental configuration fast and randomly. A delay card is used to adjust the relative time between the choice event and the other events. Independent registration is performed by individual time-tagging units on both the system and environment photon sides. The time bases on both sides are established by global positioning system (GPS) receivers (Ma, et.al. 2012, figure 5).

The authors[9] emphasize, their work „confirms that whether the correlations between two entangled photons reveal ‚which way information' or an interference pattern of one (system) photon depends on the choice of measurement on the other (environment) photon, even when all of the events on the two sides that can be space-like separated *are* space-like separated. The fact that it is possible to decide whether a wave or particle feature manifests itself long after – even space-like separated from – the measurement teaches us that we should not have any naive realistic picture for interpreting quantum phenomena".

The large space-like outer interval between entangled states in the quantum field correlated with the time reversed observer/observed relation provides a large interval of

[9] Ma, Kofler, Qarry, Tetik, Scheidl, Ursin, Ramelow, Herbst, Ratschbacher, Fedrizzi, Jennewein, and Zeilinger 2012

presence, a 'big moment' so to say within which information can be processed as if there were no arrow of time at all. This phenomenon of future events acting on past ones reoccurs in various experiments, not only in events of quantum erasure. It is not just an illusion, not just appearing as if ... These interactions can, however, take place without violating the laws of SR, as we shall work out. This fact becomes better visible, if we study the physics in the Minkowski algebra and discrete structural physics, rather than in Minkowski space and Hilbert space.

Using the technique and the results of 'weak measurement', Yakir Aharonov and coauthors (2012) have explicitly and successfully posed the question, if a future choice can affect a past measurement's outcome. An ensemble of particles is studied where each undergoes a few weak measurements of spin orientation. Each outcome is documented in the protocol. In the very last moment each particle undergoes a strong measurement in a freely chosen spin direction. Self-organization of spin directions is expected in both weak and strong measurements. However there ensues some contradictory situation: A weak measurement cannot determine the outcome of a successive strong one. Further, Bells theorem forbids definite spin to exist prior to the final collapse of the wave function. Also weak measurements do not change the entanglement. Yet, the weak measurements' outcome agrees with those of the strong measurements. In this case the considerations rest on

i) Bell's nonlocality theorem
ii) causal asymmetry between weak and strong measurement.

For any pair of entangled particles the EPR-Bell experiment proves that the outcome of a spin measurement on one particle depends on the choice of the spin-orientation to be measured on the other particle which it is an outcome of. Relativistic locality is not necessarily violated in such an experiment as there is symmetry of transaction. It can be either Alice whose choices exercised an influence over Bob's or the other way around. The authors point out that this reciprocity cannot be applied to their combination of weak and strong measurements. The strong measurement affects the weak ones, never vice versa. Thus they legitimately conclude that "when a weak measurement precedes a strong one, the only possible direction for the causal effects is from future to past". The present investigation will ratify this statement. Notably, the authors assume the only reasonable resolution would be that of the Two-State-Vector Formalism, which substantiates the fact that "the weak measurement's outcomes anticipate the experimenter's future choice, even before the experimenter himself knows what his choice is going to be".

Unfortunately the equations of motion for entangled systems, interactive quantum jumps and similar phenomena are always conceived i) as models in space-time and ii) by two or n-state vectors in Hilbert space. Regardless whether we consider a quantized local oscillator, a heterodyne detection model, a Bell inequality, a CHSH-inequality, an objective pure state system, or a nonlinear steering inequality, it is always formulated in space-time, using the x, σ_x and so on. This is a serious mistake, since proceeding in this way, we are trying to build dynamic space-time models of those events that bring about this very space-time, the 'frames', wherein they seem to move. But those 'frames' that are created in reality have a different nature. Although most of us who carry out calculations in EPR quantum-steering experiments and quantum computation are acquainted with the Kochen-Specker theorem, we are still deriving our estimates for expectation values of densities and inequalities from the implicit assumption of states in Hilbert-space. Though some of us, it appears, have cooperatively managed to close all those major loopholes, none of us has as yet realized that a closure of the locality-loophole in strong qcd-interactions is not possible. A space-like separation of hadronic events cannot easily be conceived. The reason for the weakness of our models is in the lacking of a suitable exact theory of interaction. Such a theory should be complete and phenomenologically consistent to some reasonable extent. Theoretically, both the iterant algebra of polarized entangled strings as well as the derived geometric algebra of the known space-time are both, in a way, compatible and incompatible with complete space-like separation. These instruments can show us the two sides one of which is relativistic while the other is not at all. The loophole opening up on this basis, seems to be as large and as old as that universe we pretend to know.

Chapter 4

POLARIZED BRAIDS AND LITTLE PRIMORDIAL FRAMES

MOTION BEYOND TIME

The primordial order of the process of nature is beyond space and is beyond time. Yet it reveals change, and it knows motion. There is a material system that discloses many of the features that any system has, namely signal feed and interaction, but even those properties are yet primordial. There is no definite feedback or feed forward. The orientation in systems dynamics is still chaotic. There is comparison of appearances, of characters, primordial pixels, as we shall see, entities with no definite location, as long as there is no interaction that 'measures' that location, events in a way unconscious, but not yet automatized. It is what we call the process of quantum motion. We shall have to work out how a process like this can be understood in mathematical terms. Quantum motion is not quantum mechanics, since in quantum mechanics, the dimension of time is already derived. It is introduced into the rigor and taken for granted. Quantum motion is, however, aware of the chaotic causality breaking behavior of particles when the synchronous image of a "Schrödinger wave" is being constituted. Quantum motion, first of all, is beyond time. It can even be featured as perpetual since it represents the imperishable, perennial actor of the present. We encounter this actor in the form of the entangled photons. Their correlation transcends the ordinary measures of space and time we are used to. It is interesting that a Buddhist Meditator or any human in deep meditation is able to perceive motion beyond time. Time comes in by the operation of thought. But it is difficult to construct a proper mathematical image of that motion. We shall get close to it, but I cannot promise that we solve the whole problem or provide a complete theory. We shall go together as far as possible. At any case, we do not wish to just take off and lift the whole question to the mere level of quantum informatics, because matter is more than information. Matter is extended, information is cognitive. Matter is capable of touch and emotion. Much of it is beyond cognition and analysis. 'E-motion' has to do with undivided motion, holistic dynamics. Some properties of matter seem to be ineradicably metaphysical. The most important ones concern orientation in space. The left/right-, below/above- and back/forth-dualities seem to connect inner with outer and are equally valid in the outer as in the inner worlds of matter and men, fields and cognition. The universe knows orientation even if nobody watches it. It is constantly reconstructing orientation. Another concerns the arrow of time. Physical motion, the process of nature, evolution all seem to be carried out in time. So time seems to be part of the frame where the events take place just as in space, but it may be that this is only a local affair of universal motion. Both space and time appear to be required for motion to perform. So it seems. But the real order is different. Don't we have any better idea? Some of us, the constructivists, say that time is constructed by our human minds. That is true. But even in that view there is a sensible mistake: not only us, but even nature is also constructing time. The arrow of time is constructed by both nature and human. But nature constructed time even before the human evolved. The opposite view is just anthropocentric. How come? How does the psychological illusion of the time-parameter come in? Time is thought. The process of thought is natural. Thought, as an intelligent process, goes on even in space-time without any human beings present. Space-time is there and is intelligent. It is beyond human construction. But it is not beyond construction. Quantum motion is intelligent, but beyond space-time. It has the intelligence to constitute space-time and matter at once. In this sense all three of us, nature, thought and humans represent actors in the extensive event of existence. Man is part of nature. Nature is not part of man in the same way that man is part of nature.

'POWERPOINT' ON MOTION

Any perceived element signifies a location through the act of observation. It is the creation of an bserving element that is capable to make a difference between itself and the observed element that turns observing and observed elements into active elements whereby both become locations. The relation between observer and observed bridges the interval between the two locations. Now we could make an iterant into such an observed dynamic element. But that would be almost too idealistic for physical reality. It would introduce an ideal element. For the iterant stands for an infinite iteration and alternation. As has been shown that it is in principal correlated with the ideal of a Schrödinger-wave $e^{i(kx-\omega t)}$ having infinite extension. This would not be compatible with primordial quantum events where we have perhaps only short strings that may appear and disappear within some small finite location. Yet, the mechanism of discrete *temporal shift* as well as *swap of direction* are crucially important in physics. In addition to that there are many reasons why discrete mathematics of quantum motion turns out important. But there are historically come up blocks.

Since 1926 we are seduced to give up our common sense in order to understand motion at the atomic level.

Since more than 50 years we accept, nobody can understand QED, least QCD.

Comparing QCD with QED Feynman assured "the quarks have an additional type of polarization that is not related to geometry". [1]

Each of us lives in his own small world. If we are lucky enough, we live in a beautiful house which we have built in accordance with our own construction plans. Yet we look out into the world through our own small window. The beauty of that house Richard Feynman had built, the QED, cannot be praised enough. Yet, that statement he made about quarks was

[1] With the '*powerpointed* statements' there was introduced the lecture „On Motion" held at the 10th International Conference of Numerical Analysis and Applied Mathematics, 19-25 September, 2012, Kos, Greece.

an error. It is the quarks above all that are related to geometry. We can say, the quarks make the geometry, and they make the matter. From the mathematics of strong interaction there even would follow QED. But we have put the cart before the horse long time ago. However, when we were young, the big minds were exhibited to us, and among those was Feynman. We did not meet him very often. But we liked him because of what Thirring said about his eminent work. Imagine, Feynman took it for granted, "you are not going to be able to understand it". His students could not understand it, because, as he said, "I don't understand it". What a disappointment for a young one who believes in the primacy of knowledge. Thirring used to mention the "unessential constants" as we jiggled around the complexity of ways and contributions of the coupling constant j.

Feynman argued, or should we say he confessed, we did not have a good mathematical way to describe the theory of QED. We could not include coupling distances close to zero. Calibration of constants today is a still greater problem in QCD as was shown for example by some studies at the Max Planck Institute in Aachen. Physics had to offer a lot of problems. In a way this science even had become a vanishing point for problematic minds.

- Physicists sometimes say "let us stay in close contact with empirical reality and let's not get lost in mathematical details". But
- May be we need a good double-entry bookkeeping of both quantum events and geometric realities, and, may be, some unifying insight.

How could such double-entry accountancy look like? Let us consider for simplicity two fundamentally different levels of reality:

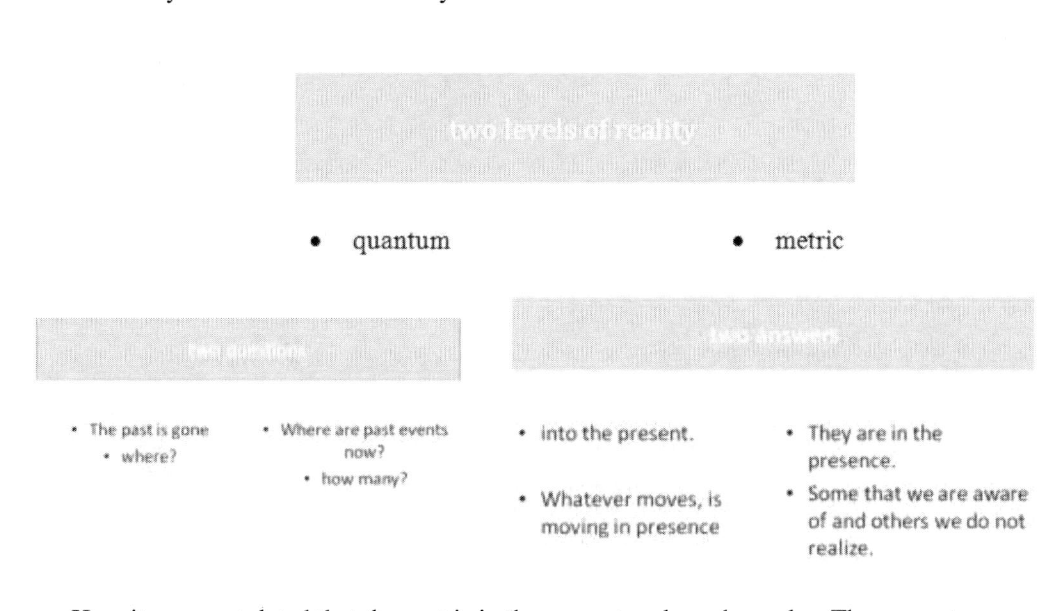

Here it was postulated that the past is in the present and nowhere else. The present

- is the platform of change
- means life
- is contact

- is interaction
- involves chaos
- decomposes the orientation

I want the reader to become aware of the rickety ground on which we are walking. We posed rather unusual questions and created a new idea, the grade 4 iterant polarity string together with a bunch of partly unusual denotations:

A grade 4 Iterant Element is a

- quad location,
- polarized string,
- constituent of quaternion
- synchronous detemporalized image
- Parsed compass
- quaternion location
- multivector in geometric algebra
- quark density component, and it looks like

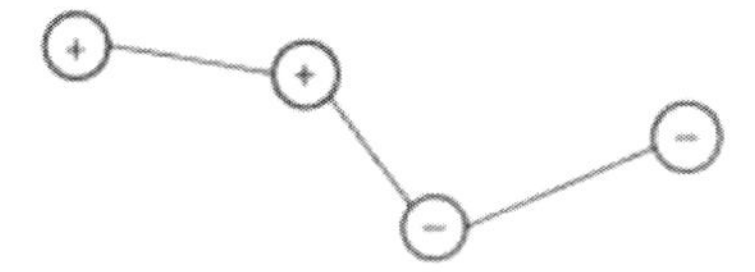

This may all look very screwy, and it is. But by the time it will become clear what it does.

Active Locations and Time-Shift

The four pucks in a polarized string have polarity + or – and are connected beyond space-time.

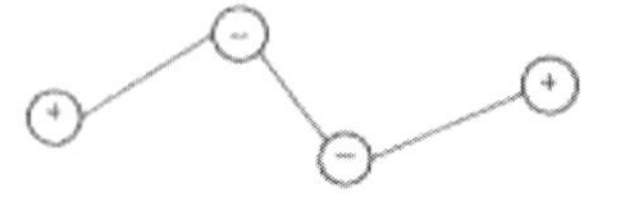

Strictly speaking, they four are one. It is utterly unimportant how far the pucks are distanced from another. Even if scientists, measuring with relativistic yardsticks, would detect a space-like distance between the places where they seem to be localized, reactions within the quad location would occur immediately as if the string was an ideal solid body. There may be no interval between the four poles or 'quadrants', as such quaternion locations represent one field. In a quad location a single entry can be changed and that may change the others too. Preferably we should look at polarized strings as iterants. The graphical connection would indicate that polarity of a neighbor is automatically flipped. So, suppose we start with

$$[+ - - +] := \ldots + - - + + - - + + - - + \ldots$$

Now we make a first move and flip any entry. Then we obtain

$$[- + + -] := \ldots - + + - - + + - - + + - \ldots$$

I do not want to fix everything at the time being. But we could regard this operation as an important possibility to handle dynamics of polarity strings. Logically it turns the 'content' of a polarized string into its negated. If we regard ourselves as observers, we may already have conceived a ready concept of space-time, say a Euclidean flat space $\mathbb{R}^3$, or a Minkowski space $\mathbb{R}^{3,1}$ or Minkowski algebra $Cl_{3,1}$ which the entangled polarity strings do not care about in the least. They just float through the universe without realizing any space-like distance between them. They do so in a manner like this

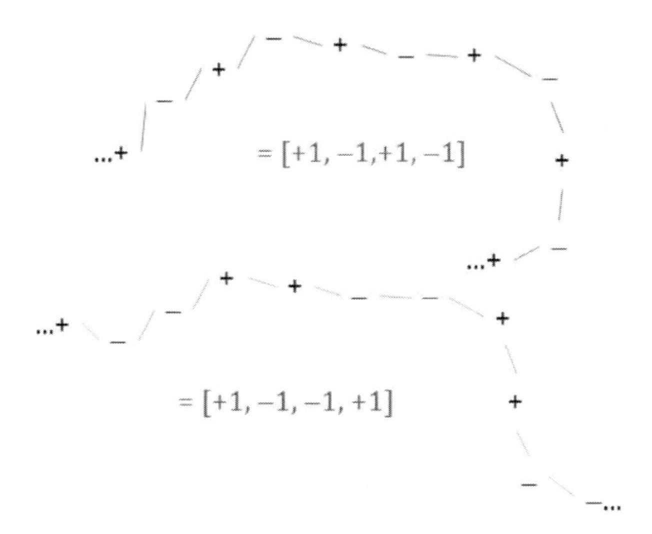

Figure 20. Floating polarity strings.

Those polarity strings have the fantastic, almost biological, capability to touch and compare each other

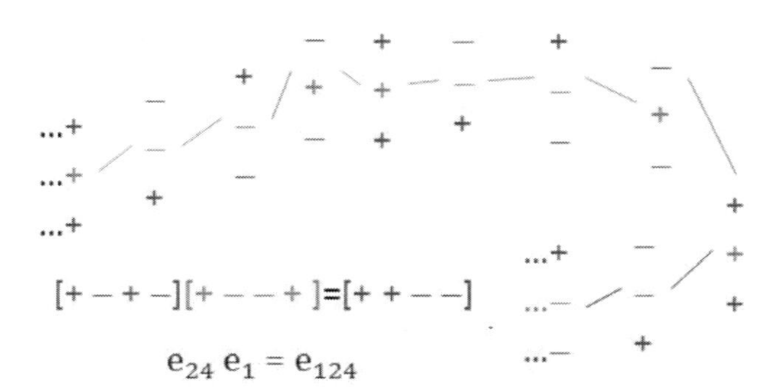

Figure 21. Touch and compare.

Here we have put something cryptic into the lower left corner, namely the formula $e_{24}e_{124} = e_1$ which actually refers to a base of a peculiar Cartan subalgebra within the Minkowski algebra. What is so special about this arrangement of locations? Well, drawing up those polarity strings we construct real associative algebras with familiar algebraic properties by some simple logic product, and as a spillover of such approach, we have obtained a peculiar temporal shift operation. Namely, as soon as the string $[+\ -\ +\ -]$ gets into touch with [+ − − +] it brings about [+ + − −], that is it turns

. . .+ − − + + − − + + − − +. . . into
. . .+ +− − + + − − + + − − . . .

This may be regarded as a time shift by one step, here represented by a shift of the first iterant by one step to the right. Having come so far, we should like to develop some of the mathematics of those peculiarly entangled strings and show if they have anything to do with the geometry of the Minkowski space.

Brief Review of Shift and Time-Shift

In what follows I am repeating a page. But the redundancy alleviates the comprehension, as it gets us back into the theme of time. In geometric algebra we use to represent the unit imaginary i by a unit bivector, say, $e_{jk} \overset{\text{def}}{=} e_j \wedge e_k$. This is the exterior product of two base units e_j, e_k each of which squared gives the positive identity Id of the algebra. After considerable progress had been made in Clifford algebra, many of us regarded that as the only appropriate introduction of the i. So the i took the form of a geometric unit. But there is that second, and in my opinion at least equally valuable, if not better, interpretation to the imaginary unit in terms of an algebraic, temporal shift operator. Kauffman had explained this very clearly in his lecture to Heinz von Foerster on his 100th birthday. We could go into this in the following way. Consider at first the binary Galois-field $GF_2 = \{0,1\}$ and the double-ring of real numbers ${}^2\mathbb{R} = \{(r,s)\}$. In both we define a 'swap' or 'swop' as an exchange of arguments:

$$\begin{aligned} &Swap \text{ acting on } GF_2: \quad (0,1) \longmapsto (1,0) \\ &Swap \text{ acting on } {}^2\mathbb{R}: \quad (r,s) \longmapsto (s,r) \end{aligned} \tag{88}$$

The swap indicates a reversion of orientation in any sense. Now consider a process $\mathfrak{R}(\ldots)$ for the eigenform i as in equation (28) defined over some numbers yet unspecified:

$$\mathfrak{R}(x) = -1/x$$

Consider first a small finite field given by $G \overset{\text{def}}{=} \{0, +1, -1\}$ constructed isomorphic with the Galois field $GF_3 = \{0,1,2\}$. Consider any multiplicative inverse $x^{-1} \overset{\text{def}}{=} 1/x$. The field G_3 is free of zero divisors, that is, the multiplicative inverse is well defined: For every non-zero element in the field there is a multiplicative inverse, namely we have $(+1)^{-1} = +1$ and $(-1)^{-1} = -1$. Letting the oscillatory process $\mathfrak{R}(x)$ operate on $+1$ iteratively, we obtain a sequence $+1, -1, +1, -1, +1, -1, \ldots$ applying it recursively on the other non-zero element, we get $-1, +1, -1, +1, -1, +1, \ldots$ Now consider two infinite series of such a type open to the left and right. Is there any way to differ between those two? The only way to distinguish between them is given by *direct confrontation*. This I called 'touch'.

$$\ldots +1, -1, +1, -1, +1, -1, \ldots$$
$$\ldots -1, +1, -1, +1, -1, +1, \ldots$$

Would we apply a backshift-operator B by half a period, we would obtain equal sequences

$$\ldots +1, -1, +1, -1, +1, -1, \ldots$$
$$\ldots +1, -1, +1, -1, +1, -1, \ldots$$

Let $\mathfrak{I}\{+1, -1\}$ stand for an undisclosed alternation or ambiguity between $+1$ and -1. The $\mathfrak{I}\{+1, -1\}$ is called 'iterant'. The undecided iterant possesses two iterant views: $[\![+1, -1]\!]$ and $[\![-1, +1]\!]$. It is obvious that the oscillatory process $\mathfrak{R}(x)$ can bring forth such iterant views when applied to real numbers. It can produce oscillations of real numbers, but it will not give us the fixed point as an eigenform of this process. Since this fixed point is in the complex domain. We have

$$\mathfrak{R}(i) = -1/i = i$$

as the i satisfies $i^2 = -1$. The iteration over the real numbers must be directly related to the idealized imaginary number i. In (29) we introduced the algebraic temporal shift operator η with familiar commutation properties

$$[\![a, b]\!]\eta = \eta[\![b, a]\!] \text{ and } \eta\,\eta = 1$$

so that concatenated observations are correlated with a time step, indicating one half period of the process. Now consider the backshift-operator B which brings forth the transformed $B[\![a, b]\!] = [\![b, a]\!]$. Backshift B has two eigenforms, namely $[\![+1, -1]\!]$ and $[\![-1, +1]\!]$. The $-B$ can be interpreted as a combination of reversion and shift or of *negation* and shift. We often take a shorthand for the backshift-operator B by the identity

$$B[\![a, b]\!] = [\![a, b]\!]' = [\![b, a]\!]$$

Addition and multiplication of iterant views can be defined as in (33). A unit is defined by $1 \stackrel{\text{def}}{=} [\![+1, +1]\!]$. The imaginary unit is $i \stackrel{\text{def}}{=} [\![+1, -1]\!]\eta$ and its conjugate $-i = [\![-1, +1]\!]\eta$, so that we have indeed $i\,i = [\![+1, -1]\!]\eta[\![+1, -1]\!]\eta = [\![+1, -1]\!][\![-1, +1]\!]\eta\,\eta = [\![-1, -1]\!] = -1$. In this way the square roots of minus one are interpreted as iterant views connected with temporal shift operators. This simple approach will now be extended so that it will be valuable for all forces and models of physical interaction.

Primordial Observation of Polarity Strings

Again we pose the question who or what observes any entity in a primordial domain of polarized locations? But now we go a little bit further and ponder over four dynamic elements. This correlates with the intuitive concept of quaternions as four locations or '4-character alternating series'. let $\mathfrak{I}\{a, b, c, d\}$ stand for an undisclosed alternation or ambiguity between a, b, c and d. The $\mathfrak{I}\{a, b, c, d\}$ is now called a '4-fold iterant'. The undecided iterant possesses 24 iterant views beginning with$[\![a, b, c, d]\!]$, $[\![b, a, c, d]\!]$ and so forth. The undecided iterant has 24 iterant views altogether in agreement with the 24 rearrangements by the symmetries of the symmetric group $\boldsymbol{S}_4$. We shall introduce two new generative units in analogy to the time shift operator that will allow us to introduce a new type of iterative motion that can be related to the quaternions and respectively to the interactive phenomenology of the Minkowski space emerging on this ground. For illustrative purposes we first need two further building blocks, namely a peculiar Cayley graph for the group $\boldsymbol{S}_4$ when generated by two specific operators of periods 2 and 4 respectively. These generators are denoted by η and t. The η will represent the time shift we are already familiar with. Both are represented by the cycles $\eta = (1\ 2)$ and $t = (1\ 2\ 3\ 4)$ as in figure 22. There the line without arrowheads represents action of η whereas the arrow stands for the action of the '4-cycle time shift' t.

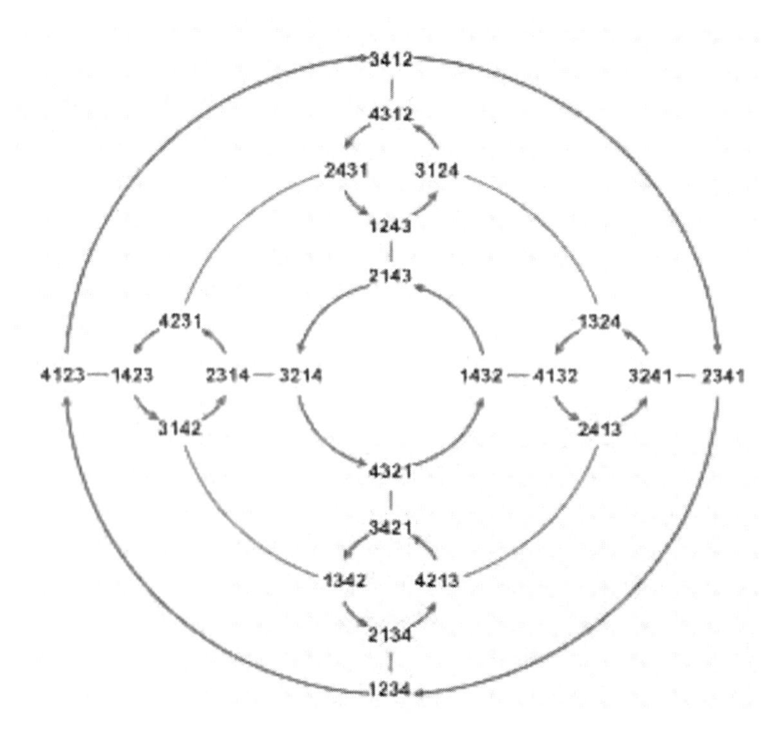

Figure 22. Cayley graph for symmetric group S_4; the 24 symmetries of $\boldsymbol{S}_4$ generated by two shift operators[2].

[2] http://en.wikiversity.org/wiki/File:Symmetric_group_4;_Cayley_graph_1,9.svg

Consider the quadruple $[\![a, b, c, d]\!]$. Then the shift operator η is supposed to bring forth a shifted iterant $[\![b, a, c, d]\!]$. It will, for example transform an alternating series $\ldots 1, -1, 0, 1, 1, -1, 0, 1, 1, -1, 0, 1, \ldots$ into $\ldots -1, 1, 0, 1, -1, 1, 0, 1, -1, 1, 0, 1, \ldots$ and generally $\ldots a, b, c, d, a, b, c, d, a, b, c, d, \ldots$ into a sequence where a is exchanged with b, that is we get $\ldots\ b, a, c, d, b, a, c, d, b, a, c, d, \ldots$ however the shift operator t should transform $\ldots 1, -1, 0, 1, 1, -1, 0, 1, 1, -1, 0, 1, \ldots$ into $-1, 0, 1, 1, -1, 0, 1, 1, -1, 0, 1, 1 \ldots$ From these generators of periods 2 and 4 respectively, we shall derive three further transpositions that will be extremely useful, namely $\varphi = t^2$, $\tau = \eta\varphi\eta$ and $\sigma = \tau\varphi$. Note that we have $\varphi^2 = \sigma^2 = \tau^2 = Id$.

The introduction of iterants of order 4 has a specific reason that can be held responsible for the emergence of the reorientation symmetries of the Minkowski space-time structure. Though the iterant views are mere sequences where, in the present case, four locations as characters are repeated or eventually permuted by the temporal shifts, they disguise a geometry we are quite familiar with.

Shaping Self Reference and Symmetry Breaking with Dihedral Group D_4

The observation of a symmetry is bound to its breakage. For consider the symbol of the quartered circle, - the compass, - with its eight symmetries:

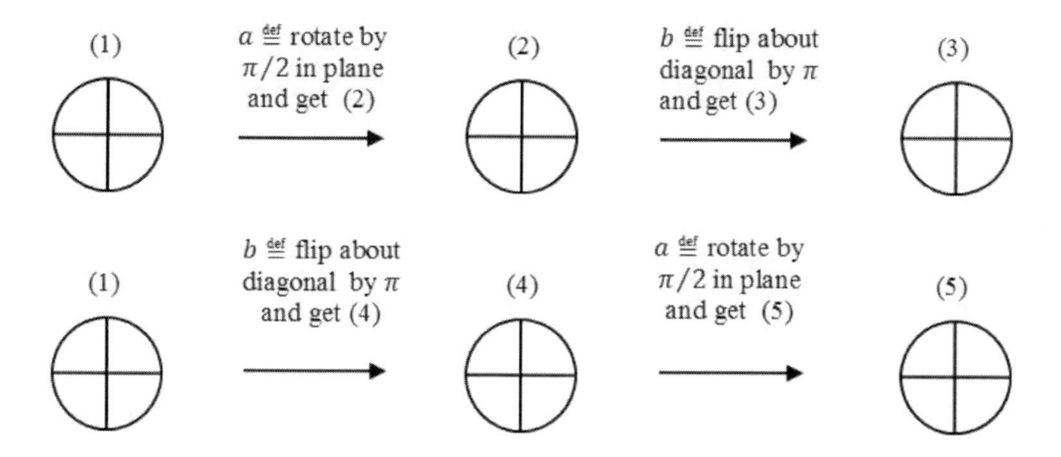

Figure 23. Symmetry without connotation.

To understand quantum motion, we got to think in a very simple, evident way. From the above figure we should see that the symmetries $a, b \in D_4$ do not commute, that is, $a\,b \neq b\,a$, and that should tell us that the dihedral group D_4 is non-abelian. But, as a matter of fact, we do not see anything and the figure alone does not tell us anything either. However, as soon as we somehow color the quadrants, enumerate them or mark them in any evident way, the non-commutativity becomes apparent:

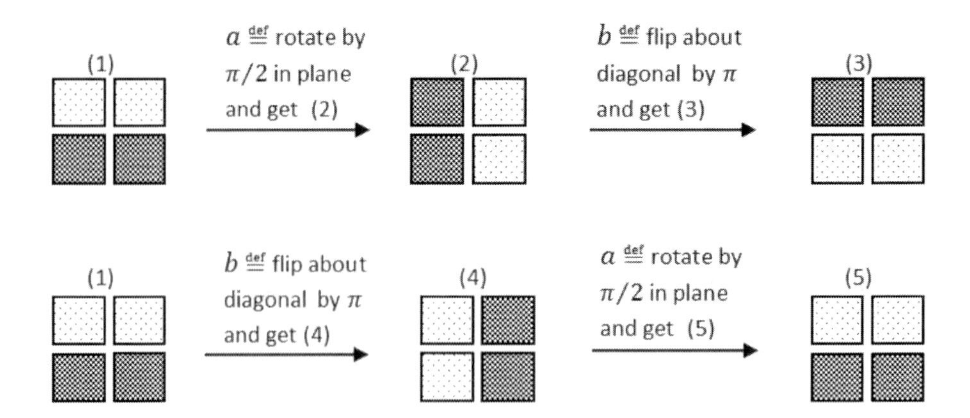

Figure 24. Symmetry breakage by connotation and coloring.

We have tinted the bottom of the start figure (1) in dark, and the upper half bright. It is only now that it becomes apparent, figures (2), (3), (4), (5) are indeed different. We detect a difference between the operations $a\,b$ and $b\,a$, that is, $a\,b \neq b\,a$. We must refer to asymmetry to detect symmetry! The product $a\,b$ 'turns the world upside-down" while $b\,a$ preserves the bottom-up direction. What can we learn from this? The inking, - the identification labels of locations, - represents a symmetry breakage. For, in a perfect symmetry we cannot make out any differences between quadrants. It's the symmetry breaking that brings upon orientation in the process. In the innermost quantum layer of real motion, the process of nature does not make a difference between orientations, that we regard as different at the macro-level, and, as we shall see, not even between different grades. It is therefore that we observe transpositions or exchanges' that we realize as 'entanglement' or information-transport beyond time and relativity. In his theory of consciousness (»Theorie des Bewußtseins«) Gerhard Frey (1980) investigated the figure of the quartered circle as an archaic thought form, the so called "Gedankenkreis". He discussed this in the pre-Socratic context, further as superposition of logic conceptional pairs and third in the context of self-reference. This morpheme represents a root form of the operational structure of thought in the structural or 'genetic' psychology of Piaget (1982). We can find this in Parmenides statement: 'being is'. Likewise 'non-being' must be self-referent. This points at both a static and a dynamic aspect when Parmenides says that 'non-being is not' and Heidegger 'das Nichts nichtet'. The *primary movens*, god or mind, must be a moving cause for itself, substance is defined as that which has its origin, its Arché, in itself (Frey p. 158).

As soon as we realize that the dihedral structure must play a much more general role for the Minkowskian phenomena of physics, namely, in the form of a direct product of different groups $D_4 \times D_4$, we enter the domain of sequences involving four characters.

Fourfold Iterants

We begin the elaboration by figuring fourfold iterants with unit entries, either negative or positive corresponding with polarity. As the iterant has four essential characters and each character can be occupied by either $+1$ or -1, there are 16 typical quadruples which we wish to denotate in the following fashion.

$$I^+ = [+1,+1,+1,+1] \stackrel{\text{def}}{=} \ldots +1+1+1+1+1+1+1+1+1+1+1+1 \ldots \quad (89)$$
$$I^- = [-1,-1,-1,-1] \stackrel{\text{def}}{=} \ldots -1-1-1-1-1-1-1-1-1-1-1-1 \ldots$$
$$P_1 = [+1,+1,+1,-1] \stackrel{\text{def}}{=} \ldots +1+1+1-1+1+1+1-1+1+1+1-1 \ldots$$
$$P_2 = [+1,+1,-1,+1] \stackrel{\text{def}}{=} \ldots +1+1-1+1+1+1-1+1+1+1-1+1 \ldots$$
$$P_3 = [+1,-1,+1,+1] \stackrel{\text{def}}{=} \ldots +1-1+1+1+1-1+1+1+1-1+1+1 \ldots$$
$$P_4 = [-1,+1,+1,+1] \stackrel{\text{def}}{=} \ldots -1+1+1+1-1+1+1+1-1+1+1+1 \ldots$$
$$N_1 = [-1,-1,-1,+1] \stackrel{\text{def}}{=} \ldots -1-1-1+1-1-1-1+1-1-1-1+1 \ldots$$
$$N_2 = [-1,-1,+1,-1] \stackrel{\text{def}}{=} \ldots -1-1+1-1-1-1+1-1-1-1+1-1 \ldots$$
$$N_3 = [-1,+1,-1,-1] \stackrel{\text{def}}{=} \ldots -1+1-1-1-1+1-1-1-1+1-1-1 \ldots$$
$$N_4 = [+1,-1,-1,-1] \stackrel{\text{def}}{=} \ldots +1-1-1-1+1-1-1-1+1-1-1-1 \ldots$$
$$I_1 = [+1,+1,-1,-1] \stackrel{\text{def}}{=} \ldots +1+1-1-1+1+1-1-1+1+1-1-1 \ldots$$
$$I_2 = [+1,-1,-1,+1] \stackrel{\text{def}}{=} \ldots +1-1-1+1+1-1-1+1+1-1-1+1 \ldots$$
$$I_3 = [-1,-1,+1,+1] \stackrel{\text{def}}{=} \ldots -1-1+1+1-1-1+1+1-1-1+1+1 \ldots$$
$$I_4 = [-1,+1,+1,-1] \stackrel{\text{def}}{=} \ldots -1+1+1-1-1+1+1-1-1+1+1-1 \ldots$$
$$I_5 = [+1,-1,+1,-1] \stackrel{\text{def}}{=} \ldots +1-1+1-1+1-1+1-1+1-1+1-1 \ldots \equiv A_1$$
$$I_6 = [-1,+1,-1,+1] \stackrel{\text{def}}{=} \ldots -1+1-1+1-1+1-1+1-1+1-1+1 \ldots \equiv A_2$$

We have arranged these 2^4 logically possible iterants in such a way that we can see the structure into which the set is partitioned by the shift operator B. Let us denote these polarity strings as PS. The polarity defines a 'signature' of the iterants. There are two *definite signatures* for I^+, I^- which are either all positive or all negative. There are six neutral signatures for I_1 to I_6 where positive and negative polarity are in balance. Of these neutral polarity strings two represent the *alternating codes* we had used in the iterant approach to the Schrödinger theory, namely the pair $I_5 = A_1, I_6 = A_2$. In the iterants P_1 to P_4 positive polarity dominates. In the iterants N_1 to N_4 the negative polarity prevails with a ratio 3:1. In the category of fourfold iterants the domains and arrows given by the shift operator are as follows

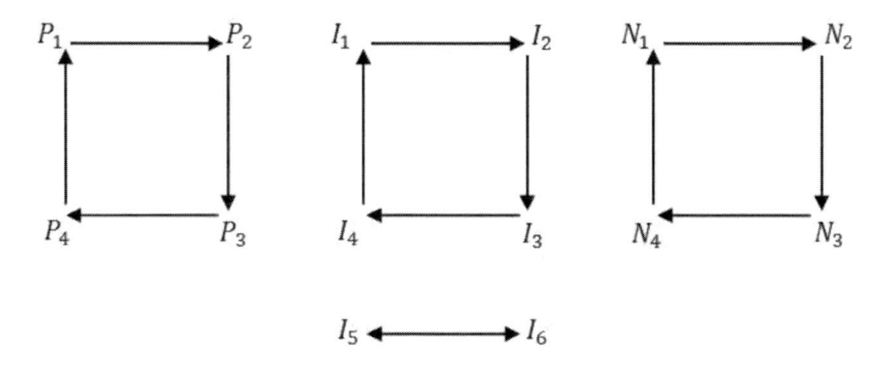

Figure 25. Partition of domain of 4-iterants by shift operator B.

So we have for example $BP_1 = P_2$, $BP_2 = P_3$, …, $BI_1 = I_2$, … et cetera. We have essentially a partition into three 4-cycles associated with a finite group structure $\mathbb{Z}_4$. The logic center of the alternating codes is of course flipped by $\mathbb{Z}_2$, while the locations with definite polarity are invariant under B. Recall that the imaginary unit generates the cyclic group $\mathbb{Z}_4$.

So the shift operator acts like the period 4 rotation generated by i within the three parts $\mathcal{P} = \{P_1, \dots, P_4\}$, $\mathcal{J} = \{I_1, \dots, I_4\}$, $\mathcal{N} = \{N_1, \dots, N_4\}$. We shall find out in a while that such a finite, primordial feature is characteristic for a quaternion structure.

The Emergence of Time Shift by Touch

All we are doing is still giving locations an activity. We give an action potential to polarity strings which touch each other. Notice that everything we were doing till now was essentially based on a handful of features of the logic binary connectives of two binary statements. From the structure we chose, there results a simple and fine

Theorem 2: To every polarity string PS which is a fundamental fourfold iterant as in (89) there exists another such iterant such that their touch causes a time shift in PS either forward or backward. There exists one definite iterant that induces a forward shift, and another definite iterant which causes a backward shift. In this way a temporal shift turns out to be resulting from interaction.
Proof: The sixteen fourfold iterants (89) denoted as PS (polarity strings) form an algebra with logic identity as multiplication. Their multiplication table is

Table 1. Symmetric multiplication table of fourfold binary iterants

	I^+	I^-	P_1	P_2	P_3	P_4	I_1	I_2	I_3	I_4	I_5	I_6	N_1	N_2	N_3	N_4
I^+	I^+	I^-	P_1	P_2	P_3	P_4	I_1	I_2	I_3	I_4	I_5	I_6	N_1	N_2	N_3	N_4
I^-		I^+	N_1	N_2	N_3	N_4	I_3	I_4	I_1	I_2	I_6	I_5	P_1	P_2	P_3	P_4
P_1			I^+	I_1	I_5	I_4	P_2	N_4	N_2	P_4	P_3	N_3	I^-	I_3	I_6	I_2
P_2				I^+	I_2	I_6	P_1	P_3	N_1	N_3	N_4	P_4	I_3	I^-	I_4	I_5
P_3					I^+	I_3	N_4	P_2	P_4	N_2	P_1	N_1	I_6	I_4	I^-	I_1
P_4						I^+	N_3	N_1	P_3	P_1	N_2	P_2	I_2	I_5	I_1	I^-
I_1							I^+	I_5	I^-	I_6	I_2	I_4	N_2	N_1	P_4	P_3
I_2								I^+	I_6	I^-	I_1	I_3	P_4	N_3	N_2	P_1
I_3									I^+	I_5	I_4	I_2	P_2	P_1	N_4	N_3
I_4										I^+	I_3	I_1	N_4	P_3	P_2	N_1
I_5											I^+	I^-	N_3	P_4	N_1	P_2
I_6												I^+	P_3	N_4	P_1	N_2
N_1													I^+	I_1	I_5	I_4
N_2														I^+	I_2	I_6
N_3															I^+	I_3
N_4																I^+

The multiplication between any pair $X, Y \in PS$ is commutative, that is $XY = YX$. For every transition within the three parts $\mathcal{P}, \mathcal{J}, \mathcal{N}$ there is indeed a definite element that carries out the shift. For example we have representation of time shift by iterant touch

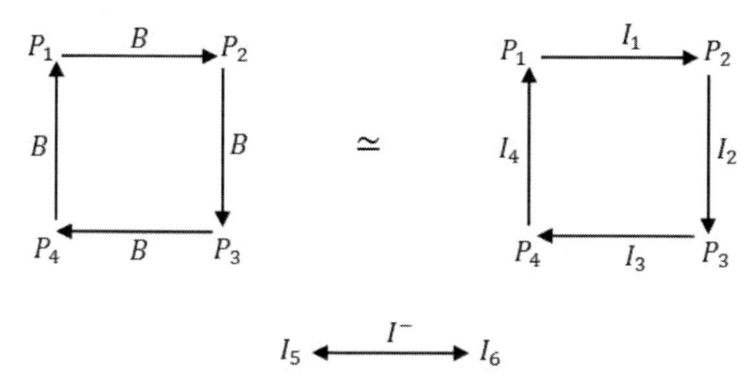

The 4-cycle of the positive iterants can be carried out by the four neutral iterants. What is interesting to observe is that the multiplication in this algebra, being logical identity of the two factors, is itself represented within PS as a polarity string, namely I_5 can be imagined to represent logic identity and I_6 exclusive disjunction, the 'XOR'. So by all means wittingly we bring here together a physical action, with some emotional meaning – the touch – the truth value of a connective, the polarity of a field and the logic coincidence. This signifies the mystery of nature, the bridgeable polarity between cognition and extension. Any element in PS is both an iterant location and a logic operation. What is so special about, say, the four neutral iterants

$$
\begin{aligned}
I_1 &= [+1,+1,-1,-1] \stackrel{\text{def}}{=} \ldots +1+1-1-1+1+1-1-1+1+1-1-1\ldots \\
I_2 &= [+1,-1,-1,+1] \stackrel{\text{def}}{=} \ldots +1-1-1+1+1-1-1+1+1-1-1+1\ldots \\
I_3 &= [-1,-1,+1,+1] \stackrel{\text{def}}{=} \ldots -1-1+1+1-1-1+1+1-1-1+1+1\ldots \\
I_4 &= [-1,+1,+1,-1] \stackrel{\text{def}}{=} \ldots -1+1+1-1-1+1+1-1-1+1+1-1\ldots
\end{aligned}
$$

The first is just the Schrödinger iterant, the *alternating code*, but with half the frequency or double wavelength. The second is obtained from the first by a single step backward of the first in relation to the informatic reference frame, that is, the four fixed character positions in every line. The third is a negation of the first, and the forth a negation of the second. If we also respect the period 2 elements, we find out, I_5 is identical with the Schrödinger iterant and I_6 is its negation. What shall turn out as surprising and important is the relation of those elements of motion with linear vector- and multivectorspaces of special relativity and high energy physics. We have already understood the meaning of a temporal shift operation η having period two, that is, $\eta^2 = Id$, but now it seems as if there were another temporal shift with period four, as soon as we study fourfold iterants and quaternion locations.

A Peculiarity of Primordial Chaos

"So all those procedures with the aid of which we erase time deserve our attention" writes Erhart Kästner in his 'greecebook'(1953, 1991p. 12): "not frighten away, indeed, not ignore, but really annihilate". Meditation means phasing out time. [...] "Fall into the landscape: pure delight, time goes out". There was war then, and people in Athens died on hunger when Kästner wrote those sentences down. What does physics have to do with it at all? Studying a temple we can go back in time and recover a story which, as Kästner pointed out, is most probably wrong. But if we permeate its inner we become aware of details that bring us closer to that which is real. The Acropolis was not a white temple. But its marble was colored. Its interior was full of objects, furniture and figurines, of different times and various social strata. Everyone wanted to take part and be part of it. It is the same in physics. We can seek the origin of matter in very old times, so that we find the Big Bang and mystify cosmic history. Or we can seek the mystery of matter by going deeply into its inner and phase out time. Then we are confronted with vivid facts of life, and the matter becomes almost a biological thing. The innermost of matter is void. It is void of time, void of scales and has no orientation. The innermost of matter, however at the same time, is a process of organization of orientation, of metric and temporal shifts. The arrow of time is its most advanced and its most illusory creation. In the inner of matter we are close to those phenomena physicists call quarks. Here we are confronted with a chaotic universe of emerging and vanishing polarities that are capable to contact each other and bring forth several phenomena which imprint a pattern of orderly process. But to create a time shift requires identification – time shift of what? – and needs memory and comparison to compare that which is identified and shifted by interaction with the outcome. All this is not yet very clear (though it is clear to me) and will need a lot of clever investigation. We are operating here at the interface between matter and mind, extension and cognition, space-time and logic. There is a very fine mystery in the simple structure of a small group – especially if we omit, for a moment, multiplication of an element with itself – and I will not yet reveal everything. Let us begin modestly with three polarity strings conceived as iterants

$$I_1 = [+1, +1, -1, -1]$$
$$I_2 = [+1, -1, -1, +1]$$
$$I_5 = [+1, -1, +1, -1] \equiv A_1$$

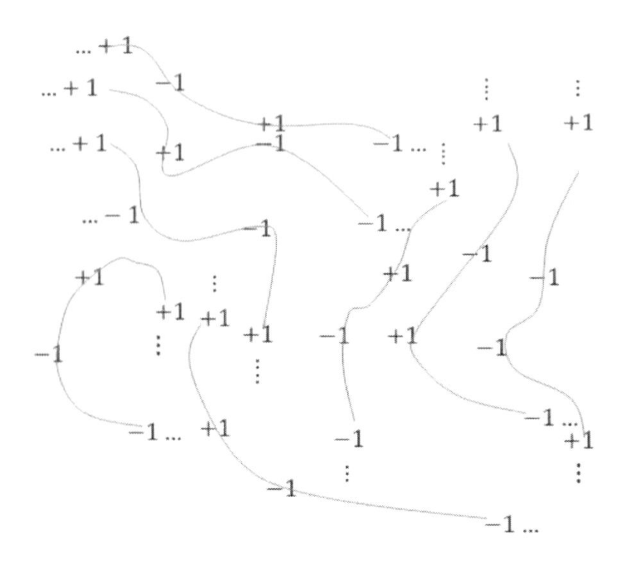

Figure 26. Iterant chaos of polarized braids in the void.

And we set these in a correspondence with three unit monomials of Clifford (or Graßmann indeed) algebra $Cl_{3,1}$where the fourth index denotes time. We must not yet understand this correspondence. The writer found it out, and you can trust him. By the time you the reader will understand it too.

$e_1 = (+1, -1, -1, +1)$ up to cyclic permutations
$e_{24} = (+1, -1, +1, -1)$
$e_{124} = (+1, +1, -1, -1)$

These polarized braids perform chaotic moves and contact each other in the void – and sure you will see, it is legitimate to call them *braids* – (The curved lines don't yet mean a thing. For now they just indicate connections.) See what those three iterants with neutral polarity are doing with each other:

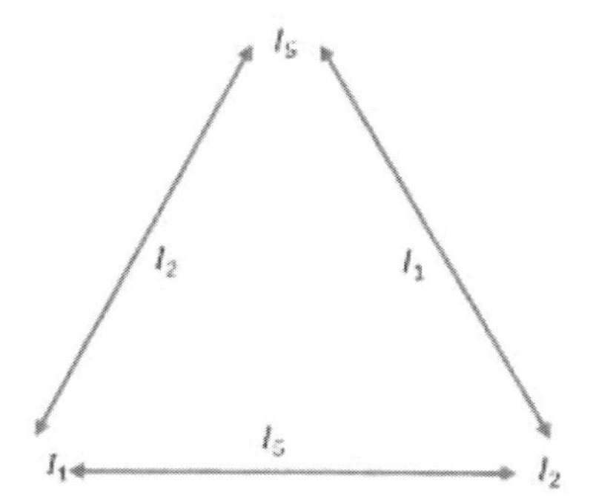

They carry out trihedral rotations thereby transmuting one into the other. It is possible to construct an oscillatory martingale such that the 3-cycle I_1, I_5, I_2 appears as an eigenform of this martingale. This is given by the recursive equation

$$Y_n = Y_{n-1}Y_{n-2} \quad \text{with logic identity as multiplication} \tag{90}$$

Beginning for example with the iterants $Y_{n-1} = I_1, Y_{n-2} = I_5$ we generate the sequence

$$I_1, I_5, I_2, I_1, I_5, I_2, I_1, I_5, I_2, \ \ldots \ \text{trihedral series of iterants} \tag{91}$$

which in the Clifford algebra $Cl_{3,1}$ shall turn out equal with the mysterious monomial sequence

$$e_{124}, e_{24}, e_1, e_{124}, e_{24}, e_1, e_{124}, e_{24}, e_1, \ \ldots$$

Let us reconsider the design of the fourfold iterants in equations 89. We can detect a doubling of 'wavelength' and respectively a halving of the frequency of recurrence of the iterant pattern as we turn from the Schrödinger iterants I_5, I_6 over to the 4-fold strings that correspond with a Clifford- and respectively a Dirac picture of synchronization. A possible wave function $\psi(x,t)$ would contain a micro-oscillatory system which would be synchronized by discrete oscillations with halved frequency.

The iterant chaos has something in common with a sea of colored fermions. And still another noteworthy feature concerning the arrow of time can be registered. While the iterant I_5 shifts the character frame forward by a quarter period

$$I_1 = [+1,+1,-1,-1] \overset{I_5}{\rightarrow} [+1,-1,-1,+1] = I_2 \tag{92}$$

the negated iterant $-I_5 = I_6$ shifts it backward

$$I_1 = [+1,+1,-1,-1] \overset{I_6}{\rightarrow} [-1,+1,+1,-1] = -I_2 = I_4$$

This simple thing carries in its heart those strange appearances of 'quantum erasure with causally disconnected choice' as we encounter in experiments where a future choice affects a past measurement's outcome. The mystery is brought in by the action of an oriented space-time area which plays a dynamic role in quantum motion without changing the laws in the generative space – the Minkowski space. The laws of Special Relativity as we are making use of today are not altered by 'spooky action'. To prove this statement which I have now only articulated in poetic ways, will require a transition from the iterant space to the Clifford algebra of the Minkowskiian 'space-time'. This is what we shall do in the next sections.

Consider the void in self-reference creating a population of iterants I_1

$$\begin{array}{ll}
\ldots +1+1-1-1+1+1-1-1+1+1-1-1\ldots & I_1 \\
\quad \textit{part of } I_1 \text{ split further into } 2\times I_5 & \\
\ldots +1 \;\vdots -1 \;\vdots +1 \;\vdots -1 \;\vdots +1 \;\vdots -1 \ldots & I_5 \\
\ldots \quad +1 \quad -1 \quad +1 \quad -1 \quad +1 \quad -1\ldots & I_5
\end{array}$$

In some unspecified area we have now two populations of iterants of the types I_1, I_5. These two have important properties, namely firstly they can interact with each other, second they form templates for material motion. While the frequency of I_1 is f the frequency of I_5 is $2f$. The cycle duration is halved by this creative splitting of the I_1. When the I_1 interacts with I_5, there is created I_2. Both I_1 and I_2 have half the frequency of I_5. We consider I_5 as an interactive element in motion that splits into a pair I_1, I_2 with double cycle duration, like signal and idler. In case that the total number of iterants I_1 minus those of them interacting with themselves is equal to twice the number of those not at all interacting, the number of types I_1 will be in balance with that of type I_2, since the I_2 stem from the I_1 that interacted with I_5. That could be a template for frequency doubling in a crystal.

Notice that the four iterants $Id = I^+$, I_1, I_2, I_5 form a vector space so that especially all 16 typical quadruples isomorphic with the logic binary connectives (89) can be generated by them. That space is further isomorphic with a small Cartan subalgebra $ch_1 = \{Id, e_1, e_{24}, e_{124}\}$ of the of Minkowski algebra, the first *chromatic space*, therefore '*ch*'. That 'color space' can be regarded as the most humble and unassuming version of commutative quaternions, as all base units squared yield $+Id$. However, as we find out next, they have a most surprising connection with Hamilton quaternions.

As we said, quantum motion is not quantum mechanics, since in quantum mechanics, the dimension of time is already derived. It is introduced into the rigor and taken for granted. Quantum motion, first of all, is constituted by polarized braids beyond any predetermined frame and so beyond time. What makes the difference of this approach to the canonic? We consider the iterants, which turn out as actual representations of geometric units, as physical realities. These strings emerging in the unmeasured void are free. They are free to interact. It is this interaction that brings upon an equilibrium state, a stable cluster of things that allow us to measure and introduce a metric for both space-time and matter. The polarity strings are alive. They ping, buzz and slosh around while being taken in by their ligature, their own attachments. It is in these liabilities that these little things are building up what we perceive as a stable portion of matter, a measurable space with a measurable arrow of time.

Entry for Dirac's Equation

Dirac's equation is like a good entry of instruments in orchestral music. In this present section it will be a re-entry to a new context. We know how cleverly this famous equation of motion makes use of base units of Clifford algebra in the form of matrices $Mat\{4, \mathbb{C}\}$, though we are not even sure if Paul Dirac knew that, as he did not mention it in his many editions of *The Principles of Quantum Mechanics* (1958). Louis Kauffman has shown how naturally the Schrödinger equation appears as an Eigenform in Quantum Mechanics, and he also has made us aware that what we called the Schrödinger iterant is indeed generative for the emergence of the hyperbolic iteration too. We can derive the Dirac equation in 'radar-coordinates' by just that same iterant which is called in informatics the alternating code. The proof is run as follows. We consider two numbers a, b as appear in the context of iterant algebra. As such they admit a rule of combination, and we assume commutativity and associativity, that is, $ab = ba$ and $(ab)c = a(bc)$. This is in accord with properties of iterant algebra. When these numbers are commutative and associative we can construct what we denote as an *invariant* $\Delta[a, b]$ by some plain equation (Kauffman 1985, p. 8)

Definition 1: invariant $\Delta[a, b] = ab$
Because of commutativity, ordering is released, so we have

$$\Delta[a, b] = \Delta[b, a] \tag{93}$$

Suppose there are numbers k that admit inverses k^{-1} so that $kk^{-1} = 1$ with left and right unit $1a = a1 = a$ for any a. Then the following identity is satisfied

$$\Delta[ka, k^{-1}b] = \Delta[a, b]. \tag{94}$$

The proving sequence of equations is

$$\Delta[ka, k^{-1}b] = (ka)(k^{-1}b) = \big((ka)k^{-1}\big)b = \big(k(ak^{-1})\big)b = \big(k(k^{-1}a)\big)b = \big((kk^{-1})a\big)b = \big((1)a\big)b = (1a)b = ab \tag{95}$$

Working with such an algebraic structure, we obtain a group of transformations having form

$$T_k[a, b] = [ka, k^{-1}b] \tag{96}$$

which preserves the magnitude $\Delta[a,b]$. Formally this is identical with the group of Lorentz transformation in one space- and one time-dimension. As we shall see now, this small structure of special relativity arises naturally by contemplating one iterant only. Namely, let us take a, b to be

$$a = t - x \qquad b = t + x \tag{97}$$

where t denotes the time-coordinate and x the space-coordinate for some observer. Let speed of light c be set equal to unity as usual. Then the quantity

$$\Delta[a,b] = c^2t^2 - x^2 = t^2 - x^2 = (t-x)(t+x) = ab \tag{98}$$

is the invariant norm for special relativity in a 2-space $\mathbb{R}^{1,1}$ with neutral signature. The term $T_k[a,b] = [ka, k^{-1}b]$ is a Lorentz transformation written in 'radar-coordinates' (Bondi 1964, 1980, Kauffman 1985).

Iterants correspond in juxtaposition and interact by touch. Any binary composition, say '∘', – if you recall logic identity and EXOR – for the ingredient numbers $(a, b, ...)$ in iterants, can be extended to the iterants themselves by defining

Definition 2: $[a,b] \circ [c,d] := [a \circ c, b \circ d]$ binary composition of iterants
Applying this to the Lorentz transformation gives iterant equations

$$[t-x, t+x] = [t,t] + [-x,x] = t[+1,+1] + x[-1,+1] \tag{99}$$

As we have $[+1,+1][a,b] = [1a, 1b] = [a,b]$ and $[0,0][a,b] = [0,0]$ we write

$$Id := [+1,+1] \text{ and } 0 := [0,0]$$

The iterant $I_6 = [-1,+1,-1,+1]$ corresponding with the 'alternating code' A_2 as a two character iterant $[-1,+1]$ is denoted by σ and called 'polarity' (Kauffman 1994).

$$\sigma := [-1,+1] \text{ polarity} \tag{100}$$

satisfying $\sigma\sigma = Id$.
So the 'Lorentz iterant' in 2D radar-coordinates takes the form

$$[t-x, t+x] = tId + x\sigma \tag{101}$$

which is an element of the real Clifford algebra $Cl_{1,0}$ spanned by units Id and σ, that is

$$Cl_{1,0} \stackrel{\text{def}}{=} span_{\mathbb{R}}\{Id, e_1\} \tag{102}$$

generated by space $\mathbb{R}^1$ with a base unit $e_1 \stackrel{\text{def}}{=} \sigma$. So this algebra is actually generated by *polarity*. Products of two elements of this Clifford algebra have the form

$$(tId + xe_1)(t'Id + x'e_1) = (tt' + xx')Id + (tx' + x't')e_1 \tag{103}$$

Or as Kauffman had put it $(t + x\sigma)(t' + x'\sigma) = tt' + xx' + (tx' + xt')\sigma$. There is only a little difference in notation, as the abstract Id allows us to lift this small structure into higher dimensional geometries without specifying the representation. Pertti Lounesto had tried to make us aware of the significance of *Study numbers* in connection with the Lorentz square norm (Lounesto 2003, p. 23f.). These numbers give us a most simple model of relativistic space-time. That is, whereas base-free representation of fermion state functions need at least a fourfold ring of real numbers, the fourfold polarity strings, the elementary swap of the 'Study numbers' proposed by Lounesto uses only pairs of reals. It can thus be seen as restricted to the phenomenology of the weak force. We can construct algebraic structures on a linear space such as $\mathbb{R}^2$ in many different ways. Component-wise multiplication $(x,y)(z,u) = (xz, yu)$ results in the double ring ${}^2\mathbb{R}$. The real algebra ${}^2\mathbb{R}$ possesses two automorphisms, namely the identity *Id* and the swap *Sw*. The swap acts as follows:

$$\text{Sw: } {}^2\mathbb{R} \rightarrow {}^2\mathbb{R}\,,\ (x,y) \rightarrow Sw(x,y) = (y,x) \tag{104}$$

Therefore it brings forth some kind of complex conjugation similar as in the complex field $\mathbb{C}$ as we have

$$Sw[a(1,1) + b(1,-1)] = a(1,1) - b(1,-1) \text{ denote the pair } (+1,-1) \text{ by } j \tag{105}$$

This is what Kauffman may have made us aware of in his 'Timespace'-paper when he says "thus the points of space-time form an algebra analogous to the complex numbers whose elements are of the form $t + x\sigma$ with the $\sigma\sigma = 1$." Note that the reflected element $j = (1,-1)$ squared gives the identity $j^2 = Id \overset{\text{def}}{=} (+1,+1)$. However, it does not give us $-Id$. Therefore, considering pairs of real numbers $(a,b) \in \mathbb{R}^2$ to form numbers

$$a + jb, \text{ with } j^2 = Id, j \neq Id \tag{106}$$

we get a peculiar conjugate $(a + bj)^- = a - bj$ together with a Lorentz squared norm $(a + jb)(a - jb) = a^2 - b^2$ and the hyperbolic polar form $a + jb = \rho(\cosh\theta + j\sinh\theta)$ for $a^2 - b^2 \geq 0$. These numbers have a matrix representation as below, this was essentially what Pertti Lounesto pointed out.

$$a + jb \cong \begin{pmatrix} a & b \\ b & a \end{pmatrix}. \tag{107}$$

In products of the Study numbers norms are preserved and hyperbolic angles are added. The linear space $\mathbb{R}^2$ endowed with the indefinite signature $\{+\,-\}$ and indefinite form $(a,b) \rightarrow a^2 - b^2$ is the hyperbolic quadratic space $\mathbb{R}^{1,1}$ which has the quadratic Clifford algebra $Cl_{1,1}$ the even part of which $Cl_{1,1}^{+}$ consists of the Study numbers. Notice, in this simple standard, the swap is represented by the first Pauli matrix.

$$j = \sigma_1 = \begin{pmatrix} 0 & 1 \\ 1 & 0 \end{pmatrix} \quad (108)$$

This can be seen as both a unit vector and a polarity in the context of iterant algebra. We chose the latter view as the whole phenomenology is made by the polarity strings and their active locations. To see how this simple frame relates to the Dirac equation, we should remember Richard Feynman. He was a master of light and a charismatic teacher. His teachings are obliging, for all we are studying is light. May the reader allow me to introduce him to a few words of Erhart Kästner who at about the same time as Feynman studied *phenomena* of light. Kästner, however, sure would have opposed if I used the word 'phenomenology' as science, analysis, for him were evidently a step away from the beauty of life, a perfidiousness to nature. He said in a chapter 'The Light of Nature': "Nature is full of signs, as we know, earth spirits there are: but how to intrude? How to suss it out? The matter is not to explore nature after all, in order to be able to better outfox her. The matter is always to advance by her to the creator. The Light of Nature. As we are a part of her there is no other way". Those people who realize the beauty of light are rare. But there are some. So we shall rely here on two other masters of light, as I see them, and do not feel ashamed to repeat their rigor here, as it is an honor to be confronted with such perfection of mathematical and phenomenological thought. As mathematicians and physicists we have no other way than to divide and follow thought, see the structure of light as related to the structure of thought, as we are part of nature. If we do it mindfully, the fruits are beyond words.

We know, it is hard to solve the general Dirac equation for any problem exactly. But there are spirits that point the way. In an exemplary investigation Louis Kauffman and Pierre Noyes (1996) rewrote in two significant forms the Dirac equation in light cone coordinates and solved them exactly in the finite difference calculus. The calculation is similar to any rigor in iterant algebra. The complex form of these equations yields 'Feynman's Checkerboard' one of the first very fine examples of path integration: a weighed sum over paths in a lattice. The second form is a real form and can be interpreted as what then was called bit-strings and what here are called polarity strings or more generally as polarized braids. These solutions are interpreted in terms of choice sequences, and the authors show how the elementary combinatorics of the imaginary unit informs the discrete physics. The imaginary unit is understood in the tradition of Hamilton's *algebra of pure time*, as an operator on ordered pairs, that is, as $i[a,b] = [-b,a]$. In this way solutions are built using only bit-strings, and no complex numbers. "Nevertheless", point out Kauffman and Noyes, "the patterns of composition of i inform the inevitable structure of negative case counting needed to build these solutions".

The Dirac equation in a space like $\mathbb{R}^{1,1}$ takes the form

$$i\hbar \frac{\partial \psi}{\partial t} = E\psi \quad (109)$$

Where the energy operator E satisfies special relativity

$$E = c\sqrt{p^2 + m^2c^2} \quad (110)$$

Where m is the mass, c the speed of light and p momentum. Dirac linearized the equation by introducing base unit elements α, β of an associative algebra commuting with c, p, m so that he obtained equation

$$E = c\alpha p + \beta mc^2 \tag{111}$$

So there follows

$$c^2(p^2 + m^2c^2) = (c\alpha p + \beta mc^2)^2 = \tag{112}$$

$$c^2p^2\alpha^2 + m^2c^4\beta^2 + c^3pm(\alpha\beta + \beta\alpha)$$

Therefore, whenever

$$\alpha^2 = \beta^2 = 1 \text{ and } \alpha\beta + \beta\alpha = 0 \tag{113}$$

these equations are satisfied. At this place let us reconsider some basic facts of Clifford algebra. We know the outstanding importance of geometric algebras for quantum physics is substantiated by their structural properties, by their idempotent lattices and their hidden Heyting algebras (Ghilardi 1992). In his celebrated book on Clifford algebras and spinors, Pertti Lounesto has worked out a substantial pictorial representation for the vector plane $\mathbb{R}^2$. To develop an appropriate picture for the Minkowski space, we shall soon have to lift Lounesto's valuable imagination. He first considered the Clifford product of a vector $\boldsymbol{r}$ with itself, which should equal the square of its length.

$$\boldsymbol{r}^2 = |\boldsymbol{r}|^2 \tag{114}$$

which is to demand

$$(x\, e_1 + y\, e_2)(x\, e_1 + y\, e_2) = x^2 + y^2. \tag{115}$$

Using the distributive rule without assuming commutativity gives

$$x^2e_1^2 + y^2e_2^2 + xy(e_1e_2 + e_2e_1) = x^2 + y^2 \tag{116}$$

which can be satisfied if the orthogonal unit vectors obey the multiplication rules

$$\begin{aligned} e_1^2 = e_2^2 &= 1 \\ e_1e_2 + e_2e_1 &= 0 \end{aligned} \quad \text{which is the same as Dirac's conditions (113)}$$

corresponding with norm and orthogonality relations $\begin{aligned} |e_1| = |e_2| = 1 \\ e_1 \perp e_2 \end{aligned}$ (117)

Calculate

$$(e_1e_2)(e_1e_2) = -e_1e_2e_2e_1 = -e_1e_1 = -1 \text{ because of (116)} \tag{118}$$

Therefore the product $e_{12} \stackrel{\text{def}}{=} e_1e_2$ squared gives the negative unit scalar from which we conclude that the quantity e_{12} is neither a scalar, nor a vector, but it is a new kind of unit, a bivector. The bivector e_{12} is said to represent the oriented plane area of the square with sides e_1 and e_2. If we compare these with the magnitudes α, β introduced by Dirac to linearize the energy operator, we realize they are the same. The α and β can be understood as geometric base units of a plane. There are even two ways to construct that plane as either a real or a complex plane. If we take the real plane $\mathbb{R}^2$, this space generates its Clifford algebra $Cl_{2,0}$. The unit vectors are $\alpha = e_1$ and $\beta = e_2$. The bivector $e_{12} \stackrel{\text{def}}{=} e_1e_2$ can be understood as a representation of the unit imaginary which usually is done so in the traditional approach of geometric algebra. If we take $\mathbb{R}^{1,1}$ as the generating space, the Clifford algebra is now $Cl_{1,1}$ with two different units; the first gives squared e_1^2=+1, the second, however $e_2^2 = -1$. The bivector is now a geometric unit with positive square norm $e_{12}^2 = 1$. So we have a correspondence between the two geometric algebras

$Cl_{2,0}$	$Cl_{1,1}$
$Id := 1$	$Id := 1$
e_1	e_1
e_2	e_{12}
e_{12}	e_2

The equation Dirac obtained can have either of two forms depending on how we interpret the units α, β. At any case we have $i\hbar\, \partial\psi/\partial t = (c\alpha p + \beta mc^2)\psi$. If we scale $c = 1,\ \hbar/m = 1$ as is usual and take for the momentum $p = (\hbar/i)\ \partial/\partial x$ the equation is

$$i\frac{\partial\psi}{\partial t} = \left(-i\alpha\frac{\partial}{\partial x} + \beta\right)\psi. \tag{119}$$

In $Cl_{2,0}$ this is the same as

$$i\frac{\partial\psi}{\partial t} = \left(-ie_1\frac{\partial}{\partial x} + e_2\right)\psi. \tag{120}$$

If we multiply this with i we obtain

$$-\frac{\partial\psi}{\partial t} = \left(e_1\frac{\partial}{\partial x} + i\, e_2\right)\psi. \tag{121}$$

Now consider that the last term, in the Clifford algebra $Cl_{2,0}$ is determined by $i\, e_2$ which squared gives $(ie_2)^2 = i^2e_2^2 = -Id$ which tells us that we are better off with the algebra $Cl_{1,1}$ in which we have $e_2^2 = -Id$. So we have cancelled the imaginary unit in the equation. This takes a most simple form in the algebra $Cl_{1,1}$:

$$-\frac{\partial\psi}{\partial t}=\left(e_1\frac{\partial}{\partial x}+e_2\right)\psi. \text{ an equation accommodated in } Cl_{1,1} \tag{122}$$

If we represent the basis in $Mat(2,\mathbb{R})$ as

$$Rep\text{:}\quad e_1=\begin{pmatrix}-1 & 0\\ 0 & +1\end{pmatrix} \text{ and } e_2=\begin{pmatrix}0 & +1\\ -1 & 0\end{pmatrix} \text{ we have } e_1^2=Id,\, e_2^2=-Id \tag{123}$$

as required. The trick to replace the complex Dirac equation by a real one was proposed probably without referring to Clifford algebra by Karmanov.[3] So we have in accord with him

$$\frac{\partial\psi}{\partial t}=\left(\begin{pmatrix}+1 & 0\\ 0 & -1\end{pmatrix}\frac{\partial}{\partial x}+\begin{pmatrix}0 & -1\\ +1 & 0\end{pmatrix}\right)\psi \tag{124}$$

This is the same as equation 8 in SLAC publication 7115 by Kauffman and Noyes (1996). If ψ is a 2-vector with two real-valued functions ψ_1,ψ_2 of x and t

$$\psi=\begin{pmatrix}\psi_1\\ \psi_2\end{pmatrix} \qquad \text{we get}$$

$$\begin{pmatrix}-\psi_2\\ \psi_1\end{pmatrix}=\begin{pmatrix}\frac{\partial\psi_1}{\partial t}-\frac{\partial\psi_1}{\partial x}\\ \frac{\partial\psi_2}{\partial t}+\frac{\partial\psi_2}{\partial x}\end{pmatrix} \tag{125}$$

The second important demand that brings us closer to polarity strings are light-cone coordinates of a point in space-time. These are defined as in equation (101) by $[r,\ell]$ $:=\left[\frac{1}{2}(t+x),\frac{1}{2}(t-x)\right]$ which yields the new 'Dirac equation' in light-cone or respectively 'radar coordinates'

$$\begin{pmatrix}-\psi_2\\ \psi_1\end{pmatrix}=\begin{pmatrix}\frac{\partial\psi_1}{\partial\ell}\\ \frac{\partial\psi_2}{\partial r}\end{pmatrix} \tag{126}$$

It is interesting to notice that Kauffman and Noyes use a first representation '*RI*' that would be the one used by Karmanov, namely

$$RI\text{: } \alpha=\begin{pmatrix}-1 & 0\\ 0 & 1\end{pmatrix},\ \beta=\begin{pmatrix}0 & -i\\ i & 0\end{pmatrix} \quad \text{in } Mat(2,\mathbb{C}) \tag{127}$$

Actually, from these units there follow the same equations (123 to 125). Because if you insert the *RI* into Dirac equation (119), you get

$$i\frac{\partial\psi}{\partial t}=\left(\begin{pmatrix}i & 0\\ 0 & -i\end{pmatrix}\frac{\partial}{\partial x}+\begin{pmatrix}0 & -i\\ i & 0\end{pmatrix}\right)\psi \text{ hence} \tag{128}$$

[3] The authors quote a private communication to H. Pierre Noyes, August 11, 1989.

$$\frac{\partial\psi}{\partial t} = \left(\begin{pmatrix}1 & 0\\0 & -1\end{pmatrix}\frac{\partial}{\partial x} + \begin{pmatrix}0 & -1\\1 & 0\end{pmatrix}\right)\psi \text{ which is} \tag{124}$$

The funny thing is that whether you start with a representation Rep as we did or with RI, in both cases you end up with equation (124). But while with R we begin with the appropriate geometric image in $Cl_{1,1}$, starting with RI we have the Clifford algebra of the Euclidean plane. For if we had $\alpha^2 = \beta^2 = Id$ and so $e_1^2 = e_2^2 = Id$ that suggests that the model is accommodated in $Cl_{2,0}$ instead of in $Cl_{1,1}$. To have a correct image of a light cone it is of course better to opt for representation Rep rather than RI.

In SLAC publication 7115 Kauffman and Noyes use a further representation

$$RII\text{: } \alpha = \begin{pmatrix}-1 & 0\\0 & 1\end{pmatrix},\ \beta = \begin{pmatrix}0 & 1\\1 & 0\end{pmatrix} \tag{129}$$

This leads to another form of the Dirac equation

$$\frac{\partial\psi}{\partial t} = \left(\begin{pmatrix}1 & 0\\0 & -1\end{pmatrix}\frac{\partial}{\partial x} + \begin{pmatrix}0 & -i\\-i & 0\end{pmatrix}\right)\psi \text{ hence} \tag{130}$$

$$\begin{pmatrix}-i\,\psi_2\\-i\,\psi_1\end{pmatrix} = \begin{pmatrix}\frac{\partial\psi_1}{\partial \ell}\\ \frac{\partial\psi_2}{\partial r}\end{pmatrix} \tag{131}$$

This is indeed correct, but again, as we can interpret the α, β as base units, we begin with a representation where $e_1^2 = e_2^2 = Id$, hence in the wrong geometric algebra $Cl_{2,0}$. If we consider however the two matrices in equation (130) which results after some multiplication with the i we seem to have a geometry of the type $Cl_{1,1}$. What we have learned from this tiny blur is that we must begin with clear images of starting tools and prerequisites of systems analysis. In our case those are geometric frames and images of temporal change. As we have today the tools of Clifford algebra and the iterant algebra, we can precisely define a dynamic system in such a way that the logic is consistent. In former times we did not know the Clifford algebraic roots of the Dirac equation(s) and made use of the i and multiplication with the i rather uncritically. But we must differ between the i as a bivector and the i as an element of time order, that is, as an iterant that can make time steps. Therefore I begun with the representation Rep in formula (123). The following rigor is closely related with the checkerboard model or 'relativistic chessboard model' which was constructed by Feynman in 1940 as a space-time approach to quantum mechanics (Feynman 1948). It was thought to give a precise sum-over-paths formulation of the kernel for a free spin ½ particle moving in one spatial dimension. It led to solutions of the Dirac equation in $(1 + 1)$-dimensional space-time in terms of discrete sums over paths in a lattice. Feynman's idea was not published in his then paper on path integrals, because it turned out difficult to generalize the result to the Minkowski space. It appeared only later in a common work with Hibbs (Feynman and Hibbs 1965). Feynman's model describes a relativistic random walk in a two-dimensional space-time lattice. Motion occurs in form of discrete time steps $c\,\Delta$ to the left or to the right, with interval Δ and speed of light usually set equal to unity. Feynman showed that each 'turn', being changes from left to right or vice versa, is weighted by the quantity $-i\Delta mc^2/\hbar$. The sum of all weighed path yields a propagator that satisfies the Dirac equation for $\mathbb{R}^{1,1}$ in the limit of vanishing checkerboard squares, $\Delta \rightarrow 0$. From this consideration there can be derived one-dimensional 'helicity' as from a cellular automaton. Feynman's creation was carrying on, and without doubt still has a value beyond words. But it has several epistemological shortcuts built in that restrict the whole discussion of freedom of motion, although the movement involves infinitely many paths. The field is thought to be a particle moving forward in (1+1)-dimensional space-time. So it gives us a forward propagator, and possible loss of micro-causality is cancelled without realizing it. There is no time-reversal in that simulation. For the iteration with a pair such as $[r, \ell] := \left[\frac{1}{2}(t+x), \frac{1}{2}(t-x)\right]$ obtrudes the illusion that the only and essential degree of freedom at any 'turn' or 'corner' is time-forward motion, that is, turnover from $+x$ to $-x$ while the t stays the same for both parts of the norm. That norm in $\{-,+\}$-Minkowski space is equal to $t^2 - x^2$. But in the 'opposite' Lorentz-metric $\{+,-\}$ it would be $x^2 - t^2$. So the appropriate iterant would rather look like $[r, l]_{opp} := \left[\frac{1}{2}(x+t), \frac{1}{2}(x-t)\right]$. That is of a special mathematical importance when we choose Minkowski space $\mathbb{R}^{3,1}$ in the opposite metric rather than $\mathbb{R}^{1,3}$ as generating space for the geometric algebra. Because in the Lorentz metric the peculiar lattice structure of primitive idempotents of $Cl_{3,1}$ provides new types of bit-strings that can give us hints as to how we could develop the idea further. So we try to handle what follows with care.

Solutions are sought in discrete calculus. For this purpose let $f(x)$ be a function of variable x and Δ a constant. Define the discrete derivative of f with respect to x by

$$D_\Delta f(x) \stackrel{\text{def}}{=} \frac{f(x+\Delta)-f(x)}{\Delta} \tag{132}$$

Consider function

$$x^{(n)} := x(x-\Delta)(x-2\Delta)\ .\ .\ .\ (x-(n-1)\Delta) \tag{133}$$

From this we obtain

$$D_\Delta\, x^{(n)} = n\, x^{(n-1)} \tag{134}$$ [4]

When Δ approaches zero, then $x^{(n)}$ approaches x^n the n'th power of x. With a binomial so called choice coefficient

$$C_n^z = \frac{z(z-1)...(z-n+1)}{n!} \tag{135}$$

$$\frac{x^{(n)}}{\Delta^n n!} = C_n^{x/\Delta} \tag{136}$$

[4] Proof is straight forward and can be found in SLAC publication 7115; arXiv:hep-th/9603202.

This is the basic equipment with which we can construct functions whose combination will give solutions to the discrete versions of the Dirac equations (126) and (131). To this end there are introduced the discrete partial derivatives $\partial_r, \partial_\ell$

$$\partial_r f := \frac{f(r+\Delta,\ell)-f(r,\ell)}{\Delta} \quad \text{and} \quad \partial_\ell f := \frac{f(r,\ell+\Delta)-f(r,\ell)}{\Delta} \tag{137}$$

Define the functions

$$\begin{aligned} \psi_R^\Delta(r,\ell) &= \textstyle\sum_{k=0}^{\infty}(-1)^k \frac{r^{(k+1)}}{(k+1)!}\frac{\ell^{(k)}}{k!} \\ \psi_L^\Delta(r,\ell) &= \textstyle\sum_{k=0}^{\infty}(-1)^k \frac{r^{(k)}}{k!}\frac{\ell^{(k+1)}}{(k+1)!} \\ \psi_0^\Delta(r,\ell) &= \textstyle\sum_{k=0}^{\infty}(-1)^k \frac{r^{(k)}}{k!}\frac{\ell^{(k)}}{k!} \end{aligned} \tag{138}$$

When Δ goes to zero these functions approach the limits

$$\begin{aligned} \psi_R(r,\ell) &= \textstyle\sum_{k=0}^{\infty}(-1)^k \frac{r^{k+1}}{k!}\frac{\ell^{k}}{k!} \\ \psi_L(r,\ell) &= \textstyle\sum_{k=0}^{\infty}(-1)^k \frac{r^{k}}{k!}\frac{\ell^{k+1}}{k!} \\ \psi_0(r,\ell) &= \textstyle\sum_{k=0}^{\infty}(-1)^k \frac{r^{k}}{k!}\frac{\ell^{k}}{k!} \end{aligned} \tag{139}$$

In a light-cone lattice the quantities r/Δ and ℓ/Δ are positive integers. In that case $\psi_R(r,\ell)$, $\psi_L(r,\ell)$ and $\psi_0(r,\ell)$ are finite sums as the quantity $x^n/n! = \Delta^n C_n^{x/\Delta}$ will vanish for sufficiently large n when x/Δ is a sufficiently large integer. Verify the identities

$$\begin{aligned} &\partial_r\psi_R^\Delta = \psi_0^\Delta,\ \ \partial_r\psi_0^\Delta = -\psi_L^\Delta, \\ &\partial_\ell\psi_L^\Delta = \psi_0^\Delta,\ \ \partial_\ell\psi_0^\Delta = -\psi_R^\Delta \end{aligned} \tag{140}$$

with $\Delta = 0$ we obtain the continuum derivatives.

Note that all calculations take the same form independent of the choice of Δ, so we can omit writing it. We are, however, aware that if r/Δ and ℓ/Δ are integers, the series bring forth discrete calculus solutions. Now as we wish to satisfy the Dirac equation in the form (126)

$$\partial_\ell\psi_1 = \psi_2,\ \ \partial_r\psi_2 = -\psi_1$$

So it follows at once that functions

$$\psi_1 = \psi_0 - \psi_L \ \ \psi_2 = \psi_0 + \psi_R \tag{141}$$

solve the Dirac equation. We can interpret the indices of the wave functions as locations and oriented moves. We have moves to the left of the light cone, moves to the right, and moves that preserve the direction. In a way we are creating space by working with these functions. Now we would like to satisfy the other Dirac equation corresponding with the equation (131)

$$\partial_\ell \psi_1 = -i\,\psi_2, \quad \partial_r \psi_2 = -i\,\psi_1$$

and solutions are now

$$\psi_1 = \psi_0 - i\,\psi_L \psi_2 = \psi_0 - i\,\psi_R \tag{142}$$

For both cases they are traced to the constituent functions ψ_0, ψ_L, ψ_R which in the lattice-case where r/Δ, ℓ/Δ are positive integers, can be rewritten as

$$\begin{aligned} \psi_R(r,\ell) &= \textstyle\sum_{k=0}^{\infty}(-1)^k \Delta^{2k+1} C_{k+1}^{r/\Delta} C_k^{\ell/\Delta} \\ \psi_L(r,\ell) &= \textstyle\sum_{k=0}^{\infty}(-1)^k \Delta^{2k+1} C_k^{r/\Delta} C_{k+1}^{\ell/\Delta} \\ \psi_0(r,\ell) &= \textstyle\sum_{k=0}^{\infty}(-1)^k \Delta^{2k+1} C_k^{r/\Delta} C_k^{\ell/\Delta} \end{aligned} \tag{143}$$

We revert the historic way of development and decompose the radar iterant $[(t+x)/2, ((t-x))/2]$ into its light cone coordinates $r = (t+x)/2$ and $\ell = ((t-x))/2$, and we construct a lattice in nicely located plane sections of the light cone in Minkowski space. The combinatorics does not change, if we set the speed of light c and Δ equal to 1. If motion were in a Minkowski space with signature $\{+1,-1\}$ we had such a plane lattice as in figure 27; any pair of numbers r, ℓ determines a rectangle with sides of length ℓ and r on the left and right pointing sections of the light cone. For better contemplation we can turn the rectangle by 45° and consider locus $A = (0,0)$ as the origin of motion for a particle that travels from the 'left' to the 'right' light cone, that is, to locus $B = (r,\ell)$. The particle is thought to make steps to the left or right on the lattice path, and it travels every possible path but forward time. It is better to imagine the *field*, whatever it may be, to make a series of decisions if it goes left or right. These decisions constitute the path. To represent any path under consideration we sketch the oriented corners by which it is constituted by the most basic 'elements of orientation'. These are the elementary images of spin

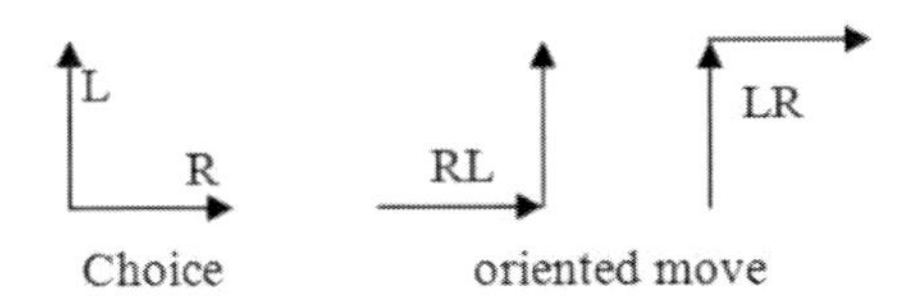

We denote field motion which preserves the direction by RR and respectively LL. A path may take off at R and terminate in L. Or it may begin in L and terminate in R. Then we call it an RL-path. When a choice is made, we imagine an oriented move RL or LR. These moves where the path increments are the turns Feynman counted on the checkerboard. So we have altogether 4 possibilities of 'two step' elements of a path of the types RR, RL, LR and LL. These elements are counted. Any path of type RL takes k locations from the R axis and $k+1$ locations from the L axis. On the other hand any path of type LR involves $k+1$ locations from the R axis and k locations from the L axis. Paths that preserve orientation take equal numbers of k locations from each axis. An example is portrayed in Figure 28.

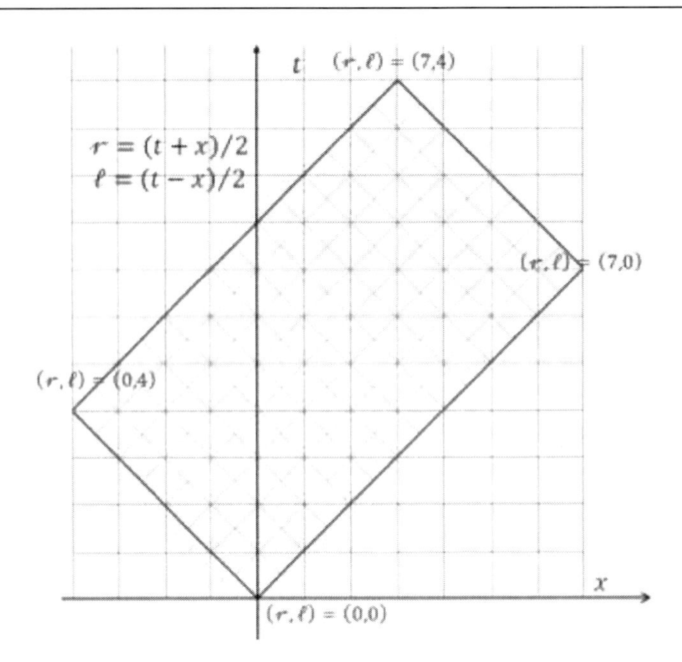

Figure 27. Checkerboard light cone lattice frame.

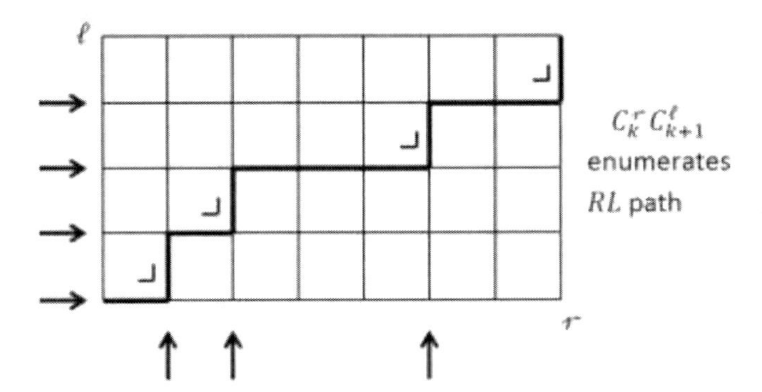

Figure 28. Enumeration of paths in the light cone lattice frame.

In the example we have paths with 4 elementary oriented moves of type RL and $r = 7, \ell = 4$. We obtain

$$C_k^r = C_3^7 = \frac{7\times6\times5}{3\times2} = 35; C_{k+1}^\ell = C_4^4 = 1$$

a total of 35 such paths.

Let $\|XY\|$ denote the counted number of paths from the initial location (0,0) to the terminating location (r, ℓ) in the lattice, then

$$\|RL\| = \sum_k C_k^r C_{k+1}^\ell \qquad (144)$$
$$\|LR\| = \sum_k C_{k+1}^r C_k^\ell$$

$$\|RR\| = \|LL\| = \sum_k C_k^r C_k^\ell$$

So the functions $\psi_R(r,\ell), \psi_L(r,\ell), \psi_0(r,\ell)$ turn out to be weighted sums over the different types of lattice paths, where the weights are given by magnitudes $(-1)^k$, that is, the authors interpret the weights by the number of choices and respectively corners within the paths. This corresponds with what Feynman did (Feynman and Hibbs 1965). Consider the re-interpretation

Paths types $RR \Longrightarrow 2k$ corners
$LR \Longrightarrow 2k+1$ corners
$RL \Longrightarrow 2k+1$ corners
$LL \Longrightarrow 2k$ corners

Hence if $N_c(XY)$ denotes the number of paths with c 'corners' of type XY then

$$\psi_0 = \Sigma_c(-1)^{\frac{c}{2}} N_c(LL) = \Sigma_c(-1)^{\frac{c}{2}} N_c(RR) \tag{145}$$
$$\psi_R = \sum_c (-1)^{\frac{c-1}{2}} N_c(LR)$$
$$\psi_L = \sum_c (-1)^{\frac{c-1}{2}} N_c(RL)$$

The authors realized then that "from the point of view of the solution to the RI Dirac equation ($\psi_1 = \psi_0 - \psi_L$, $\psi_2 = \psi_0 + \psi_R$) it is an interesting puzzle in discrete physics to understand the nature of the negative case counting that is entailed in the solution". It is indeed possible to answer this important question, if we go deeper into the nature of a corner. Obviously this involves a dynamic element that is related to the oriented areas in Minkowski algebra.

The solution (142) to the *RII* Dirac equation provides a different point of view. Being aware that we have $i^{2k} = (-1)^k$ whereas $i^{2k+1} = (-1)^k i$ we can represent the solutions by formulas

$$\psi_1 = \Sigma_c(-i)^c N_c(R), \qquad \psi_2 = \Sigma_c(-i)^c N_c(L) \tag{146}$$

where $N_c(R)$ denotes the number of paths that go to the right and have c crossings, while $N_c(L)$ is the number of paths that go to the left with c crossings. This is essentially equal to the solution presented by Feynman and Hibbs in their checkerboard model of the Dirac propagator. The problems that remained are related to the difference between dynamic elements that come in at the very beginning: the meaning of the 'i', the meaning of an oriented area and the meaning of an oriented space-time area. There are important differences.

The Concept of Pre-Spinor in Clifford Algebra

What is a crossing in the checkerboard model, what is a corner? If we consider this small figure once more

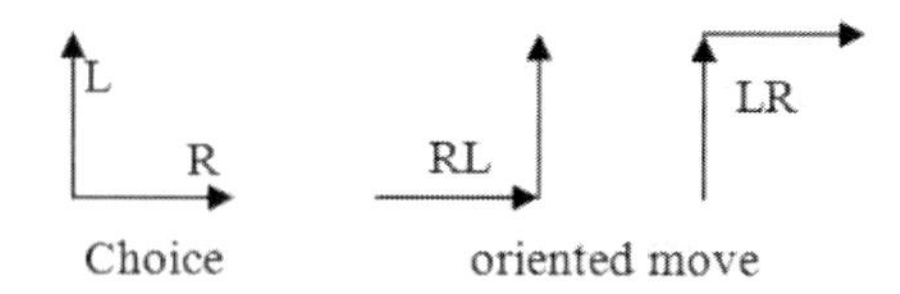

and look at it relaxed and a little bit beyond context. We find out, the first partition looks like two different lines originating in the same point and starting off into different directions. The second figure seems to represent two different lines with different points of departure, but where the 'first' ends the second begins. The same holds for the third partition of the figure. But is a turnover from the second to the third image a self-evident thing? Is it easy to understand? As the lines are directed arrows, the second graph seems to indicate a counterclockwise rotation, while the third graph sketches a clockwise direction. There is a little illusion involved: the images look like line sequences. But they are oriented areas. In the Clifford algebra of the Euclidean plane $Cl_{2,0}$ the second graph would indicate the oriented area $e_{12} = e_1 \wedge e_2$, namely the exterior product of the two unit vectors e_1, e_2, the arrows. As this exterior product does not commute, but anti-commute, that is,

$$\{e_{12}, e_{21}\} = e_{12} + e_{21} = 0. \tag{147}$$

The oriented area e_{12} is different from e_{21}, namely its negative, that is, the 'corner' RL is really *another corner* than LR. First of all, the difference is geometric, but second it is dynamic. This cannot immediately be seen and requires a little meditation. Namely, the base units e_1, e_2 generate by Clifford multiplication the discrete multivector group of the Clifford algebra $Cl_{2,0}$. This has eight elements, namely $\{Id, e_1, e_2, e_{12}, -Id, -e_1, -e_2, -e_{12}\}$. This small algebraic group is classified as the dihedral group D_{2d} for $d = 2$ diagonals as below

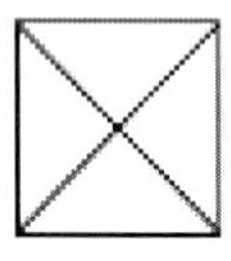

also called the space congruence group D_4 of the square. Why space congruence? Well, the group has two elements with period 4. Call them S_4 and S_4^{-1}. The first turns the square, say, counter clockwise, then the inverse turns it clockwise. Now, how can S_4, a counter clockwise rotation in the plane, be transformed into a clockwise rotation S_4 in the same plane? This can only be performed by a spatial operation that goes beyond the operations in the plane, namely by a *spatial* diagonal flip. By some authors in space groups and crystallography (Belger, Ehrenberg 1981) the eight elements of the dihedral group D_{2d} have been listed as $\{E, C_2, C_2', C_2'', S_4, S_4^{-1}, \sigma_d', \sigma_d''\}$. The eight elements are graphically represented in figure 29. These symmetry elements of the space group can be represented by permutation cycles, by

Schönfließ-symbols (Schmeikal 2000, p. 272ff.) and in some infinitely more ways. I give here the irreducible faithful representation in the real matrix algebra $Mat(2,\mathbb{R})$

$$E = \begin{pmatrix} 1 & 0 \\ 0 & 1 \end{pmatrix} \simeq Id,\ C_2 = S_4^2 = \begin{pmatrix} -1 & 0 \\ 0 & -1 \end{pmatrix} \simeq -Id,\ C_2^{'} = \begin{pmatrix} -1 & 0 \\ 0 & 1 \end{pmatrix},\ C_2^{''} = \begin{pmatrix} 1 & 0 \\ 0 & -1 \end{pmatrix} \tag{148}$$
$$S_4 = \pi_1 = \begin{pmatrix} 0 & -1 \\ 1 & 0 \end{pmatrix} \simeq i,\ S_4^{-1} = S_4^3 = \begin{pmatrix} 0 & 1 \\ -1 & 0 \end{pmatrix},\ \sigma_d^{'} = \begin{pmatrix} 0 & 1 \\ 1 & 0 \end{pmatrix},\ \sigma_d^{''} = \begin{pmatrix} 0 & -1 \\ -1 & 0 \end{pmatrix}$$

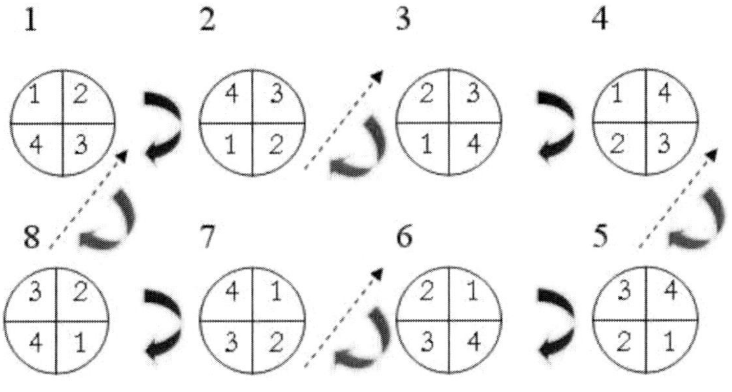

Figure 29. Elements of D_4.

The preferred direction of the arrow of time is connected with the non-commutative structure of the dihedral reorientation group. This is a natural source of asymmetry of time, since the sequential carrying out of a turn by a 4-cycle S_4, generated by the i, and followed by a swap, in the form of a diagonal flip or reflection on a diagonal, is not abelian, that is, we have $S_4 \times flip \neq flip \times S_4$. Now, let us go slowly and consider these facts very seriously. There is a sequence of analogies that we have to ponder over. If we consider integer powers of the imaginary unit, we indicate a directed area with clockwise orientation.

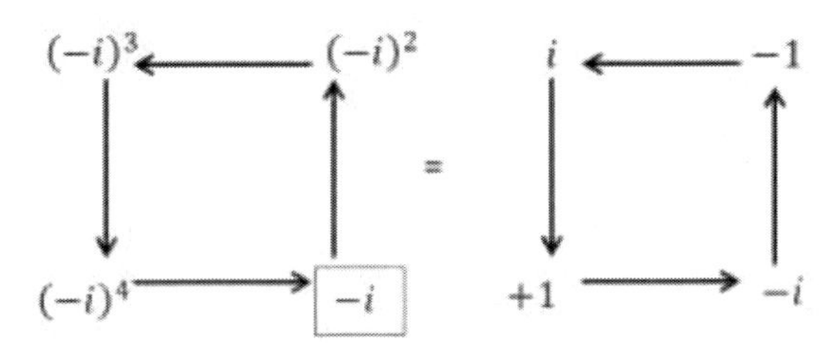

Figure 30. Powers of the imaginary unit and its inverse.

If we consider powers of $-i$ we obtain a directed area with opposite orientation (figure 30). Clearly, these observations are the same, if we consider powers of the spatial bivector e_{12}, and compare this with powers of the reverted bivector e_{21}. The rolling direction inverts. The same holds for iterants $[-1,+1]\eta$ that represent the imaginary unit. The same is true for the grade 4 iterant time t which we may represent for short as a permutation 4-cycle:

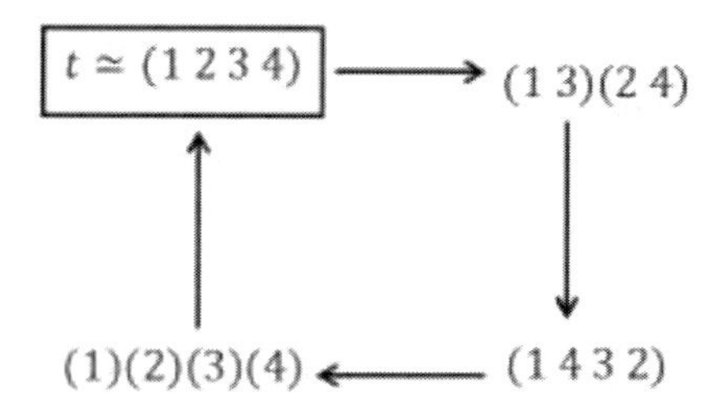

Figure 31. Powers of bivectors, the unit imaginary iterant, iterant time and their inverses.

A 'choice-sequence' such as $RLRRLRRLLRRRL$ involves the elementary transitions from $r = (t + x)/2$ to $\ell = \big((t - x)\big)/2$. But additionally there occurs reversion in the orientation of directed areas. Which directed areas? Recall our takeoff to the Dirac equation with representation R in Clifford algebra $Cl_{1,1}$ (123) which is indeed the proper representation.

$$R: \quad e_1 = \begin{pmatrix} -1 & 0 \\ 0 & +1 \end{pmatrix} \text{ and } e_2 = \begin{pmatrix} 0 & +1 \\ -1 & 0 \end{pmatrix} \text{ we have } e_1^2 = Id,\, e_2^2 = -Id$$

Any transition of the type RL involves a turnover from $r = (t + x)/2$ to $\ell = \big((t - x)\big)/2$. This is corresponding with a reversion of the space-time area $e_{12} \to e_{21} = -e_{12}$. This is 'reversion' in the proper sense of geometric algebra. And within this picture of the Dirac equation of motion the real solutions (126) follow quite naturally. But why is there no $'i'$ coming out in the calculation of the paths integral? Simply because bivector $e_{12} \in Cl_{1,1}$ contributes to positive signature, we have $e_{12} \in Cl_{1,1} \Rightarrow e_{12}^2 = Id$ as opposed to $Cl_{2,0}$. We can draw a picture to show a most essential difference between $Cl_{1,1}$ and $Cl_{2,0}$.

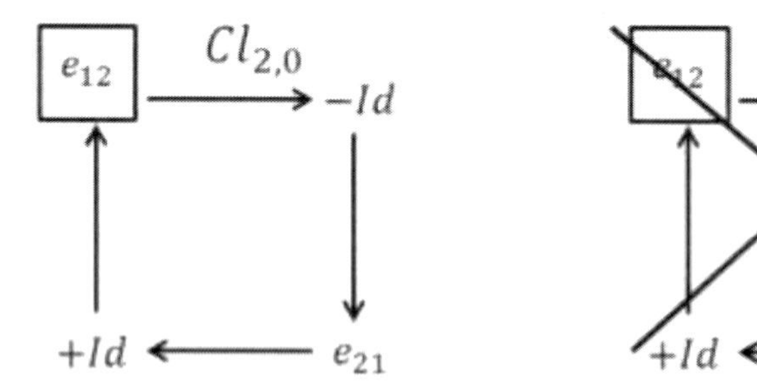

Figure 32. In signature $\{+,-\}$ the $e_{12} \not\simeq i$.

Geometric Action of the Time Shift Operator

Although the unit bivector does not give us an "imaginary square" with a cyclic group structure $\mathbb{Z}_4$, we nevertheless have a directed unit space-time area e_{12}, and we dare say that we have

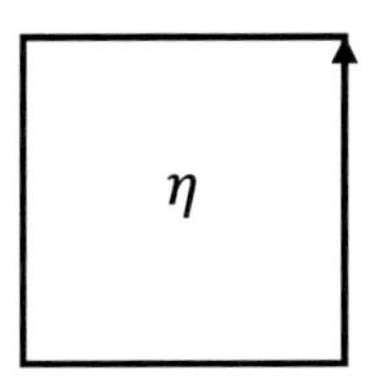

Kauffman and Noyes discerned a *pre-spinor* as they wrote: "In our *RI* formulation, no complex numbers appear and none are needed if we take a combinatorial interpretation of i as an operator on ordered pairs $i[a,b]$=$[-b,a]$. Then we can think of a "pre-spinor" in the form of a labeled $\pi/2$ angle associated to each corner:

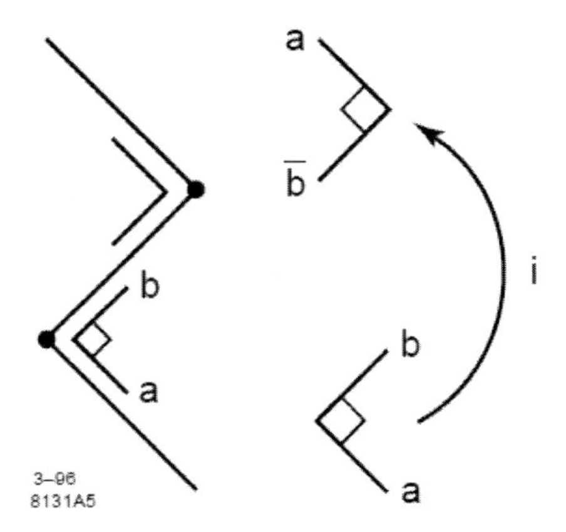

As the particle moves from corner to corner its pre-spinor is operated on by i. There is a combination of one sign change and one change in order. The total sign change from the beginning of the path to the end documents the positivity or negativity of the count".

Theorem 3: The grade 2 iterant algebra of pre-spinors is a Clifford algebra $Cl_{1,1}$ of spinors

Lemma 1: A directed unit space-time area squared gives the identity $+Id$. Proof: The spatial unit squared $e_1^2 = Id$, whereas the temporal unit squared gives $e_2^2 = -Id$. Therefore, given that $e_{12} := e_1e_2$, there follows $e_{12}^2 = e_1e_2e_1e_2 = -e_2e_1e_1e_2 = -e_2Ide_2 = -Ide_2e_2 = -Id(-Id) = Id$. Therefore a unit directed space-time area cannot replace the imaginary unit. Consequently we can ask what could be the counterpart of a directed space-time area in the iterant context?

Lemma 2: The iterant algebra of degree 2 is a Clifford algebra of the type $Cl_{1,1}$

Proof: take as base units $e_1 = [-1,+1]$ and $e_2 = [-1,+1]\eta$. We have

$e_1e_1 = [-1,+1][-1,+1] = [+1,+1] = Id$ and also

$e_2e_2 = [-1,+1]\eta[-1,+1]\eta = [-1,+1][+1,-1]\eta\eta = [-1,-1]Id = -Id$.

The unit bivector is $e_{12} := e_1e_2 = [-1,+1][-1,+1]\eta = [+1,+1]\eta = Id\,\eta = \eta$, hence $e_{12}^2 = \eta^2 = Id$. So we finally found out that the temporal shift operator η is a directed unit space-time area. The η does not generate the cyclic group $\mathbb{Z}_4$, but instead $\mathbb{Z}_2$. Yet, it represents faithfully the director or "oriented volume" in $Cl_{1,1}$. The Clifford algebra $Cl_{1,1}$ is isomorphic with the matrix algebra $Mat(2,\mathbb{R})$. It is a simple algebra with the squared oriented volume equal to $+Id$. The primitive idempotents in the standard basis are equal to $f_{\pm} = (Id \pm e_{12})/2$. Automorphism groups of spinors, for reversion is $O(1,1)$ and for conjugation $Sp(2,\mathbb{R})$. So if we take the $f_{+} = (Id + \eta)/2$, then the space $\mathfrak{S} \stackrel{\text{def}}{=} f_{+}Cl_{1,1}$ is our prespinor-space. In which manner does the time shift operator act on the Lorentz iterants? We first consider the alternating code of the iterant of any grade $2n$ and its associate time shift of order two, that is we have iterants essentially given by the short term $[a,b]$ and a time shift η that acts like

$$\eta[a,b] = [b,a]\eta \text{ with } \eta^2 = Id \tag{149}$$

We showed in theorem 3, the grade 2 iterant algebra of pre-spinors is a Clifford algebra $Cl_{1,1}$ of spinors. The time shift turned out to be an oriented space-time area of typical signature $+1$. It would be interesting to know more about the geometric effect of the time shift on an arbitrary element of the iterant algebra.

The effect of the action of a unit space-time area η on a Lorentz iterant $L(t,x) = [t-x, t+x]$ is that of a rotation. A rotation R carries an element L to an element RLR^{-1}. As the η is its own inverse, we have

$$\eta L(t,x)\eta = \eta(t\,Id + x\,e_1)\eta = t\,Id - x\,e_1 = [t,t] - x[-1,1] = [t+x, t-x] = L(t,-x)$$

Hence

$$\eta\colon\ [t-x, t+x] \longrightarrow [t+x, t-x] \tag{150}$$

or in terms of the checkerboard lattice coordinates

$$\eta\colon\ (r,\ell) \longrightarrow (\ell, r) \tag{151}$$

which is a clockwise rotation of the frame by $\pi/2$. If we let η act on this result

$$\eta L(t,-x)\eta = \eta(t\,Id - x\,e_1)\eta = t\,Id + x\,e_1 = [t,t] + x[-1,1] = [t-x, t+x] = L(t,x)$$ hence

$$\eta\colon\ (\ell, r) \longrightarrow (r,\ell) \tag{152}$$

The transformation works in both directions which corresponds to clockwise and counter clockwise rotations of the lattice.

Lusona: Plane Dynamic Oriented Area and Motion which Creates a Frame

As we saw, the features of orientation, the left-right business and the logic we use have to do with extension and cognition. They are not mere objective predicates of space-time. Rather the concepts of space and time were constructed by us in such a way that these images correspond with the process of nature. The laws of shifts, corners and turns that accompany motion mean a structural givenness that is intrinsic to both thought and nature. This fact is familiar to various degrees in civilizations and pre-civilized populations. There are some cultures where social life and history are combined with geometry in a special way. Interestingly our discrete considerations in quantum space time can learn from them. I will give us two examples of *(Lu)Sona* that will improve our knowledge about the role of the *'i'* in motion and story-telling without giving a complete theory of the mathematics of Lusona. The interested reader may find those riches in a book on ethno-mathematics by Paulus Gerdes (1997).

The Chokwe people of northeastern Angola, and adjacent countries, are well known for their sand drawings. Their drawings in the sand are called Sona (singular lusona). These drawings are used both in their decorative artwork and in their story-telling tradition. To draw lusona, the Chokwe smooth an area of sand, and then set out with their fingertips a net of points. These are then used in drawing the figure. The distances between horizontal or vertical neighboring points should be the same. The number of points depends on the story. When telling their stories through these pictures, the drawing process often parallels the development of the story, and these points may represent characters in the stories. What is so peculiar about lusona is its monolinearity. Gerdes (1997, p. 36) pointed out that monolinearity can represent a cultural value. Fontinha (1983) collected and published 141 drawings of ten different types, and 61 of them were monolinear, that is, drawn in one line. They were drawn by eight different masters, the *akwa kuta sona.* One of them, named Samesa, drew 13 sona of which 12 were monolinear. To draw a lusona without interruption requires a high degree of attention. In high attention the feeling of time disappears. The following sona is called the fleeing cock.

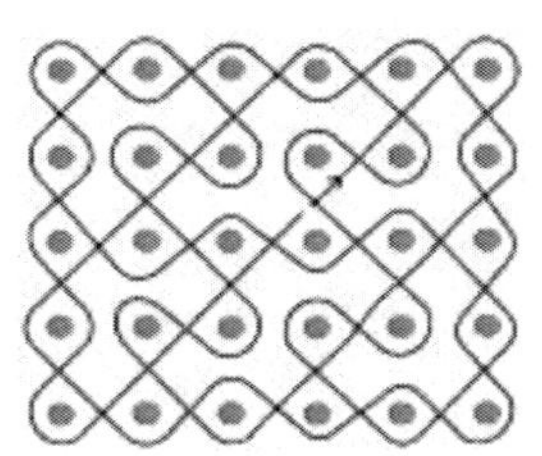

Figure 33. LuSona 'Fleeing Cock'.

Gerdes (1997, p. 280) had discovered that this pattern essentially represents a magic array modulo 8. For any row, sums taken modulo eight give a remainder of 2.

106	105	69	70	102	101	33	34	98	97	5	6
107	68	104	103	71	32	100	99	35	4	96	7
67	108	76	75	31	72	40	39	3	36	8	95
66	109	77	30	74	73	41	2	38	37	9	94
110	65	29	78	62	61	1	42	58	57	93	10
111	28	64	63	79	120	60	59	43	92	56	11
27	112	84	83	119	80	48	47	91	44	12	55
26	113	85	118	82	81	49	90	46	45	13	54
114	25	117	86	22	21	89	50	18	17	53	14
115	116	24	23	87	88	20	19	51	52	16	15

826	2
826	2
650	2
650	2
666	2
786	2
802	2
802	2
626	2
626	2

Figure 34. Framing and natural counting of the route segments in the Sona 'Fleeing Cock'.

If, however, we replace the natural counting by a modulo 4 enumeration, we obtain the following pattern. This can be seen as a sequence of oriented unit areas, like $e_{12}, e_{21}, e_{12}, e_{21}, e_{12}, e_{21}$ for each row.

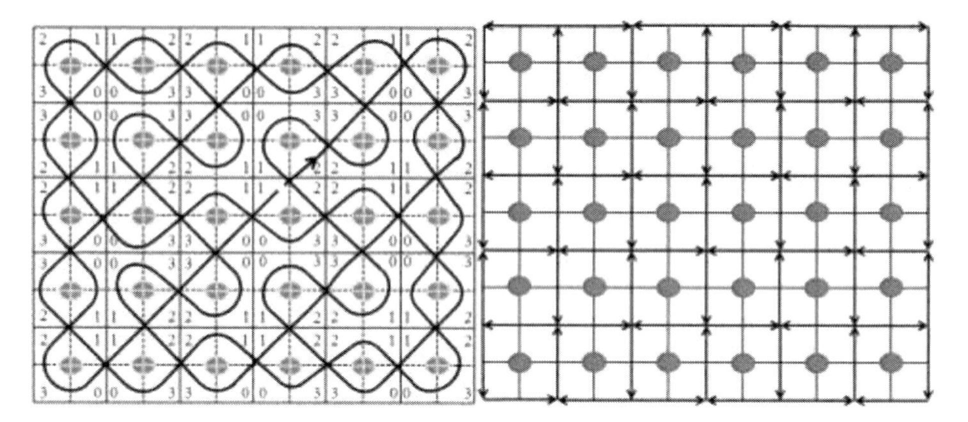

Figure 35. How motion in Lusona creates an oriented Peano Fractal Frame.

We began with a fleeing cock, analyzed its motion and obtained an orderly pattern of whirls in the form of a Peano curve. There are important lessons we learn from this 'natural process'.

- Motion creates the frame
- The imaginary unit orders the pattern
- Every advanced path has a retarded double
- Time reversion is a transition $e_{12} \to e_{21}$ and respectively $[+1,-1]\eta \to [-1,+1]\eta$

This could be a lesson for discrete quantum motion: it is motion that brings about space-time. Our concept of space-time is resulting from the process of motion.

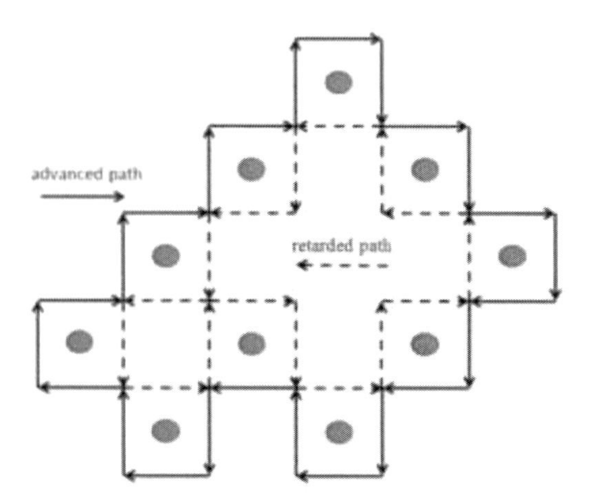

Figure 36. The Peano fractal frame results from advanced and retarded paths.

In many processes, as in positrons travelling backward in time, the noncommutativity of the dihedral reorientation group can be seen as the origin of time reversion. But it is not a general fact. An alternating series such as $e_{12}, e_{21}, e_{12}, e_{21}, e_{12}, e_{21}, \ldots$ may very well represent a mere noise pattern. As a *choice sequence of orientation*, the forward path in figure 36 can be seen as a series of ten directed areas: $e_{12}, e_{12}, e_{12}, e_{12}, e_{12}, e_{12}, e_{12}, e_{12}, e_{12}, e_{12}$. The backward path is resulting from Clifford reversion in the pure algebraic sense: $e_{21}, e_{21}, e_{21}, e_{21}, e_{21}, e_{21}, e_{21}, e_{21}, e_{21}, e_{21}$. This can just as well be written in the form $-e_{12}, -e_{12}, -e_{12}, -e_{12}, -e_{12}, -e_{12}, -e_{12}, -e_{12}, -e_{12}, -e_{12}$ which is telling us that the rotational direction has changed. Such change of orientation can indeed be traced back to the difference between the particular order in which two operations are carried out. Linear propagation, in that context, seems rather boring:

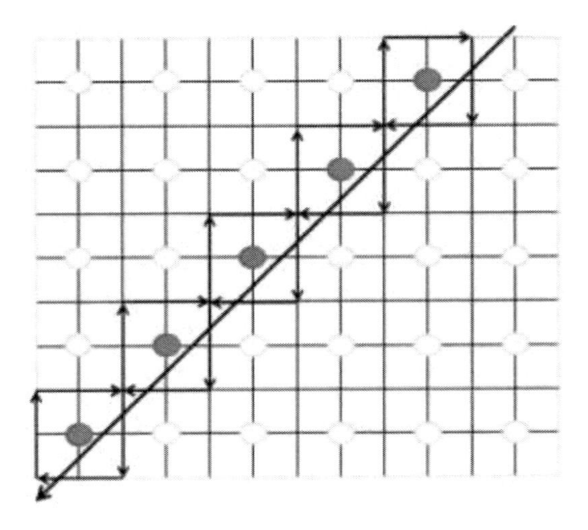

Figure 37. Linear propagation creating sequence $e_{12}, e_{12}, e_{12}, e_{12}, e_{12}$.

Let us consider those generative symmetries of planar reorientation:

$C_2' := (1\ 2)(3\ 4)$ a flip or 'period 2 rotation about a main (say vertical) axis
$\sigma_d' := (2\ 4)$ a diagonal flip

Both symmetry operations are represented as permutations of quarters in cycle format. Now carry out the diagonal flip first and the vertical flip thereafter, this gives

$$\sigma_d' C_2' = (1\ 2\ 3\ 4) \simeq t \qquad (153)$$

which is isomorphic with what we introduced as period 4 iteration time for fourfold iterants. If we revert the order of multiplication, we get

$$C_2' \sigma_d' = (1\ 4\ 3\ 2) \simeq t^{-1} \qquad (154)$$

isomorphic with reverse iteration time. A pretty graphic representation for such performance of operations stems from chemistry. See the effect of noncommutativity on the chirality of an Allen-molecule C_2H_4.

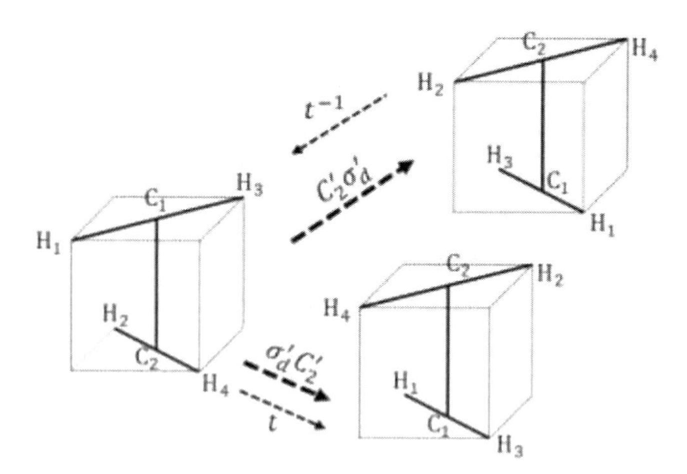

Figure 38. Effect of non abelian symmetries of Clifford algebra on the helicity of a molecule.

Both operations $\sigma_d^{'}C_2^{'} \simeq t$ and $C_2^{'}\sigma_d^{'} \simeq t^{-1}$ turn the molecule upside down, but relative to one another the two have opposite helicity. Now, please, recall what we said in the section 'Shaping self reference and symmetry breaking with dihedral group D_4'! Both molecules look absolutely equal. We would not be able to make out any difference, if we would not introduce some kind of symmetry breakage that makes the difference visible at all. To quote Gregory Bateson, here it is really the difference that makes the difference. We could not realize the symmetry operation without introducing some asymmetry.

Now, if it were possible that this simple difference between spatial operations can make the difference between iteration time and inverse iteration time, again there must be something that realizes that difference, in the sense that it makes that difference part of reality. Usually we assume that it is the human cognition that perceives and constructs the difference. But there must be some memory built into the process itself in order that the image of what has been and how that differs from what is now can be realized. If it is true that the whole process goes on in presence, and in presence only, if the Eleates are right, and if I am right, there is some structure and some phenomenon always going on that allows the present (bit string, polarized braid etc.) to realize its appearance in relation to what it was before. Both the what is and the what was are built into the present structure which includes memory and comparison. This does not mean that the presence has no temporal extension, on the contrary. But the installation of an arrow of time in that movement is not self-evident and remains a bit of a mystery.

Chapter 5

EMERGENCE OF PRIMORDIAL MINKOWSKI FRAMES

EXTENSION OF THE ITERANT MODEL

Richard Feynman established a very close connection between quantum mechanics and space-time. One can get a very convincing impression of this alliance reading the book on Quantum Mechanics and Path Integrals coauthored by Hibbs (Feynman and Hibbs 1965, Problem 2-6, pp. 34-36). It took twenty years until Feynman's original idea on path integrals was published therein. The model had not been included in his original paper on path-integrals since a generalization of the method to dimension 4 could not be found then. One of the first interpretations of the propagator in dimension 1+1 in terms of Feynman's "amplitudes" was worked out by Narlikar (1972). The denotation 'Feynman chessboard model' was coined by Gersch (1981) who saw the relation to the small Ising model involving 2 neighboring spins. Gaveau (1984) investigated a relationship between the checkerboard model and the Brownian motion in telegraph equations due to Kac (1974) by analytic continuation. Ord (1992) showed that the model could be derived without analytical continuation as a classical system. And at about the same time Kauffman and Noyes gave the fully discrete version in terms of polarity strings. So it took more than fifty years until we could be sure of this matter. It is known from his archived notes that Feynman wanted to establish a link between the imaginary unit and his path integral for a relativistic particle travelling in one dimension. He wanted to apply this to his discovery with Wheeler that antiparticles are travelling backwards in time. His notes contain several sketches of paths with space-time loops and spiral models. Further extensions of the original model involve generalized lattices. However in those complex pictures the property of a single isotropic speed of light is lost. It is well known meanwhile that the generalization of the discrete checkerboard path integral method to Minkowski space is a very difficult thing. The reason for this much misunderstood difficulty is in the lack of awareness that motion is a matter of the algebra that is generated by the space wherein the field moves. The quantum mechanics is taking place in indefinite space-time locations with an iterant structure such that the outcome is the Clifford algebra generated by a space-time of the type Minkowski. It would be straight forward to take as a hint that the finite multivector group of the Minkowski space is a product of two dihedral groups. As $Cl_{3,1} \simeq Cl_{2,2}$ we would guess the following

Theorem 4: The grade 4 iterant algebra of pre-spinors is a Clifford algebra $Cl_{2,2}$ of spinors and give the proof a little later. This is actually true. But the situation becomes much more complicated than it is suggested by the apparent simplicity of the theorem, since there are entirely new degrees of freedom coming in as we investigate the reorientation lattice within $Cl_{2,2}$. This is of a much higher order than in the iterant algebra involving two grades only.

Quaternions, Quad Locations and Tangle Time

In accordance with the idea of primordial observation and the symmetries of the octahedral permutations S_4 we shall now define some time shift operators η and t, that will turn out fundamental for the iteration of fourfold locations in a relativistic space-time. In this way we shall also establish a relation between polarity strings and braids of grade 4 with the quaternion subspaces of the Minkowski space.

Definition 2: quad-locations are fourfold locations given by iterant views having form $[a, b, c, d]$.

Examples:
$P_1 = [+1, +1, +1, -1] \stackrel{\text{def}}{=} ... +1+1+1-1+1+1+1-1+1+1+1-1...$
$I_2 = [+1, -1, -1, +1] \stackrel{\text{def}}{=} ... +1-1-1+1+1-1-1+1+1-1-1+1...$

Definition 3: of time shift operators. Sequences are iterated by *iterant time* t which satisifies

$$t\,[a, b, c, d] = [b, c, d, a]\,t \text{ with } t^4 = Id$$

and by *tangle-time* η which is the period-2 iterant time shift that we already know but now applied to iterants of degree 4 and satisfying

$$\eta[a, b, c, d] = [b, a, c, d]\eta \text{ with } \eta^2 = Id$$

Examples:
t acting on P_1 brings forth $... +1+1-1+1+1+1-1+1+1+1-1+1...$
tangle time η acting on I_2 results in $... -1+1-1+1-1+1-1+1-1+1-1+1...$

Theorem 5: Iterant views plus two time-shifts according to definitions 2, 3 generate quaternions.

Proof: Using the iterant units e_1, e_2, e_3 [1]

$$e_1 := I_1 = [+1, +1, -1, -1]; \quad (155)$$
$$e_2 := I_5 = [+1, -1, +1, -1];$$

[1] We change notation in order to obtain Hamilton's quaternions from unit iterants.

$e_3 := I_2 = [+1, -1, -1, +1]$ we calculate

$$[+1, +1, -1, -1]\, t\, t\, [+1, +1, -1, -1]\, t\, t = [+1, +1, -1, -1]\, t\, [+1, -1, -1, +1]\, t^3 = [+1, +1, -1, -1]\, [-1, -1, +1, +1]\, t^4 = [-1, -1, -1, -1] = -Id$$

Summing it up we write for this equation

$$e_1\varphi = e_1 t^2 = I \text{ with } I^2 = -Id = I^- \qquad (156)$$

indicating thereby that applying the 4-cycle t twice we obtain permutation $\varphi = (1\,3)(2\,4) \in \boldsymbol{S}_4$. Recall, we permute characters in the iterants. Next consider the term $e_2\eta t^2\eta$ which is the same as $e_2\tau$ and calculate

$$[+1, -1, +1, -1]\, \eta t^2 \eta\, [+1, -1, +1, -1]\, \eta t^2 \eta =$$
$$= [+1, -1, +1, -1]\, \eta t^2\, [-1, +1, +1, -1] Id\, t^2 \eta =$$
$$= [+1, -1, +1, -1]\, \eta\, [+1, -1, -1, +1]\, t^4 \eta =$$
$$= [+1, -1, +1, -1]\, [-1, +1, -1, +1] Id\, \eta^2 = -Id$$

We write $e_2\tau = e_2\,\eta\varphi\eta = J$ with $J^2 = -Id$ (157)

Last we construct the term $e_3\eta t^2 \eta t^2 = e_3\tau\varphi$ in agreement with the Cayley graph of $\boldsymbol{S}_4$ (figure 22) with the period-2 cycle $\sigma \stackrel{\text{def}}{=} \tau\varphi = (1\,2)(3\,4)$. The shift σ acts similar like η. That is, we can briefly calculate

$$[+1, -1, -1, +1]\, \sigma\, [+1, -1, -1, +1]\, \sigma = [+1, -1, -1, +1][-1, +1, +1, -1]\, \sigma^2 = -Id$$

So we put $e_3\sigma = e_3\tau\varphi = K$ with $K^2 = -Id$ (158)

These quantities derived from the iterant views

$$I = [+1, +1, -1, -1]\, \varphi$$
$$J = [+1, -1, +1, -1]\, \tau$$
$$K = [+1, -1, -1, +1]\, \tau\varphi$$

satisfy the multiplication table of the quaternions. One can easily verify

$$I^2 = J^2 = K^2 = IJK = -Id,\ IJ = K \text{ and so on.} \qquad (159)$$

What can we learn from this?

We can construct quaternions from quad-iterant views by two temporal shift operators. Now consider those quad-locations coherent like in a solid object, rather compact and connected. We have four cohesive locations. As soon as we apply the tangle-time operator, the object is tangled and decomposed. However, if we apply the shift-operator t locations are merely rotated, but not disaggregated. It is therefore that the η was called '*tangle-time*'. Tangle-time preserves the quaternion I and of course the Id, but it changes quaternions J, K. It is on the basis of these temporal shifts and the existence of fourfold locations that the relativistic space-time and respectively its Minkowski algebra can be thought to be brought about. Notice that until to this paragraph everything works without metric. We can visualize the quartered disc as a rolled up iterant. The infinite length of the iterant is reduced to 4 because of the modulo 4 repetition of the characters. In this way the iterant, by closing it, obtains a simple topological structure. The Schrödinger iterant when closed represents the local synchronous image of a harmonic oscillation. Then there arises the question how such a local repetition of cyclic motion can be memorized by the system itself, without overmuch outlay and complexity. If we allow that the iterants traverse their own structure, we obtain general discrete topological and pre-topological spaces with even more general pre-topological neighborhood systems (Schmeikal 2012, Part IV). I am saying this here just to make us aware how many new questions arise in that context of polarized braids. We cannot answer them all in this treatise, but I can promise that all the questions I asked until now could be answered. This is a context in which the right questions can be posed.

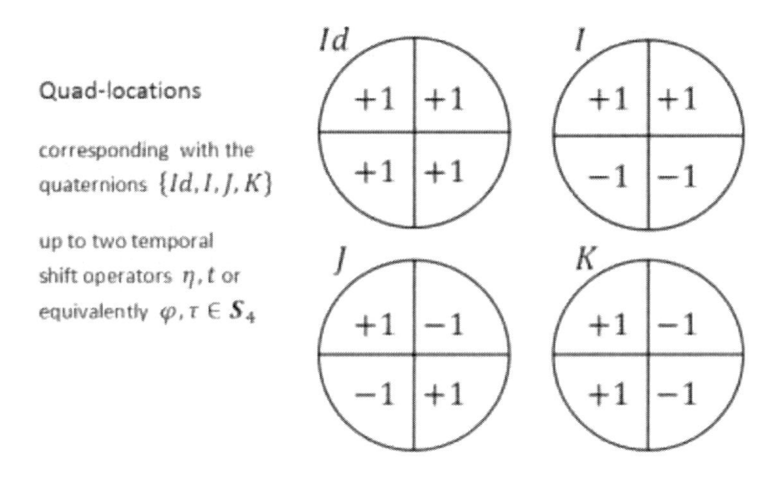

Figure 39. Quad-locations.

DEGREES OF FREEDOM

We are so used, so conditioned by our habits that we correlate in our imagination a line with dimension one, a square with 2 and a volume with dimension 3, so that practically every new theory is blocked by our conditioning. Looking at the figures we immediately project out tokens $\mathbb{R}, \mathbb{R}^2, \mathbb{R}^3$ as below

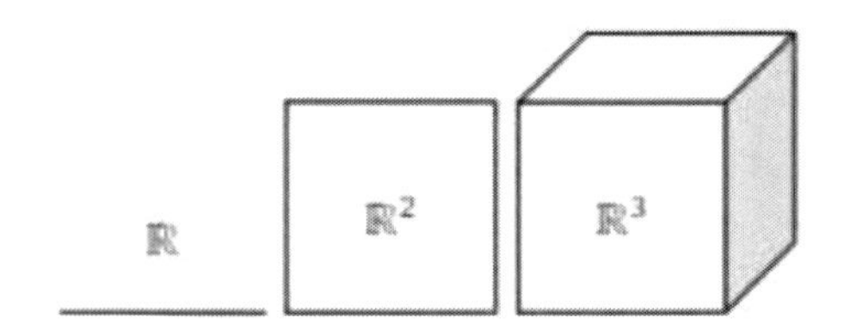

though we already know that this Peano fractal

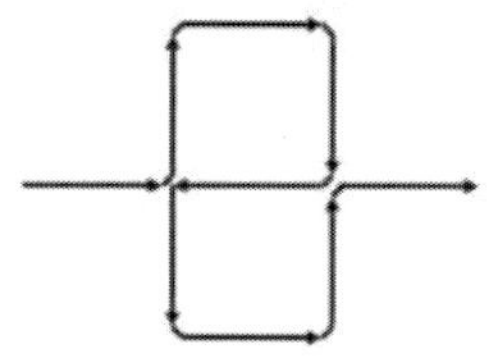

facilitates a parquetting of the plane $\mathbb{R}^2$ by a line, that is, by a Peano curve, like in figure 35, or here

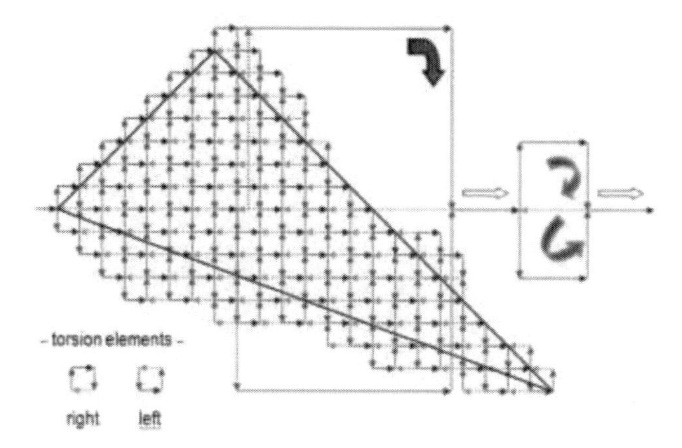

Or if we like to go further, we can even find a line that allows for 'parquetting' Euclidean space by a fractal structure. If we replace in the below figure every straight line that is enumerated by a small version of that figure, we can fill up the 3-space. Now, what is surprising for the newcomer in geometric algebra, is the fact that this 'dimensional equivalence' of subspaces that seem to have different dimension, is a natural thing in geometric Clifford algebra.

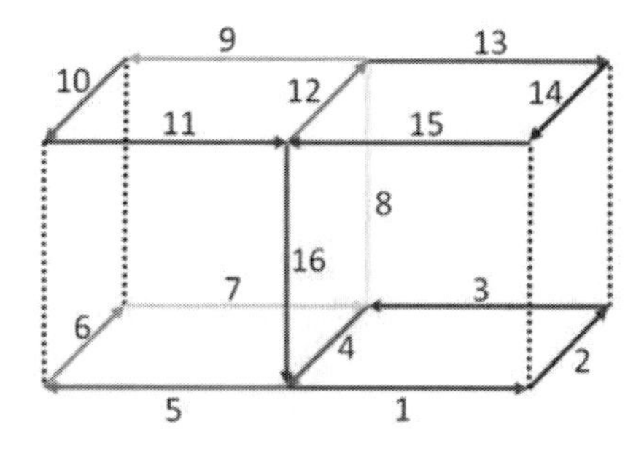

We shall work this out for the six isomorphic subspaces that we have to recognize as Cartan subalgebras of the Minkowski algebra. If we consider the Clifford algebra $Cl_{3,1}$ in the Lorentz metric, with the forth index indicating a temporal base unit e_4 with $e_4e_4 = -Id$, then the three units e_1, e_{24}, e_{124} denote what I decided to call the first chromatic space ch_1 of $Cl_{3,1}$. In accordance with the features of a Cartan subalgebra, these three unit multivectors commute, that is, we have

$$e_1 e_{24} - e_{24} e_1 = 0$$
$$e_{24} e_{124} - e_{124} e_{24} = 0$$
$$e_1 e_{124} - e_{124} e_1 = 0 \tag{160}$$

Proof: As we have $e_2 e_2 = Id$ and $e_4 e_4 = -Id$ the bivector e_{24} is a directed space-time unit area with $e_{24} e_{24} = Id$, thus $e_1 e_{24} = e_1 e_2 e_4 = e_{12} e_4 = -e_2 e_1 e_4 = -e_2 e_{14} = -e_2(-e_{41}) = e_2 e_{41} = e_{24} e_1$. No one would guess at first encounter that the following pattern of polarity strings could represent such line-, area-, and volume elements of three different grades. If we add the string $I^+ = ...+1+1+1...$, we have even four different grades, namely $0, 1, 2, 3$. Would you believe that?

$$...+1+1-1-1+1+1-1-1+1+1-1-1...$$
$$...+1-1+1-1+1-1+1-1+1-1+1-1...$$
$$...+1-1-1+1+1-1-1+1+1-1-1+1...$$

$$...+1-1+1-1+1-1+1-1+1-1+1-1...$$
$$...+1-1-1+1+1-1-1+1+1-1-1+1...$$
$$...+1+1-1-1+1+1-1-1+1+1-1-1...$$

$$...+1-1-1+1+1-1-1+1+1-1-1+1...$$
$$...+1+1-1-1+1+1-1-1+1+1-1-1...$$
$$...+1-1+1-1+1-1+1-1+1-1+1-1...$$

Figure 40. 4-fold polarity string pattern in unspecified domain.

Actually the above bit strings have the capability to represent elements of different grade as elements having different dimension in the algebra of space-time. This will give us a new imagination of scalars, lines, areas and volumes in space-time.

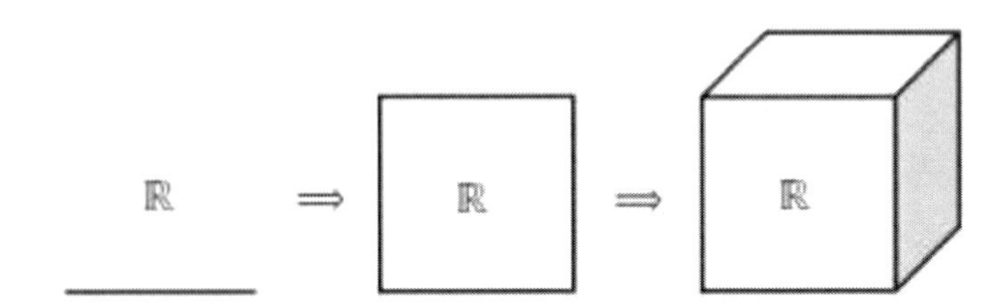

Figure 41. Looking beyond the edge of the plate.

As this concerns a real degree of freedom for a field's motion in relativistic space-time algebra, we shall, for a moment, go into the structure of that algebra which is by no means totally unknown to us. We may try to imagine a natural episode as unfolding from a line.

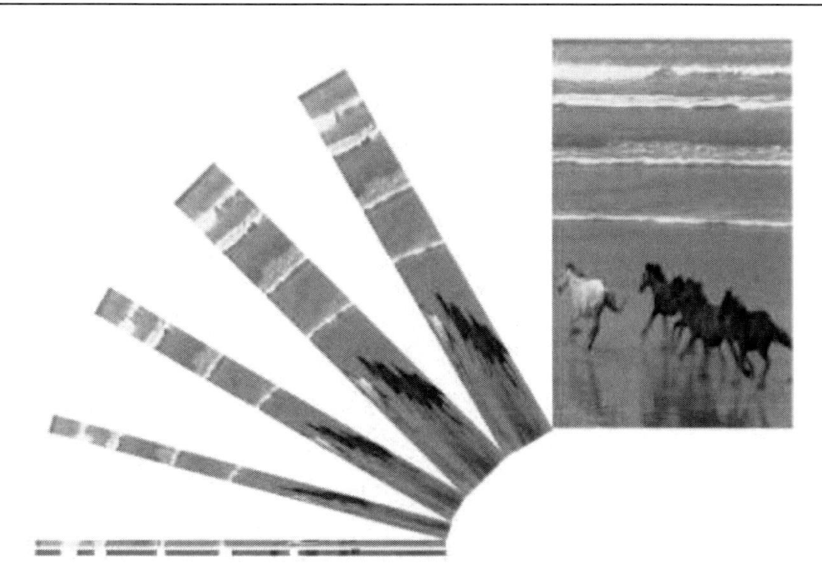

Figure 42. Nature unfolding.

Now, how is this actually working in quantum motion? The matter is a bit more complicated and requires a fine rigor in algebra. We shall now introduce the metric Minkowski space-time just in order to get a hint about the important relation between the large motion-groups of this Clifford algebra and the discrete location approach to polarity strings with the time-shift operators. We verify at first

Theorem 6: $Cl_{3,1} \Rightarrow Lie(Cl_{3,1})$

The Minkowski algebra can be conceived as a Lie algebra. The proof is easy. I give an abbreviated version. There can be found more in my second volume on Primordial Space. We disaggregate a mathematics object from the predicate $Lie(Cl_{3,1})$, namely a graded differential Lie algebra $(Cl, [\,,], D)$ with bracket and derivative element D. We have to show that $(Cl, [\ ,\ ])$ exists. Consider, therefore, three elements $a, b, c \in Cl$ having the general form

$$\begin{aligned} a &= x_2\epsilon_1 + x_3\epsilon_2 + \ldots + x_{15}\epsilon_{234} + x_{16}\epsilon_{1234} \\ b &= y_2\epsilon_1 + y_3\epsilon_2 + \ldots + y_{15}\epsilon_{234} + y_{16}\epsilon_{1234} \\ c &= z_2\epsilon_1 + z_3\epsilon_2 + \ldots + z_{15}\epsilon_{234} + z_{16}\epsilon_{1234} \end{aligned} \tag{161}$$

Define $\epsilon_k \stackrel{\text{def}}{=} e_k/2$ as half the Clifford base units and eliminate the scalar Id. So, when $Cl(Q) = Cl_{3,1}$ we get a Lie group $\mathfrak{L}_3$ having rank 3 and dimension 15, that is, 1 less than the Clifford algebra of the Minkowski space. Indeed, the bracket is skew symmetric and the magnitudes $a, b, c \in Cl$ satisfy the Jacobi identity

$$[[a,b],c] = [a,[b,c]] - [b,[a,c]] \quad \text{q.e.d.} \tag{162}$$

This is independent of the grade-mix of any element a, b, c. The calculation was carried out with MAPLE-Clifford, a program developed by Rafal Ablamowics and Bertfried Fauser

(2005). According to a theorem by Elié Cartan, all maximal abelian subalgebras of a semi-simple Lie algebra are mutually isomorphic. In the Minkowski algebra we obtain a 6-star of maximal Cartan subalgebras. We are using this to get an impulse towards new polarity strings.

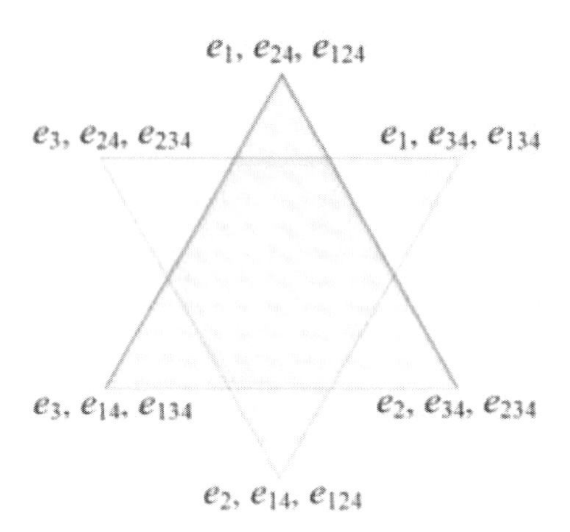

Figure 43. The seal of space-time – **Cartan' subalgebras of the motion**-group.

It has been shown that the minimum number of dimensions for a primitive idempotent in the algebra $Cl_{3,1}$ is equal to 4. Pertti Lounesto in his book (2001) uses all sign combinations for the 4 mutually annihilating primitive idempotents which generate a 16-element lattice.

$$f_1 = \frac{1}{2}(Id + e_1)\frac{1}{2}(Id + e_{24}) \quad f_2 = \frac{1}{2}(Id - e_1)\frac{1}{2}(Id - e_{24}) \tag{163}$$
$$f_3 = \frac{1}{2}(Id - e_1)\frac{1}{2}(Id + e_{24}) \quad f_4 = \frac{1}{2}(Id + e_1)\frac{1}{2}(Id - e_{24})$$

These span a commutative subspace, the 'color-space' ch_1. Let me mention a little prematurely, the idempotent primitive in $Cl_{3,1}$ and represented by f_1 generates a 1-norm. That is, the primitive f_1 endows ch_1 with a 1-norm. For take any $X = aId + be_1 + ce_{24} + de_{124}$ $\in ch$ and verify that

$$f_1X = (a + b + c + d)f_1 \quad \text{with 1-norm} \tag{164}$$
$$L = a + b + c + d.$$

Therefore, we say that the f_1X provides an *eigenform* for a *1-norm* (Kauffman 2009, Heinz von Foerster 1981). Take $L = a + b + c + d = 1$ to obtain $f_1X = f_1$. It can be shown (Schmeikal 2004) that color spaces $ch_1, ch_2, \ldots$ are isomorphic with the 4-fold real ring ${}^4\mathbb{R} = \mathbb{R}\oplus\mathbb{R}\oplus\mathbb{R}\oplus\mathbb{R}$. The proof can be run as follows. Consider the idempotent

$$f = \frac{1}{2}(Id - e_1) \tag{165}$$

not primitive in the Minkowski algebra and the isospin

$$\Lambda_3 = \frac{1}{2}(e_{24} - e_{124}) \tag{166}$$

Both are elements in ch_1. For any natural number $n \in \mathbb{N}$ we verify the identities

$$\Lambda_3^{2n} = f \text{ and } \Lambda_3^{2n-1} = \Lambda_3 \tag{167}$$

Therefore, we can state that the Clifford number Λ_3 represents a swap (or swop), similar as in the equations (88, 104). The color space can now be decomposed into two ideals according to the equations

$$ch_1 = ch_1 f \oplus ch_1 \hat{f} = \mathcal{G}_1 \oplus \hat{\mathcal{G}}_1 \tag{168}$$

with main involuted $\hat{f}$, and spaces $\mathcal{G}_1 \stackrel{\text{def}}{=} span\,\{f, \Lambda_3\}$ and $\hat{\mathcal{G}}_1 \stackrel{\text{def}}{=} span\,\{\hat{f}, \widehat{\Lambda}_3\}$. According to a theorem by Elié Cartan, all maximal abelian subalgebras of a semi-simple Lie algebra are mutually isomorphic. The equations (167) imply that both $\mathcal{G}_1$ and $\hat{\mathcal{G}}_1$ are isomorphic with the small Clifford algebra $Cl_{1,0} := \{Id, e_1\} \simeq {}^2\mathbb{R} = \mathbb{R}\oplus\mathbb{R}$ - the double ring of real numbers. Therefore, due to equation (168) we end up with a fundamental decomposition

$$ch_1 \simeq ch_\chi \simeq \mathbb{R}\oplus\mathbb{R}\oplus\mathbb{R}\oplus\mathbb{R} \quad \text{for each color space in the seal} \tag{169}$$

Clearly, any color space can be spanned either by its orthogonal primitive idempotents or by its base units. If we consider the top of the seal, we represent the base units of ch_1 by the following quadruples:

$$Id = (+1, +1, +1, +1);\; e_1 = (+1, -1, -1, +1); \tag{170}$$

$$e_{24} = (+1, -1, +1, -1);\; e_{124} = (+1, +1, -1, -1)$$

These numbers correspond with the primitive idempotents $f_1 = (+1,0,0,0)$; $f_2 = (0, +1,0,0)$; $f_3 = (0,0, +1,0)$; $f_4 = (0,0,0, +1)$; (equations 163) and $\mathcal{G}_1$ spanned by $f = (0, +1, +1,0)$ and $\Lambda_3 = (0, -1, +1,0)$.

See the analogy between the quad-locations and the ${}^4\mathbb{R}$ representation of the base units of color space $ch_1 = span\,\{Id,\ e_1, e_{24}, e_{124}\}$. Due to the peculiar construction of the iterant algebra, we can identify the iterant views with units having different grades: a spatial unit, a space-time area, a space-time volume.

$$e_1 := [+1, -1, -1, +1];\; e_{24} := [+1, -1, +1, -1];\; e_{124} := [+1, +1, -1, -1] \tag{171}$$

As we know it is the trigonal transition among those iterants that brings upon discrete colors satisfying the unitary symmetry of the motion. On the other hand, the color space being a commutative Cartan subalgebra of the $Cl_{3,1}$ is derived from the quaternion algebra by abstracting from the temporal order imposed on the iterants correlated with space ch_1 by the permutations φ, σ and τ. In this sense each color space ch_χ $(\chi = 1, \ldots, 6)$ turns out to be a *contemporalized synchronous image* of the quaternion iterant temporal structure of relativistic quantum motion.

The Iterant Flavor Rotation

It seems that the trigonal rotation that connects the 4-fold polarity strings I_1, I_2, I_5 as in equation (90) is also qualified to connect the commuting base units in a Cartan subalgebra of the Minkowski algebra. This movement signifies a real degree of freedom, a rotation of grade. I like to call such a 'trans grade' rotation that carries lines into directed space-time areas and those into space-time volumes as 'pulsation' or 'breath'. We need not take the word to serious. It is rather a means to discern several degrees of freedom of motion in a sufficiently strong and meaningful way. Similar like in the case of the Hamilton quaternions we can now introduce a trigonal rotation without any immediate reference to the Minkowski space and its 'colored' motion groups and without using a multivector base and metric. Yet, we show how that is made possible by abusing the Clifford algebra as a crutch to iterants with high internal structural complexity.

Definition 4 Pulsation: In agreement with the Cayley graph of the symmetric group $\boldsymbol{S_4}$ we define the permutation 3-cycle $C_3 \stackrel{\text{def}}{=} (2\ 3\ 4)$ acting on fourfold bit strings. It is called color-shift operator. Acting on the iterants I_1, I_2, I_5 we verify

$$\begin{aligned}
[+1,+1,-1,-1]\,C_3 &= [+1,-1,+1,-1] \\
[+1,-1,+1,-1]C_3 &= [+1,-1,-1,+1] \\
[+1,-1,-1,+1]C_3 &= [+1,+1,-1,-1]
\end{aligned} \tag{172}$$

This is the iterant trigon we have already discovered in the section "a peculiarity of primordial chaos". It is a discrete object that does not yet have to do with a metric space. But later it will turn out that it has a multivector analogue within the Minkowski algebra. So, if there is any reasonable way in which these polarized strings interact, that force should ultimately bring forth or reproduce the metric of space-time.

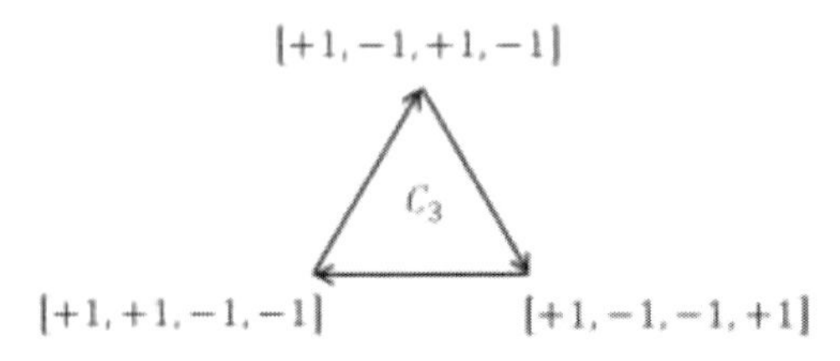

Figure 44. Action of the color shift operator on 3 strings which are polarized braids.

As we can see from the representation (171) the color-shift operator actually transposes a line element onto a directed spatio-temporal area and the latter to a directed spatio-temporal volume. Intuitively we may already have a presentiment that fourfold strings will not be enough to understand the whole of quantum motion. But that prejudice is unsubstantiated. So let us begin with the inquiry. We have six isomorphic color spaces and we got to differ between color and flavor which is indeed possible as soon as we understand why we need polarity strings. Now that we have defined 'touch' and 'pulsation' the matter becomes rather biological which it actually is.

Theorem 7: As the sixteen fourfold iterants (89) denoted as PS (polarity strings) form an algebra with logic identity as multiplication, we find among them four primitive idempotents. These are the strings (163) in the representation (171). The first primitive idempotent is invariant under C_3.

Proof: $f_1 \stackrel{\text{def}}{=} \frac{1}{2}(Id + e_1)\frac{1}{2}(Id + e_{24})$

$$\frac{1}{2}(Id + e_1) := \frac{1}{2}([+1,+1,+1,+1] + [+1,-1,-1,+1]) = [+1,0,0,+1] \tag{173}$$
$$\frac{1}{2}(Id + e_{24}) := \frac{1}{2}([+1,+1,+1,+1] + [+1,-1,+1,-1]) = [+1,0,+1,0]$$

$[+1,0,0,+1][+1,0,+1,0] = [+1,0,0,0]$ hence $f_1 = [+1,0,0,0]$
$f_1 C_3 = [+1,0,0,0](2\;3\;4) = [+1,0,0,0]$ so we have altogether
$f_1 f_1 = f_1$ and $f_1 C_3 = f_1$q. e. d. (174)

In contrast to that we define $f_2 = \frac{1}{2}(Id - e_1)\frac{1}{2}(Id - e_{24})$. This is a peculiar calibration which guarantees that the second entry will become unity in both the iterant and the diagonal representation matrix in the Clifford algebra. We have

$$\frac{1}{2}(Id - e_1) = \frac{1}{2}([+1,+1,+1,+1] + [-1,+1,+1,-1]) = [0,+1,+1,0]$$
$$\frac{1}{2}(Id - e_{24}) := \frac{1}{2}([+1,+1,+1,+1] + [-1,+1,-1,+1]) = [0,+1,0,+1]$$
$$[0,+1,+1,0][0,+1,0,+1] = [0,+1,0,0]$$

hence $f_2 = [0,+1,0,0]$ and $f_2 f_2 = f_2$ and further (175)
$f_3 = [0,0,+1,0]$ and $f_3 f_3 = f_3$
$f_4 = [0,0,0,+1]$ and $f_4 f_4 = f_4$

So we have a single invariant polarity string and three of them which form a flavor trigon. In a while we shall understand these peculiar primitive idempotents as energy densities of fermions having one definite color and different flavor.

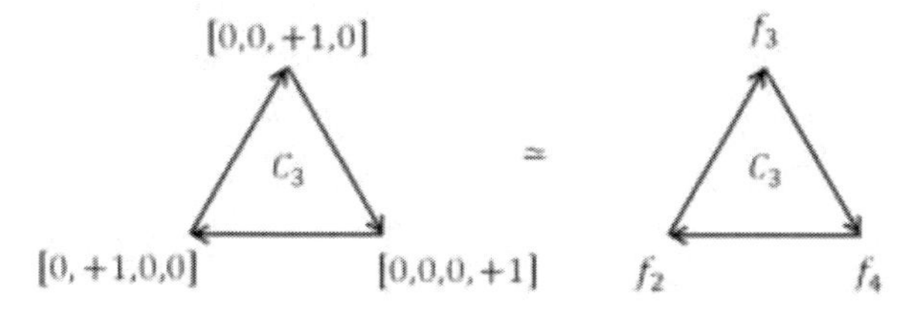

Figure 45. Action of the flavor shift operator on primitive idempotents.

Notice that the polarized strings taken alone do not yet require or suggest any specific metric as their location is unspecified. But in terms of representation theory we can already see their specific effects as elements in a Minkowski space with Lorentz metric. They are indeed a bit of a hybrid form. As sequences of symbols they are mere mathematical objects, the iterants. But as polarity strings they are more like physical objects, spin chains, isospin networks or similar such material things. Polarity strings turn out as braids by mere mathematical reasons which I will show in a while. But in physical imagination they float around as material fields as in figures 20, 21 or 27. They are logistic animals. As such they may change their nature after a while. As not every of the 6x4 = 24 fundamental classes of primitive idempotents of the Minkowski algebra have such nice representations as in (173) to (175), it may seem that we need a large bracket $[b_1, b_2, \ . \ . \ . \ , b_{16}\,]$ to represent a general 'primitive idempotent iterant'. But this is not the case. Fourfold strings are sufficient. So it will turn out that the concept of touch as logic identification can easily be upheld as a main donator of algebraic multiplication, at least not unless we introduce some further 'topological' rules for which elements have to touch where. For such rule exists for example in matrix multiplication. (Lay the first row of a $Mat(4, \mathbb{R})$ matrix on the first column of a second element of $Mat(4, \mathbb{R})$ and go on in this manner with the second column. If rows and columns are polarity strings, then such a rule is a topological statement about which bits are supposed where to touch.) As long as we have a line instead of a matrix and multiplication is mere logic identification, no problem comes up. But if we want to have all flavors and colors and leptons within the constitution of space-time algebra, those problems may seem to become severe. But surprisingly, they do not. Our tools in discrete mathematics are just to be developed a little bit further. The slew may dissolve if we can show that the proper iterant algebra of polarized strings as prespinors is isomorphic with some Clifford algebra, as was the case with the checkerboard pre-spinors. The most serious challenge that arises in the task of generalizing the original path integral method stems from the new degrees of freedom that come in by necessity as soon as we turn from a grade 1 element e_α to a unit involving 4 different grades. Let us investigate this kinship between geometric and iterant algebra by first looking at the Clifford numbers involved in our last problem. Consider some peculiar standard representation of the Minkowski algebra, namely the base units

$$e_1 = \begin{bmatrix} 1 & & & \\ & -1 & & \\ & & -1 & \\ & & & 1 \end{bmatrix}, \quad e_2 = \begin{bmatrix} & 1 & & \\ 1 & & & \\ & & & 1 \\ & & 1 & \end{bmatrix}, \quad e_3 = \begin{bmatrix} & & 1 & \\ & & & -1 \\ 1 & & & \\ & -1 & & \end{bmatrix}, \quad e_4 = \begin{bmatrix} & -1 & & \\ 1 & & & \\ & & & -1 \\ & & 1 & \end{bmatrix} \tag{176}$$

With these we obtain the oriented unit space-time area and unit space-time volume in diagonal form.

$$e_{24} = \begin{bmatrix} 1 & & & \\ & -1 & & \\ & & +1 & \\ & & & -1 \end{bmatrix} \quad e_{124} = \begin{bmatrix} 1 & & & \\ & 1 & & \\ & & -1 & \\ & & & -1 \end{bmatrix} \tag{177}$$

Now we can immediately realize that some of those important polarity strings – we shall later call them fermion bit strings – provide the diagonal representations of the first Cartan subalgebra of the Minkowski algebra. As we have postulated that the trigonal rotation shown in figure 45 is a flavor rotation, we should expect that there is another degree of freedom which should be color. Flavor will be related to the triangles in Solomon's seal (figure 43). The three bit strings

$$\begin{aligned} I_2 &\simeq e_1 = [+1,-1,-1,+1] \\ I_5 &\simeq e_{24} = [+1,-1,+1,-1] \\ I_1 &\simeq e_{124} = [+1,+1,-1,-1] \end{aligned} \tag{178}$$

form one of six corners of the seal. Each corner is one Cartan subalgebra of the rank 3 Lie group correlated with the Minkowski algebra. Within each corner there occur the flavor rotations. For example in the color space ch_1 we encounter the flavor rotation between f_2, f_3 and f_4 with one invariant 'lepton' f_1. Suppose we denote the upper corner in figure 43 the *fermion corner for a red u-quark*. Then it would be natural to understand how we can pass over to the corner given by the elements $\{e_2, e_{34}, e_{234}\}$, call it the 'green u-quark corner'. To carry out such color rotation, and find out in addition how the flavor of such green u-quark can be altered, confronts us with the fundamental degrees of freedom that are given in the Clifford algebra for the Minkowski space(-time). We have learned that parity, time reversal and charge conjugation are the essential degrees of freedom for relativistic motion. But already Feynman's path integral makes it clear that the dynamic operation of a directed space-time area such as e_{24} is another important contribution to freedom of motion in quantum dynamics. Note that the e_{24} as an iterant representative of the alternating code is relevant not only for the stochastic process of the Schrödinger equation, but also for the iterant Dirac equation in signature $\{+,-\}$. And it seems, we are just learning that this basic iterant plays an active role in strong interacting fields too. The most general degrees of freedom stem from a unitary transformation that knows pulsation, and thereby transposes units of different grade onto each other. Such non grade preserving transformations are necessary as the basic constituents of a spinor space, forming the minimal ideal, are not primitive idempotents derived from a single unit such as e_1, that is, $f = \frac{1}{2}(Id + e_1)$ but from a period 2 element involving at least four different grades: :

$$\mathbb{e}_1 = Id - 2f_1 = \frac{1}{2}(Id - e_1 - e_{24} - e_{124}) \tag{179}$$

Note, we also have $e_1 = Id - 2f$ in the theory of Pauli spinors. The idempotent primitive in the algebra of the underlying space is always derived from a period 2 element by the form $f_{()} := (Id \pm \mathbb{e})/2$.

Example: $\mathbb{e}_1$ has period 2.

Proof: $\mathbb{e}_1\mathbb{e}_1 = \frac{1}{2}(Id - e_1 - e_{24} - e_{124})\frac{1}{2}(Id - e_1 - e_{24} - e_{124}) =$ (180)

$$\frac{1}{4}(Id - e_1 - e_{24} - e_{124} - e_1 + Id + e_{124} + e_{24} -$$
$$-e_{24} + e_{124} + Id + e_1 - e_{124} + e_{24} + e_1 + Id) = Id$$

There exist 24 'reflections' of the form $\mathbb{e}_1$, since the seal has six corners and in each there are four primitive idempotents. These reflections can be imagined to symbolize the *new degrees of freedom* that are brought into the space-time by the peculiar Clifford algebra $Cl_{3,1}$. These are capable to carry out the important non grade preserving transpositions within the basis. Now, even if we are sure that the process of nature is a translocal chaotic process with many polarized strings, appearing and disappearing in the void, that would not change the fact that macro physics shows us a relativistic Minkowski space which, right from the start, is a geometric algebra. It turns out that it is therefore that the phenomenon of entanglement is not disturbing principles of relativistic motion at all, rather it is to be expected as a natural feature of that geometry.

We want to represent the iterant that shifts flavor. For this purpose we consider the Clifford algebra representations we are by now familiar with. We need two reflections to bring on this trigonal rotation.

$$s_{2,3} = Id - 2f_{2,3} = \frac{1}{2}(Id + e_1 - e_{34} + e_{134}) \quad \text{and} \qquad (181)$$
$$s_{6,4} = Id - 2f_{6,4} = \frac{1}{2}(Id - e_3 + e_{24} - e_{234})$$

Then the flavor rotation within the first color space is the element

$$Tf_1 = s_{6,4}s_{2,3} = \frac{1}{4}(Id + e_1 + e_2 - e_3 + e_4 + e_{12} + e_{13} + e_{14} - e_{23} + e_{24} - e_{34} + \qquad (182)$$
$$+e_{123} + e_{124} + e_{134} - e_{234} + e_{1234}) \quad \text{with the inverse}$$
$$Tf_1^{-1} = s_{2,3}s_{6,4} = \frac{1}{4}(Id + e_1 + e_2 - e_3 - e_4 - e_{12} - e_{13} + e_{14} + e_{23} + e_{24} - e_{34} +$$
$$-e_{123} + e_{124} + e_{134} - e_{234} - e_{1234})$$

In the basis (176) these are the familiar matrices

$$T_1 = \begin{bmatrix} 1 & 0 & 0 & 0 \\ 0 & 0 & 0 & 1 \\ 0 & 1 & 0 & 0 \\ 0 & 0 & 1 & 0 \end{bmatrix},\quad T_1^{-1} = \begin{bmatrix} 1 & 0 & 0 & 0 \\ 0 & 0 & 1 & 0 \\ 0 & 0 & 0 & 1 \\ 0 & 1 & 0 & 0 \end{bmatrix}$$

Terms $T_1 f_{1\alpha} T_1^{-1}$ and $T_1^{-1} f_{1\alpha} T_1$ rotate the flavor of $f_{1\alpha}$ within the first color in opposed directions. They likewise rotate the basis of the first color space, preserving the identity, that is, we shift in a 3-cycle $e_1 \to e_{24} \to e_{124} \to e_1$. We give a first representation Lemma for the flavor rotation in color one.

Lemma 3: The flavor rotation for color space ch_1 is the iterant

$$T_{f,1} = [+1,0,0,0] + [0,0,0,-1]\sigma + [0,-1,0,0]\varphi + [0,0,+1,0]\tau \quad \text{with the inverse}$$
$$T_{f,1}^{-1} = [+1,0,0,0] + [0,0,-1,0]\sigma + [0,0,0,-1]\varphi + [0,+1,0,0]\tau$$

Where the indices mean f =flavor and 1 =first color (e.g. red), ... $_{f,1}$ flavor rotation is in first color space. To prove this lemma we need a theorem.

ISOMORPHY THEOREM FOR GRADE 4 ITERANT ALGEBRA

Even to the beginner it may seem obvious that there is some prudent relation between the algebra of fourfold iterants and the Clifford algebra. To investigate this connection, some experience is certainly of advantage. In the section on quaternions, quad locations and tangle time we have seen that there is some reasonable relationship between the fourfold polarity strings and the Hamilton quaternions I, K, J. For someone who sees the proper domain of the calculations involved, namely iterants having grade four and hypercomplex numbers that can essentially be represented by four real numbers and their various combinations due to the iterations, it is a challenge to find out if the iterant algebra is perhaps isomorphic with the Majorana algebra $Mat(4, \mathbb{R})$, the 4×4 matrix algebra with real entries which again is isomorphic with the quadratic Clifford algebra $Cl_{3,1}$ of the Minkowski space in the Lorentz metric. After having gone through the rigor, we can formulate

Theorem 8: The iterant algebra with four grades is isomorphic with the Clifford algebra $Cl_{3,1}$

Proof: Consider the four real iterants e, f, g we are already familiar with

$$e = [+1, +1, -1, -1],\ f := [+1, -1, -1, +1],\ \ g := [+1, -1, +1, -1], \tag{183}$$

Together with the permutation operators $\sigma := (1\ 2)(3\ 4)$, $\varphi := (1\ 3)(2\ 4)$, $\tau :=$ $:= (1\ 4)(2\ 3)$ which are generated by iteration time t and tangle time η. Sequences are iterated by *iterant time* t which satisifies

$$t\,[a, b, c, d] = [b, c, d, a]\,t \text{ with } t^4 = Id \tag{184}$$

and by *tangle-time* η which is the period-2 iterant time shift that we know from quad locations, applied to iterants of degree 4 and satisfying

$$\eta[a, b, c, d] = [b, a, c, d]\eta \text{ with } \eta^2 = Id \tag{185}$$

The iterant time t can be represented by a permutation 4-cycle $(1\ 2\ 3\ 4)$ and the tangle time by a 2-cycle $(2\ 1)$. These two generate the symmetric group S_4 of which we need the three operators σ, φ, τ which satisfy similar equations like the tangle time, namely

$$\begin{aligned} \sigma[a, b, c, d] &= [b, a, d, c]\sigma \text{ with } \sigma^2 = Id \\ \varphi[a, b, c, d] &= [c, d, a, b]\varphi \text{ with } \varphi^2 = Id \\ \tau[a, b, c, d] &= [d, c, b, a]\tau \text{ with } \tau^2 = Id \end{aligned} \tag{186}$$

These can be derived from the generating iterant- and tangle-time operators in the following manner

$$\begin{aligned} \varphi &= t^2 = (1\ 2\ 3\ 4)(1\ 2\ 3\ 4) = (1\ 3)(2\ 4), \\ \tau &= \eta\varphi\eta = (2\ 1)\big((1\ 3)(2\ 4)\big)(2\ 1) = (1\ 4)(2\ 3), \\ \sigma &= \tau\varphi \end{aligned} \tag{187}$$

Now there exist nine possibilities to let any permutation operator act on the unit iterants. Among these nine products there are six quaternions. Among those there are the three we already know from the analysis of quad locations. Three of the nine squared give the identity Id. The nine terms are

$$e\sigma, e\varphi, e\tau, f\sigma, f\varphi, f\tau, g\sigma, g\varphi, g\tau \tag{188}$$

Calculating the iterant product of the first three terms with themselves we obtain

$$\begin{aligned}
&(e\sigma)^2 = [+1,+1,-1,-1]\sigma[+1,+1,-1,-1]\sigma = \\
&= [+1,+1,-1,-1][+1,+1,-1,-1]\sigma^2 = [+1,+1,+1,+1]Id = +Id \\
&(e\varphi)^2 = [+1,+1,-1,-1]\varphi[+1,+1,-1,-1]\varphi = \\
&= [+1,+1,-1,-1][-1,-1,+1,+1]\,\varphi^2 = [-1,-1,-1,-1]Id = -Id \\
&(e\tau)^2 = [+1,+1,-1,-1]\tau[+1,+1,-1,-1]\tau = \\
&= [+1,+1,-1,-1][-1,-1,+1,+1]\,\tau^2 = [-1,-1,-1,-1]Id = -Id
\end{aligned} \tag{189}$$

We know the first quaternion. We may call it K. There are two quaternions among the three terms and one monomial with square equal Id. Analogues formulae are obtained for the other six terms. We get

$$\begin{aligned}
&(e\sigma)^2 = +Id, \\
&(e\varphi)^2 = -Id, \text{ bivector} \\
&(e\tau)^2 = -Id, \text{ quaternion } K \\
&(f\sigma)^2 = -Id, \text{ bivector} \\
&(f\varphi)^2 = -Id, \text{ quaternion } J \\
&(f\tau)^2 = +Id, \\
&(g\sigma)^2 = -Id, \text{ quaternion } I \\
&(g\varphi)^2 = +Id, \\
&(g\tau)^2 = -Id \text{ bivector}
\end{aligned} \tag{190}$$

The proof can proceed once we understand why we have six instead of three quaternions. Namely, if we consider the Clifford algebra $Cl_{3,1}$ in a standard basis, we realize that we have a triple of bivectors which represent quaternions, the $\{e_{12}, e_{23}, e_{13}\}$ and we have a further triple of time-like quaternions with different grades, the time-space $\{e_4, e_{123}, e_{1234}\}$. If we pose these two sets of quaternions in a proper way, we can see

$$\begin{bmatrix} e_{12} & e_4 \\ e_{23} & e_{123} \\ e_{13} & e_{1234} \end{bmatrix} \Rightarrow \begin{matrix} e_{124} \\ e_1 \\ e_{24} \end{matrix} \tag{191}$$

How quaternions are carried to the Cartan subalgebra, that is to the color space of logic units. The Clifford product in each row gives a component of the first color space, each of which squared gives the Identity. Therefore it is reasonable to assume that the six quantities $e, f, g, \sigma, \varphi, \tau$ generate a geometric algebra that includes even more than just two sets of quaternions. This could be the Clifford algebra $Cl_{3,1}$ of the Minkowski space. We can use the above relation to get the base units, one after the other. To abbreviate the proof, let us factor in how the six quantities $e, f, g, \sigma, \varphi, \tau$ interact. We formulate as

Lemma 1: The polarity strings e, f, g constitute the commutative algebra of a Klein-4 group; all the same the permutations σ, φ, τ satisfy the same algebra. The mixed products of polarity strings and permutations commute or anti-commute according to table 2.

Table 2. Multiplication tables and commutation relations among polarity strings and permutations

	e	f	g
e	Id	g	f
f	g	Id	e
g	f	e	Id

polarity strings

	σ	φ	τ
σ	Id	τ	φ
φ	τ	Id	σ
τ	φ	σ	Id

permutations

	e	f	g
σ	$e\sigma$	$-f\sigma$	$-g\sigma$
φ	$-e\varphi$	$-f\varphi$	$g\varphi$
τ	$-e\tau$	$f\tau$	$-g\tau$

Commutation relations

Direct component-wise multiplication gives the first table, the second is well known property of permutation group $\boldsymbol{S}_4$, the third part must be verified:

Use $e = [+1, +1, -1, -1]$, $\sigma := (1\,2)(3\,4)$, and rule (68) $\sigma[a, b, c, d] = [b, a, d, c]\sigma$ to get $\sigma e = (1\,2)(3\,4)[+1, +1, -1, -1] = [+1, +1, -1, -1]\sigma = e\sigma$, the matrix element in first row, first column; further use $f := [+1, -1, -1, +1]$, $\sigma := (1\,2)(3\,4)$, and rule (68) $\sigma[a, b, c, d] = [b, a, d, c]\sigma$ to get $\sigma f = (1\,2)(3\,4)[+1, -1, -1, +1] = [-1, +1, +1, -1]\sigma = -f\sigma$, the matrix element in first row, second column; use $g := [+1, -1, +1, -1]$ and σ to verify $\sigma g = -g\sigma$, the matrix element in first row, third column; and so on until to $\tau g = -g\tau$, the last matrix element in third row, third column. Now it is clear how we place the elements e, f, g at the positions e_1, e_{24}, e_{124} in the 16 element basis of $Cl_{3,1}$ and if we put $e_2 = -\varphi$, $e_3 = f\tau$ and $e_4 = f\varphi$, we get unit vectors with the appropriate signature $(+ + + -)$ of the Minkowski space in the opposite (Lorentz) metric. These satisfy the commutation relations of this Clifford algebra. The result of exterior multiplication gives us the following representation of the Clifford algebra of Minkowski space.

$$
\begin{array}{llll}
Id & e_1 = e & e_2 = -\varphi & e_3 = f\tau \\
e_4 = f\varphi & e_{12} = \varphi e & e_{13} = g\tau & e_{14} = g\varphi \\
e_{23} = f\sigma & e_{24} = f & e_{34} = \sigma & e_{123} = g\sigma \\
e_{124} = g & e_{134} = e\sigma & e_{234} = -\tau & e_{1234} = \tau e
\end{array} \tag{192}
$$

These quantities satisfy the multiplication table of the base units of Clifford algebra $Cl_{3,1} \simeq Cl_{2,2}$ such that the table 2 of (anti)commutation relations is satisfied. This can be verified by MAPLE Clifford. q.e.d. In this way we get a color space basis of iterants. The diagonals appear as cyclically rotated within the brackets.

$$e = [+1,+1,-1,-1] \simeq e_1 = \begin{bmatrix} +1 & & & \\ & -1 & & \\ & & -1 & \\ & & & +1 \end{bmatrix} \tag{193}$$

$$f = [+1,-1,-1,+1] \simeq e_{24} = \begin{bmatrix} +1 & & & \\ & -1 & & \\ & & +1 & \\ & & & -1 \end{bmatrix}$$

$$g = [+1,-1,+1,-1] \simeq e_{124} = \begin{bmatrix} +1 & & & \\ & +1 & & \\ & & -1 & \\ & & & -1 \end{bmatrix}$$

together with the temporal permutation operators

$$\varphi \simeq \begin{bmatrix} & -1 & & \\ -1 & & & \\ & & & -1 \\ & & -1 & \end{bmatrix}; \ \tau \simeq \begin{bmatrix} & & +1 & \\ & & & +1 \\ +1 & & & \\ & +1 & & \end{bmatrix}; \ \sigma \simeq \begin{bmatrix} & & & -1 \\ & & -1 & \\ & -1 & & \\ -1 & & & \end{bmatrix} \tag{194}$$

Now we can prove lemma 3. The flavor rotation $T_{f,1}$ takes the form

$$Tf_1 = s_{6,4}s_{2,3} = \frac{1}{4}(Id + e_1 + e_2 - e_3 + e_4 + e_{12} + e_{13} + e_{14} - e_{23} + e_{24} - e_{34} + \tag{195}$$
$$+e_{123} + e_{124} + e_{134} - e_{234} + e_{1234})$$
$$\simeq \frac{1}{4}(Id + e - \varphi - f\tau + f\varphi - e\varphi + g\tau + g\varphi - f\sigma + f - \sigma + g\sigma + g + e\sigma + \tau -$$
$$e\tau) = \frac{1}{4}((Id + e + f + g) + (-Id + e - f + g)\sigma + (-Id - e + f + g)\varphi + (Id - e -$$
$$f + g)\tau)$$

We calculate these four iterant terms, the binary series up to the factor $1/4$ and

$$\begin{array}{llll}
\{Id = [+1,+1,+1,+1]; & \{-Id = [-1,-1,-1,-1]; & \{-Id = [-1,-1,-1,-1]; & \{+Id = [+1,+1,+1,+1] \\
+e := [+1,+1,-1,-1]; & +e := [+1,+1,-1,-1]; & -e := [-1,-1,+1,+1]; & -e := [-1,-1,+1,+1] \\
+f := [+1,-1,-1,+1]; & -f := [-1,+1,+1,-1]; & +f := [+1,-1,-1,+1]; & -f := [-1,+1,+1,-1] \\
+g := [+1,-1,+1,-]\}; & +g := [+1,-1,+1,-]\}\}; & +g := [+1,-1,+1,-]\}\}; & +g := [+1,-1,+1,-]\} \\
\hline
= [+1,0,0,0] & = [00,0,-1] & = [0,-1,0,0] & = [0,0,+1,0]
\end{array}$$

$$\text{Hence } T_{f,1} = [+1,0,0,0] + [0,0,0,-1]\sigma + [0,-1,0,0]\varphi + [0,0,+1,0]\tau \tag{196}$$

The inverse can be obtained in the same way

$$Tf_1^{-1} = s_{2,3}s_{6,4} = \frac{1}{4}(Id + e_1 + e_2 - e_3 - e_4 - e_{12} - e_{13} + e_{14} + e_{23} + e_{24} - e_{34} +$$
$$-e_{123} + e_{124} + e_{134} - e_{234} - e_{1234})$$
$$T_{f,1}^{-1} = \frac{1}{4}(Id + e - \varphi - f\tau - f\varphi + e\varphi - g\tau + g\varphi + f\sigma + f - \sigma - g\sigma + g + e\sigma + \tau +$$
$$e\tau)$$
$$= \frac{1}{4}((Id + e + f + g) + (-Id + e + f - g)\sigma + (-Id + e - f + g)\varphi + (Id + e - f -$$
$$g)\tau)$$

Carrying out the rigor gives:

$$T_{f,1}^{-1} = [+1,0,0,0] + [0,0,-1,0]\sigma + [0,0,0,-1]\varphi + [0,+1,0,0]\tau \qquad (197)$$

We test how this transforms the second primitive idempotent. We have to compute $T_{f,1}f_{1,2}T_{f,1}^{-1}$:

$$T_{f,1}\, f_{1,2} = (\emptyset + [0,+1,0,0][+1,0,0,0] + [0,0,0,-1]\sigma + [0,-1,0,0]\varphi + [0,0,+1,0]\tau) =$$
$$\emptyset + \emptyset + \emptyset + [0,0,+1,0]\tau,$$

with the zero iterant $\emptyset \stackrel{\text{def}}{=} [0,0,0,0]$. Therefore

$$T_{f,1}\, f_{1,2}\, T_{f,1}^{-1} = [0,0,+1,0]\tau([+1,0,0,0] + [0,0,-1,0]\sigma + [0,0,0,-1]\varphi + [0,+1,0,0]\tau) =$$
$$= [0,0,+1,0][0,0,0,+1]\tau + [0,0,+1,0][0,-1,0,0]\tau\sigma + [0,0,+1,0][-1,0,0,0]\tau\varphi +$$
$$+[0,0,+1,0][0,0,+1,0]\tau\tau = \emptyset + \emptyset + \emptyset + [0,0,+1,0]Id = f_{1,3} \qquad (198)$$

since we have $\emptyset + \emptyset + \emptyset + [0,0,+1,0]Id = [0,0,+1,0]$ due to (186),

This is the primitive idempotent $f_{1,3}$. Thus the $T_{f,1}$ applied in this way turns the flavor clockwise

$$f_{1,2} \to f_{1,3} \to f_{1,4} \to f_{1,2}.$$

Considering the iterants $[0,1,0,0], [0,0,1,0], [0,0,0,1]$ as fermion densities, the quantity $[1,0,0,0]$ is supposed to be preserved. The proof just carries out the conjugate form:

from left side:

$$T_{f,1}\, f_{1,1} = ([+1,0,0,0] + [0,0,0,-1]\sigma + [0,-1,0,0]\varphi + [0,0,+1,0]\tau)[+1,0,0,0] =$$
$$[+1,0,0,0][+1,0,0,0] + [0,0,0,-1][0,+1,0,0]\sigma +$$
$$+[0,-1,0,0][0,0,+1,0]\varphi + [0,0,+1,0][0,0,0,+1]\tau = [+1,0,0,0] + \emptyset + \emptyset + \emptyset = f_{1,1}$$

and likewise from the right with $T_{f,1}^{-1}$

$$f_{1,1}([+1,0,0,0] + [0,0,-1,0]\sigma + [0,0,0,-1]\varphi + [0,+1,0,0]\tau) =$$
$$[+1,0,0,0][+1,0,0,0] + [+1,0,0,0][0,0,-1,0]\sigma + [+1,0,0,0][0,0,0,-1]\varphi$$
$$+ [+1,0,0,0][0,+1,0,0]\tau =$$
$$[+1,0,0,0] + \emptyset + \emptyset + \emptyset = f_{1,1} \qquad (199)$$

making sure that $f_{1,1}$ is preserved under a flavor rotation in space ch_1 and can represent a lepton. What is surprising in this approach that we could transform the whole matrix mechanics of the strong force so to say to linear writing. Namely the flavor rotation turns out to be representable by a fourfold iterant that can act on any subspace of the whole Minkowski algebra, as for example it can transform a time unit e_4 into thermodynamic magnitudes namely space-time volumes. So we summarize:

$T_{f,1} = [+1,0,0,0] + [0,0,0,-1]\sigma + [0,-1,0,0]\varphi + [0,0,+1,0]\tau$
$T_{f,1}^{-1} = [+1,0,0,0] + [0,0,-1,0]\sigma + [0,0,0,-1]\varphi + [0,+1,0,0]\tau$
$T_{f,1}$ replaces the 4×4 –matrix

$$T_1 = \begin{bmatrix} 1 & 0 & 0 & 0 \\ 0 & 0 & 0 & 1 \\ 0 & 1 & 0 & 0 \\ 0 & 0 & 1 & 0 \end{bmatrix}$$

in the iterant algebra. It represents a multivector involving all 16 base units and 5 grades of the Clifford algebra $Cl_{3,1}$. The inverse flavor rotation $T_{f,1}^{-1}$ has a linear design with comparable symmetric appearance seen parallel to $T_{f,1}$.

The Iterant Color Rotation

In order to make a difference between the six isomorphic color spaces, we give the primitive idempotents a first index $i = 1, \dots, 6$ and use the denotation f_{ij} with $i = 1, \dots 6; j = 1, \dots, 4$. The first color space has now the primitive idempotents

$$f_{11} = \tfrac{1}{2}(Id + e_1)\tfrac{1}{2}(Id + e_{24}) \quad f_{12} = \tfrac{1}{2}(Id - e_1)\tfrac{1}{2}(Id - e_{24}) \qquad (200)$$
$$f_{13} = \tfrac{1}{2}(Id - e_1)\tfrac{1}{2}(Id + e_{24}) \quad f_{14} = \tfrac{1}{2}(Id + e_1)\tfrac{1}{2}(Id - e_{24}).$$

These are supposed to represent one lepton and three quarks of different flavor. We want to carry these to the second color space, that is, to the second corner in the first triangle of Solomon's seal. That second corner is constituted by the primitive idempotents

$$f_{21} = \tfrac{1}{2}(Id + e_2)\tfrac{1}{2}(Id + e_{34}) \quad f_{22} = \tfrac{1}{2}(Id - e_2)\tfrac{1}{2}(Id - e_{34}) \qquad (201)$$
$$f_{23} = \tfrac{1}{2}(Id - e_2)\tfrac{1}{2}(Id + e_{34}) \quad f_{24} = \tfrac{1}{2}(Id + e_2)\tfrac{1}{2}(Id - e_{34}).$$

The elements carrying out such *permutations of grade 1 base units* e_i, e_k *with positive signature* are Weyl reflections typically represented by the six roots from the root lattice A_2 corresponding with the $su(3)$; those are the terms

$$\tfrac{1}{\sqrt{2}}(\pm e_1 \mp e_2), \tfrac{1}{\sqrt{2}}(\pm e_2 \mp e_3), \tfrac{1}{\sqrt{2}}(\pm e_1 \mp e_3), \qquad (202)$$

Suppose, you have any Clifford monomial X, and you want to carry out an exchange of e_i with e_k – we call such non grade preserving exchanges transpositions and we denote them by the word $\tau_{ik}(X)$ – then the Weyl transposition is given by $\tau_{ik}(X) := u_{ik}\hat{X}u_{ik}^{-1}$ [2] where $\hat{X}$ is the grade involuted monomial. Now, let us apply this to the $f_{1\alpha}$ and $f_{2\alpha}$. We are carrying colored u-quarks to colored d-quarks. Define the rotation

[2] The mathematics of not grade preserving transpositions in the Clifford basis is dealt with in Schmeikal (2004).

$$\tau_{(123)} := \frac{1}{\sqrt{2}}(e_1 - e_2)\frac{1}{\sqrt{2}}(e_2 - e_3) = \frac{1}{2}(Id + e_{12} + e_{23} - e_{13}) \tag{203}$$

This is an even element of the algebra. Its inverse is given by the reversion

$$\tau_{(123)}^{-1} = \tau_{(132)} = \tilde{\tau}_{(123)} = \frac{1}{2}(-Id - e_{12} - e_{23} + e_{13}) \tag{204}$$

Now one can verify by some Clifford calculator that $f_{1\alpha}$ by the rotation $\tau_{(123)}$ are carried to the $f_{2\alpha}$.

$$\tau_{(123)} f_{11} \tau_{(123)}^{-1} = f_{21} \text{ and generally } \tau_{(123)} f_{1\alpha} \tau_{(132)} = f_{2\alpha}. \tag{205}$$

These trigonal rotations are color rotations in the proper sense, exact symmetries beyond special relativity. Contrasting these, the flavor rotations imply symmetries of special relativity.

Hence, in the standard representation (176)

$$\tau_{(123)} = \frac{1}{2}\begin{bmatrix} 1 & 1 & -1 & -1 \\ -1 & 1 & 1 & -1 \\ 1 & -1 & 1 & -1 \\ 1 & 1 & 1 & 1 \end{bmatrix} \quad \tau_{(132)} = \frac{1}{2}\begin{bmatrix} 1 & -1 & 1 & 1 \\ 1 & 1 & -1 & 1 \\ -1 & 1 & 1 & 1 \\ -1 & -1 & -1 & 1 \end{bmatrix} \tag{206}$$

The rotation $\tau_{(123)}$ transforms the first idempotent from color space ch_1 to the second color space ch_1; in this representation

$$\tau_{(123)} f_{1,1} \tau_{(132)} = \frac{1}{2}\begin{bmatrix} 1 & 1 & -1 & -1 \\ -1 & 1 & 1 & -1 \\ 1 & -1 & 1 & -1 \\ 1 & 1 & 1 & 1 \end{bmatrix}\begin{bmatrix} 1 & 0 & 0 & 0 \\ 0 & 0 & 0 & 0 \\ 0 & 0 & 0 & 0 \\ 0 & 0 & 0 & 0 \end{bmatrix}\frac{1}{2}\begin{bmatrix} 1 & -1 & 1 & 1 \\ 1 & 1 & -1 & 1 \\ -1 & 1 & 1 & 1 \\ -1 & -1 & -1 & 1 \end{bmatrix} = f_{2,1}$$

This is the matrix

$$f_{2,1} = \frac{1}{4}\begin{bmatrix} 1 & -1 & 1 & 1 \\ -1 & 1 & -1 & -1 \\ 1 & -1 & 1 & 1 \\ 1 & -1 & 1 & 1 \end{bmatrix} \tag{207}$$

which satisfies indeed $f_{2,1} f_{2,1} = f_{2,1}$.

The trigonal color rotations have the iterant representations

$$\tau_{c,123} \simeq \frac{1}{2}(Id - e\varphi + f\sigma - g\tau) \tag{208}$$
$$\tau_{c,123}^{-1} \simeq \frac{1}{2}(-Id + e\varphi - f\sigma + g\tau)$$

A color rotation carries lepton $f_{1,1}$ to

$$f_{2,1} = \frac{1}{2}(Id - \varphi)\frac{1}{2}(Id + \sigma) = \frac{1}{4}(Id + \sigma - \varphi - \varphi\sigma) = \frac{1}{4}(Id + \sigma - \varphi - \tau) \tag{209}$$

And transforms, say, a red u-quark

$$f_{1,2} = \frac{1}{2}(Id - e)\frac{1}{2}(Id - f) = \frac{1}{4}(Id - e - f + g) \tag{210}$$

Into, say, a green u-quark

$$f_{2,2} = \frac{1}{2}(Id + \varphi)\frac{1}{2}(Id - \sigma) = \frac{1}{4}(Id + \varphi - \sigma - \varphi\sigma) = \frac{1}{4}(Id - \sigma + \varphi - \tau)$$

By table 2 and representation (192) these equations can easily be verified. The iterant approach is absolutely qualified to describe the entirety of amplitudes in high energy physics.

Flavor Rotations

The flavor group is generated by some complex elements with reference to the fixed point $f_{1,1}$, the density matrix of $|\overline{\nu_e}\rangle$. So we have an iterant flavor $SU(2)$ that operates on the unitary states of $SU(3)$.

$$\dot{\lambda}_2 = i[0,0,+1,-1]\sigma,\ \dot{\lambda}_5 = i[0,+1,-1,0]\tau\ ,\ \dot{\lambda}_7 = i[0,-1,0,+1]\varphi \tag{211}$$

We shall now go a little into this group because it is responsible for the topological entanglement that occurs through the action of flavor transformations. After all, this is not all too surprising, as we are able to make pigtails in Euclidean 3-space. Just as we can put a plaited loaf into the furnace, so we can put the three strand Artin braid group B_3 into the special unitary group $SU(2)$. It has been shown that this can even be done densely, that is, in such a way that we obtain representations of a manifold of braid groups that lie dense in the unitary group $U(2)$. A proof of this conjuncture has been given by Kauffman and Lomonaco (2009). The groups $SU(2)$ and $U(2)$ are playing a special role in quantum informatics. Their symmetry elements are regarded as operating on a single qubit, the unitary two spin eigenstates $|\pm 1/2\rangle$. We are used to represent an element of the symmetric unitary group $SU(2)$ by four complex numbers arranged in a 2×2 −matrix. In this small section I shall show how we can stress the matrix aspect of the idea. With that in mind the reader can go further all by himself. We denote these matrices by $Mat(2,\mathbb{C})$ in the manner of our deceased friend Pertti Lounesto. We demand that a matrix $M \in Mat(2,\mathbb{C})$ has a unit determinant and its inverse is equal to the conjugate transpose, and we usually write for this

$$det(M) = 1 \tag{212}$$

$$M^{\dagger} = M^{-1}$$

In our case the mathematics is a little bit more complicated as our so called motion group is given by symmetric unitary matrices $S \in SU(3)$, however, represented by matrices in $Mat(4,\mathbb{C})$. In accordance with formulas yet to be derived, (257) and (259), we shall have

$$\lambda_2 = \begin{bmatrix} 0 & 0 & 0 & 0 \\ 0 & 0 & -i & 0 \\ 0 & i & 0 & 0 \\ 0 & 0 & 0 & 0 \end{bmatrix} \quad \lambda_5 = \begin{bmatrix} 0 & 0 & 0 & 0 \\ 0 & 0 & 0 & -i \\ 0 & 0 & 0 & 0 \\ 0 & i & 0 & 0 \end{bmatrix} \quad \lambda_7 = \begin{bmatrix} 0 & 0 & 0 & 0 \\ 0 & 0 & 0 & 0 \\ 0 & 0 & 0 & -i \\ 0 & 0 & i & 0 \end{bmatrix} \tag{213}$$

The exponential map brings forth the group elements having form

$$S_2(x) := e^{\lambda_2 x} = \begin{bmatrix} 1 & 0 & 0 & 0 \\ 0 & \frac{1}{2}e^{x} + \frac{1}{2}e^{-x} & -\frac{i}{2}e^{x} + \frac{i}{2}e^{-x} & 0 \\ 0 & \frac{i}{2}e^{x} - \frac{i}{2}e^{-x} & \frac{1}{2}e^{x} + \frac{1}{2}e^{-x} & 0 \\ 0 & 0 & 0 & 1 \end{bmatrix} \text{ a. s. o. with} \tag{214}$$

$$det\big(S_2(x)\big) = e^{x}e^{-x} = 1$$

This matrix satisfies $S_2(x)^{\dagger} = S_2(x)^{-1}$ q. e. d. From here we can go further.

Freedom of Motion

Nature is free. She takes every possible course to escape the restrictions we pose on her. She takes the opportunity to move freely along every possible path, even if she starts off at a definite origin, as we believe, and ends the journey at a definite destiny, imposed by us. Such are the protocols. Nature takes the liberty to move in a way, we humans can never make use of. We are not free to move that way. Quantum motion is a no go for macro physics, it seems. I do say 'it seems' because it may be that someday psychology or cosmology teach us the contrary. - Unfolding the path and thus bringing forth its integral, the fields touch in a trans-local manner and transact polarity across space and time. Such is their life. *Touch* of strings and *translocal polarity flips* are nature's freedom to move. When Feynman said "If we interrupt the course of the event before its conclusion with an observation of the state of the particles involved in the event, we disturb the construction of the overall amplitude", he sensed the restrictions matter experiences in her interaction with the human observer. Clearly he projects what a human experiences in a macro experiment, but do we have any other way? "Thus if we observe the system of particles to be in one particular state, we exclude the possibility that it can be in any other state, and the amplitudes associated with the excluded states can no longer be added in as alternatives in computing the overall amplitude." We cannot restrict nature at all. She just takes other paths, and we have to change the integrals. If we recall Feynman's summary of probability concepts given in section 1-4 of 'Quantum Mechanics and Path Integrals', there he investigates the paths of a particle going from a to b by putting several screens with several holes drilled in them between the source at a and some detector at b. The particle can take every possible route through the holes, and for every route there is some amplitude. The amplitude of the resulting event requires the addition of the amplitudes, one for each path. We can go on drilling holes and adding more screens between a and b. So the arrangement becomes more and more complicated. Bu still the principle of superposition applies and we have to take the sum of amplitudes over all possible path. And Feynman and Hibbs (p. 21) go further, saying "we can make a still finer specification of the motion". In addition to specifying particular paths $x(y)$ we can specify the time at which each point is passed.

Feynman had discovered the surprising feature of light to touch every location of space-time between source location a and destiny b while it leads us to believe in a particle travelling on a straight line connecting a with b. That led to the exact checkerboard model with its image of a plane lattice of pathways and to our late understanding of the action of oriented space-time area η

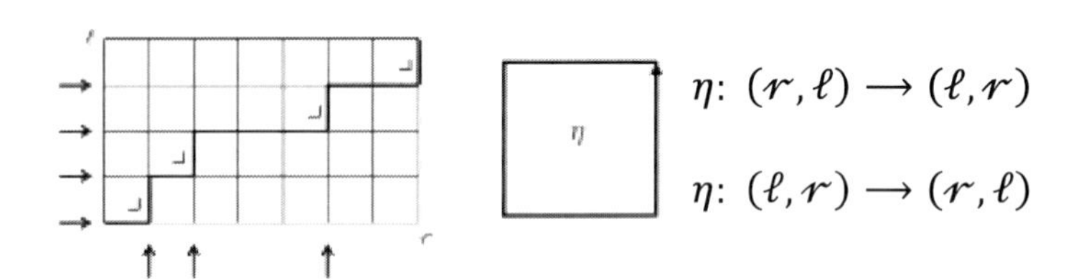

which performs rotations by $\pi/2$, that is, clockwise and counter clockwise rotations of the lattice. This concept of a pre-spinor had an exact model in the Clifford algebra $Cl_{1,1}$. It is exactly here, where the new degrees of freedom come in. It is not only that we can, but rather that we have to introduce much finer specification of the motion. Light has still more possibilities to touch the whole of space-time. In the plane with signature $\{+,-\}$ we have just one single 'pre-spinorial' shift operator η which indeed represents one single directed unit space-time area. But in the Clifford algebra of Minkowski space there are three oriented space time areas with signature $\{+,-\}$, namely $e_{14} = g\varphi$, $e_{24} = f$, $e_{34} = \sigma$. Each of these has the power to perform pre-spinor flips in the geometry of Minkowski space. This is the reason, why the problem of generalizing the checkerboard model could not be solved till now. For, not only are there three fundamental shift operators of grade 2, but they imply further degrees of freedom, as they generate the corresponding oriented unit space-time volumes with grade 3. These are the three quantities

$$e_{124} = g \quad e_{134} = e\sigma \quad e_{234} = -\tau\ ,$$

So that we have altogether six discrete operators that perform shifts analogous to the 'corner' moves in the checkerboard model. It is really surprising how serious the consequences were that the incompleteness of our insight into the geometric algebra of quantum motion brought forth. As we have pointed out, the iteration with a pair such as $[r,\ell] := \left[\frac{1}{2}(t+x),\frac{1}{2}(t-x)\right]$ obtrudes the illusion that the only and essential degree of freedom at any 'turn' or 'corner' is time-forward motion, that is, turnover from $+x$ to $-x$ while the t stays the same for both parts of the norm. That norm in $\{-,+\}$-Minkowski space is equal to $t^2 - x^2$. But in the 'opposite' Lorentz-metric $\{+,-\}$ it would be $x^2 - t^2$. So the appropriate iterant would rather look like $[r,l]_{opp} := \left[\frac{1}{2}(x+t),\frac{1}{2}(x-t)\right]$ which takes into account the reversion of time in the motion. Already Pertti Lounesto was aware of this matter when he investigated the smallest building blocks of quantum mechanics. The linear space $\mathbb{R}^2$ endowed with the indefinite signature $\{+\ -\}$ and indefinite quadratic form $(a,b) \to a^2 - b^2$ is the hyperbolic quadratic space $\mathbb{R}^{1,1}$ which has the quadratic Clifford algebra $Cl_{1,1}$ the even part of which $Cl_{1,1}^{+}$ consists of the Study numbers. That could have given us enough motivation for the creation of something like an iterant $[x+t, x-t]$. The idea is not new. It also goes back to the very first diagrams drawn by Richard Feynman. It will be very important to factor in time reversion and symmetry breakage in the appropriate way.

We can identify the space-time area η in the plane with e_{24} in $Cl_{3,1}$. But then there immediately follows the existence of a further degree of freedom having grade 3, namely the oriented space-time volume e_{124}. That is, considering a 3-dimensional Minkowski space $\mathbb{R}^{2,1}$ with its Clifford algebra $Cl_{2,1}$ we already had to factor in two directed unit space-time areas, e_{14}, e_{24} and one directed unit space-time volume e_{124}. As we know, the triple $\{e_1, e_{24}, e_{124}\}$ forms a Cartan subalgebra of the $SU(4)$-Lie group that we derive from the Clifford algebra $Cl_{3,1}$. Now we saw that in $Cl_{3,1}$ this immediately brings in and respectively confronts us with the phenomenology of a flavor rotation in the first color space, that is, the element $T_{f,1}$. Now we could say, well what, and does that mean anything in $Cl_{2,1}$? And the answer is 'yes'. Because there exists a projected image of this rotation in the smaller Clifford algebra. We saw, the rather complicated, graded operator $T_{f,1}$carries out a rotation of flavor in the Cartan algebra ch_1. But the three polarized strings $\{e_1, e_{24}, e_{124}\}$ have the capacity to do the same by mere touch, by logic interaction, as was depicted by formula (91).

The triple $\{e_1, e_{24}, e_{124}\}$ span a commutative subspace, the 'color-space'ch_1. The idempotent primitive in $Cl_{3,1}$ and represented by $f_{1,1}$ generates a 1-norm. That is, the primitive $f_{1,1}$ endows ch_1 with a 1-norm. Consider any $\phi = \varphi Id + xe_1 + ye_{24} + z \in ch_1$ and verify that

$$f_{1,1}\phi = (\varphi + x + y + z)f_{1,1} \quad \text{with 1-norm} \qquad (215)$$
$$L = \varphi + x + y + z.$$

Therefore, in second order cybernetics we say that the $f_{1,1}\phi$ provides an *eigenform* for a *1-norm*. Take $L = \varphi + x + y + z$ to obtain $f_{1,1}\phi = \phi f_{1,1} = f_{1,1}$. As this Clifford multiplication represents the interaction of two polarity strings by touch, this has a special meaning. Let us go into this. We observe a mix of polarity strings in some translocal arrangement of locations.

So we have three types of strings

$$e_1 \simeq e = [+1, +1, -1, -1]$$
$$e_{24} \simeq f = [+1, -1, -1, +1]$$
$$e_{124} \simeq g = [+1, -1, +1, -1]$$

which form the elementary unity of some mix of grade 4 bit-strings. A standard color field thus can be seen as a fermionic field brought forth by interacting polarized strings. It is given by a quadruple of real numbers

$$\phi = \{\varphi, x, y, z\} \quad \text{and equivalently by some multi-vector} \qquad (216)$$

$$\phi = \varphi + xe_1 + ye_{24} + ze_{124} \in ch_1 \qquad (217)$$

It can be written as a linear combination of the orthogonal primitive idempotents, as a mixture of what some of us called pure states

$$\phi = a_1 f_{1,1} + a_2 f_{1,2} + a_3 f_{1,3} + a_4 f_{1,4} \text{ with numbers} \tag{218}$$
$$a_1 = \varphi + x + y + z$$
$$a_2 = \varphi - x - y + z$$
$$a_3 = \varphi - x + y - z$$
$$a_4 = \varphi + x - y - z$$

So the weights of the strings determine the weights of the pure fermion states. These four numbers reproduce the 1-norm $L = \varphi + x + y + z$ for the bit string mix. Assuming that these numbers constitute the relative frequencies of the polarity strings, we may consider L to be equal to 1. Hence the mix interacting with a pure state $f_{1,1}$ just reproduces that pure state, as we have $f_{1,1}\phi = f_{1,1}$. So the fermion field may have a special freedom of motion, namely the reproduction of a lepton. But for the sake of the 'whole', every string that represents a primitive idempotent has in its associated color space a 1-norm. Also $f_{1,2}, f_{1,3}, f_{1,4}$ provide 1-norms, however, with the appropriate signs as are determined by (218). This clever arrangement nature is confronting us with, will become a prudent pidgin once we see it operates like a phase gate in a quantum computer. We just have to find the natural combinatorial gate that performs the entanglement.

Freedom of Entanglement

Most of us say they do not understand what is going on in nature as it is disclosed in their experiments or they prefer to accommodate to the current viewpoint: "we cannot understand it". Where does the entanglement stem from? All we described in the chapter "phenomenology of immediacy" where we investigated delayed choice, orientation and quantum erasure, is indeed connected with a natural property that is indicated by the structure of Clifford algebra of the Minkowski space. Entanglement is a necessary equipment of the real geometry and thus of the grade 4 iteration of space. As in the Anti-Physics of motion, matter is space, it is natural that entanglement in quantum motion turns out as topological entanglement. To see that point more clearly, for most of us it would be sufficient to take the following hint: Consider the three idempotent iterant constituents not primitive in the algebra

$a := \frac{1}{2}(Id + e),\ \ b := \frac{1}{2}(Id + f),\ \ c := \frac{1}{2}(Id + g)$ representing Clifford monomials

$$a \simeq \frac{1}{2}(Id + e_1), b \simeq \frac{1}{2}(Id + e_{24}), c \simeq \frac{1}{2}(Id + e_{124}) \tag{219}$$

These quantities satisfy a quasi braid algebra characteristic for the commutative Klein 4 group:

$$aba = bab \tag{220}$$
$$aca = cac$$
$$bcb = cbc$$

To see the consequences, consider at first the 'braid operator'

Figure 46. The braiding operation.

Such an operator can be seen as an element of quantum information. It represents an entry to a certain viewpoint of geometry, namely to the attitude that a location, an element of space, or a whole space act like operators on spaces. This idea is central to Clifford algebra and I have variously used it, for example when I showed it is meaningful to consider locations as rotators. It is also a fundamental module in topology where the Artin braid group and braids in general can be seen as operators on spaces. Mathematicians speak of a shift that carries elements of a topological category to morphisms in an associated algebraic category (Kauffman and Lomonaco 2002). Consider two sets of locations in an unspecified domain, with locations in each set arranged vertically for convenience, so it appears in our imagination those two sets are somehow neighboring each other and can be connected. The connecting strands constitute a one-one correspondence, so each location in the first set is connected with a location in the neighboring set. Some strand will pass over some other or it will pass under it, which makes a difference. Such a pattern of connections is called a braid. In general terminology braids are patterns of entangled strands. For example

Figure 47. Unequal braids with two strands.

A braiding operator can be conceived like a 'topological transformer'. It can transform the neutral element into a non-trivial braid

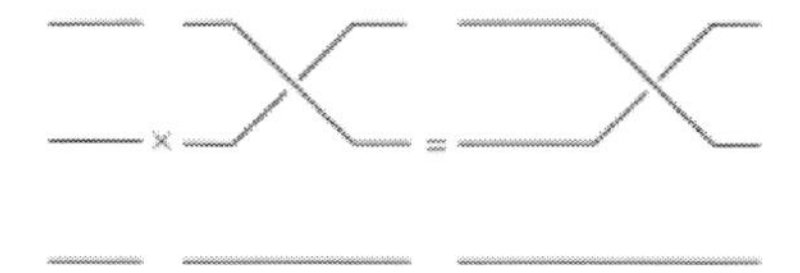

Figure 48. Concatenation of braids.

Like the finite Coxeter groups, so braid groups have generators and relations. For example every braid in the group B_4 can be written as a composition of a number of three generating braids:

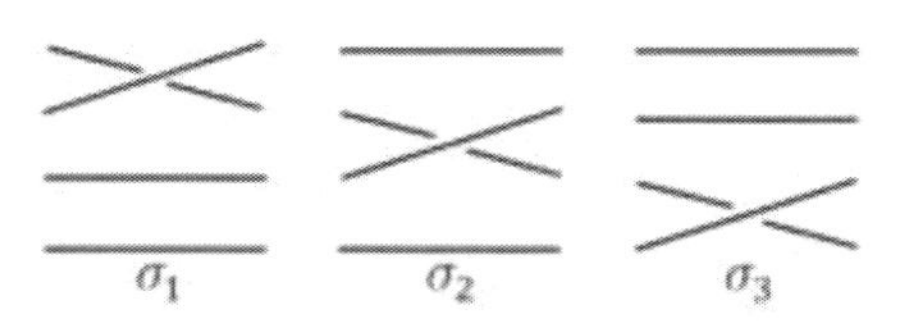

Figure 49. Generators of the braid group B_4.

The braid group B_3 is isomorphic to the knot group of the trefoil knot. It is an infinite non-abelian group. The braids of B_3 satisfy the diagram of the Young-Baxter equation

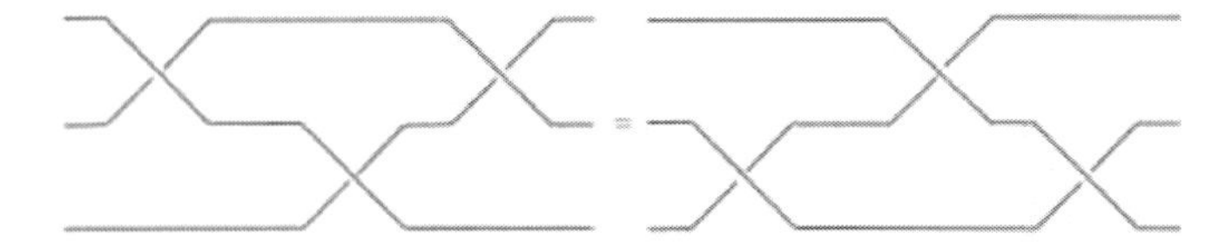

Figure 50. Young-Baxter equation.

The construction of a relation between topological and quantum entanglement implies the possibility to associate a unitary operator with a braid. So we are looking for unitary representations of the Artin braid group. Kauffman and Lomonaco (2002) have investigated such a representation of the knot group, annotating that it was not the purpose of their paper to give an exhaustive account of such representations. But interestingly their case characterizes the whole entanglement phenomenology of quantum motion that brings forth the motion groups of the Minkowski algebra. Suppose V is a two-dimensional complex vector space. The braiding operator of an elementary 2 strand braid is a mapping

$$R\colon V \otimes V \longrightarrow V \otimes V \tag{221}$$

The input and output lines in figure 48 should be thought to represent this map of tensor products. For the braid group B_3 we have to consider a mapping

$$V^{\otimes 3} \longrightarrow V^{\otimes 3}. \tag{222}$$

The algebraic form of the diagram depicted in Figure 49 is the algebraic equation

$$(R{\otimes}I)(I{\otimes}R)(R{\otimes}I) = (I{\otimes}R)(R{\otimes}I)(I{\otimes}R). \tag{223}$$

Generally, a representation of the Artin braid group to the automorphisms $V^{\otimes n} \longrightarrow V^{\otimes n}$ is constructed by terms

$$\sigma_k^{rep} := I \otimes \ldots I \otimes R \otimes I \otimes \ldots I \tag{224}$$

where R occupies the entries k and respectively $k+1$ in the tensor product. If R is invertible and satisfies the Young-Baxter equation, then the σ_k^{rep} represent the generators of the braid group B_n. If R is unitary, we have constructed a unitary representation. A peculiar matrix R has been examined that is a unitary matrix and satisfies the Young-Baxter equation. As R is a unitary matrix it can be considered as a quantum gate and since it can represent the braiding operation R, it was presumed that it performs both topological and quantum entanglement. This matrix has four scalar entries on the unit circle in the complex plane

$$R := \begin{bmatrix} a_1 & 0 & 0 & 0 \\ 0 & 0 & a_2 & 0 \\ 0 & a_3 & 0 & 0 \\ 0 & 0 & 0 & a_4 \end{bmatrix} \tag{225}$$

From the point of view of quantum information, we can consider any superposition of states in the first color space $\phi = a_1 f_{1,1} + a_2 f_{1,2} + a_3 f_{1,3} + a_4 f_{1,4}$ as a phase gate, since the $f_{1,\alpha}$ are primitive idempotents. Note that in the representation (176) these are diagonal matrices with just one non vanishing entry in row α, column α equal to one. So we consider a phase gate represented by the matrix

$$\phi = \begin{bmatrix} a_1 & 0 & 0 & 0 \\ 0 & a_2 & 0 & 0 \\ 0 & 0 & a_3 & 0 \\ 0 & 0 & 0 & a_4 \end{bmatrix} \tag{226}$$

Now consider a flavor rotation $T_{f,1} = s_{6,4} s_{2,3}$ composed by two natural units of the Minkowski space, reflections, or flips in the iterant algebra

$$s_{2,3} = \frac{1}{2}(Id + e - \sigma + e\sigma) \tag{227}$$

In the Majorana representation this is the matrix

$$\dot{s}_{2,3} = \begin{bmatrix} 1 & 0 & 0 & 0 \\ 0 & 0 & 1 & 0 \\ 0 & 1 & 0 & 0 \\ 0 & 0 & 0 & 1 \end{bmatrix} \tag{228}$$

indicating the $Mat(4, \mathbb{R})$ representation by an over-dot, and the other reflection

$$s_{6,4} = \frac{1}{2}(Id + f + \tau - f\tau)$$

is a swap matrix too

$$\dot{s}_{6,4} = \begin{bmatrix} 1 & 0 & 0 & 0 \\ 0 & 0 & 0 & 1 \\ 0 & 0 & 1 & 0 \\ 0 & 1 & 0 & 0 \end{bmatrix} \tag{229}$$

First, recall, what are these operators doing? What do they effect in the Minkowski algebra? They carry out flips in the color space ch_1. But physically these are no trivial flips, but they change grade. Consider the directed unit space-time area e_{24}. The swap $s_{2,3}$ takes it to the directed unit space-time volume element, as we have

$$s_{2,3}\, e_{24}\, s_{2,3} = e_{124} \tag{230}$$

And it also has the potential to flip it back:

$$s_{2,3}\, e_{124}\, s_{2,3} = e_{24}, \tag{231}$$

and what about the $s_{6,4}$? It maps the directed unit space-time volume element onto the line element e_1

$$s_{6,4}\, e_{124}\, s_{6,4} = e_1 \tag{232}$$

and as we would expect from the braiding phenomena, the other way around too

$$s_{6,4}\, e_1\, s_{6,4} = e_{124}. \tag{233}$$

It is due to the action of these two peculiar swaps that their product $T_{f,1}$ operates as a trigonal rotation on the Minkowski algebra. Now some might still wonder why we are moving in circles in a commutative space spanned by four mysterious units, namely $\{Id, e_1, e_{24}, e_{124}\}$, the iterants $\{Id, e, f, g\}$. Physicists are asking me this repeatedly since many years, and the answer is simply: we obtain spinors from a spinor space, and spinor spaces from primitive idempotents, and primitive idempotents from a directed spatial element of physical motion which squared gives the identity. In Euclidean space this unitary element of space is simply a line element, say, e_1 or e_2 or e_3, and the idempotent primitive in its Clifford algebra $Cl_{3,0}$ (the prominent Pauli algebra) are the 'paravectors' $(Id \pm e_1)/2$, $(Id \pm e_2)/2$ and $(Id \pm e_3)/2$. From these we obtain as minimal ideals the spinor spaces, say, $fCl_{3,0}$ with $f = (Id + e_3)/2$. You see, the whole unitary business, in the Pauli algebra of Euclidean space, is connected with this natural unitary line element of space along which particles seem to move. But this is not the case in the Clifford algebra of the Minkowski space, and this is not a fact for relativistic motion. In relativistic quantum mechanics there is no single unitary line element of the form e_1 for motion! No e_1 which can be considered as a unit element of paths! The most simple such directed unitary element of spatial motion, in Minkowski space-time algebra has the form $Id - 2f$ where f is an idempotent primitive in this algebra. This, and only this, is analogous in this concern. In the Clifford algebra of Euclidean 3-space we also have a directed element of spatial motion having form $Id - 2f$. And what is it? It is $Id - 2((Id \pm e_3)/2) = e_3$. In Euclidean space such a line element makes sense in motion. But as soon as relativity comes in, quantum motion has another minimal structure for motion, namely a quantity of the form, say, $s_{1,1} = (1/2)(Id - e_1 - e_{24} - e_{124})$ or $s_{1,2} = (1/2)(Id + e_1 + e_{24} - e_{124})$ or else $s_{2,3} = (1/2)(Id + e_1 - e_{34} + e_{134})$. There are $6 \times 4 = 24$ such 'minimalistic elements' of motion in the standard representation. They form equivalence classes of period 2, graded swaps, permutations of base unit monomials.

Now, let us consider quantum informatics and $\dot{s}_{2,3}$ as a swap gate for quantum motion. So we have a phase gate ϕ and a swap gate $\dot{s}_{2,3}$ and we verify the equation

$$\phi = R\dot{s}_{2,3} \tag{234}$$

$$R\,\dot{s}_{2,3} := \begin{bmatrix} a_1 & 0 & 0 & 0 \\ 0 & 0 & a_2 & 0 \\ 0 & a_3 & 0 & 0 \\ 0 & 0 & 0 & a_4 \end{bmatrix} \begin{bmatrix} 1 & 0 & 0 & 0 \\ 0 & 0 & 1 & 0 \\ 0 & 1 & 0 & 0 \\ 0 & 0 & 0 & 1 \end{bmatrix} = \begin{bmatrix} a_1 & 0 & 0 & 0 \\ 0 & a_2 & 0 & 0 \\ 0 & 0 & a_3 & 0 \\ 0 & 0 & 0 & a_4 \end{bmatrix} = \phi$$

Here the iterant R restores the superposition of states in color space ch_1. But R has the potential to entangle quantum states. Note that it is itself a composition of a period 2 generator of the reorientation group, in this notation a mere swap, and a phase gate. For multiply equation (234) from the right by $\dot{s}_{2,3}$, we get

$$\phi\dot{s}_{2,3} = R\dot{s}_{2,3}\dot{s}_{2,3} = R\, Id \text{ hence } R = \phi\dot{s}_{2,3} \tag{235}$$

R is a Clifford product of a phase gate and swap gate. This swap gate is better denoted as that what it is, namely a transposition in the graded basis of the Minkowski algebra: a reorientation. Therefore I would like to introduce a 'space time informatic' denotation.

A braiding multivector R is composed by a phase gate ϕ and a unitary reorientation gate $s_{\chi,\alpha}$.

For the special design of the matrix (225) it has been shown that such a braiding operator can entangle quantum states. Once we realize that we have indeed 24 equivalence classes of such transposition elements, the proof given by Kauffman and Lomonaco gains enormous generality.

Hence, consider the action of the unitary transformation R on quantum states, and use the customary 'bra'-'ket' notation. Notice however that this tensor decomposition is not at all needed. It serves mere illustrative purposes to recall the old context, how experimenters imagined their entangled states.

(i) $R|0\ 0\rangle = a_1|0\ 0\rangle$ (236)
(ii) $R|0\ 1\rangle = a_3|1\ 0\rangle$
(iii) $R|1\ 0\rangle = a_4|0\ 1\rangle$
(iv) $R|1\ 1\rangle = a_2|1\ 1\rangle$

Lemma 4: Let $\psi := |0\rangle + |1\rangle$. If R is chosen so that $a_1a_2 \neq a_3a_4$, then the state $R\ \psi\otimes\psi$ is entangled.

Proof: In accord with (236)

$$\Psi = R(\psi\otimes\psi) = R\big((|0\rangle + |1\rangle)\otimes(|0\rangle + |1\rangle)\big) = a_1|0\ 0\rangle + a_3|1\ 0\rangle + a_4|0\ 1\rangle + a_2|1\ 1\rangle \tag{237}$$

If this state is not entangled then there are constants A, B, C, D such that

$$\Psi = (A|0\rangle + B|1\rangle)\otimes(C|0\rangle + D|1\rangle) \tag{238}$$

This implies that

(i) $a_1 = AC$ (239)
(ii) $a_3 = BC$

(iii) $a_4 = AD$
(iv) $a_2 = BD$

which means that $a_1 a_2 = a_3 a_4$. Therefore, whenever $a_1 a_2 \neq a_3 a_4$ we can conclude that if Ψ is a quantum state, it is an entangled state. Beware of believing that this would be just a special case. We shall work out step by step the proof for its generality. Begin by considering the second component of the flavor rotation $T_{f,1} = s_{6,4} s_{2,3}$, that is, we keep the ball moving by investigating the other reorientation gate $s_{6,4}$.
We have a new braid operation U

$$U\,\dot{s}_{6,4} := \begin{bmatrix} a_1 & 0 & 0 & 0 \\ 0 & 0 & 0 & a_2 \\ 0 & 0 & a_3 & 0 \\ 0 & a_4 & 0 & 0 \end{bmatrix} \begin{bmatrix} 1 & 0 & 0 & 0 \\ 0 & 0 & 0 & 1 \\ 0 & 0 & 1 & 0 \\ 0 & 1 & 0 & 0 \end{bmatrix} = \begin{bmatrix} a_1 & 0 & 0 & 0 \\ 0 & a_2 & 0 & 0 \\ 0 & 0 & a_3 & 0 \\ 0 & 0 & 0 & a_4 \end{bmatrix} = \phi \tag{240}$$

which is also composed by a phase gate and a reorientation gate, as we have

$$U = \phi \dot{s}_{6,4} \tag{241}$$

We are repeating the argumentation. Henceforth, consider the action of the unitary transformation U on quantum states, and use the customary 'bra'-'ket' notation.

(i) $U|0\,0\rangle = a_1|0\,0\rangle$ (242)
(ii) $U|0\,1\rangle = a_3|1\,0\rangle$
(iii) $U|1\,0\rangle = a_4|0\,1\rangle$
(iv) $U|1\,1\rangle = a_2|1\,1\rangle$

Lemma 4: Let $\psi := |0\rangle + |1\rangle$. If U is chosen so that $a_1 a_2 \neq a_3 a_4$, then the state $U\,\psi \otimes \psi$ is entangled.
Proof: In accord with (236)

$$\Psi = U(\psi \otimes \psi) = U\big((|0\rangle + |1\rangle) \otimes (|0\rangle + |1\rangle)\big) = a_1|0\,0\rangle + a_3|1\,0\rangle + a_4|0\,1\rangle + a_2|1\,1\rangle \tag{243}$$

If this state is not entangled then there are constants A, B, C, D such that

$$\Psi = (A|0\rangle + B|1\rangle) \otimes (C|0\rangle + D|1\rangle) \tag{244}$$

This implies that

(v) $a_1 = AC$ (245)
(vi) $a_3 = BC$
(vii) $a_4 = AD$
(viii) $a_2 = BD$

which means that $a_1 a_2 = a_3 a_4$. Therefore, whenever $a_1 a_2 \neq a_3 a_4$ we can conclude that if Ψ is a quantum state, it is an entangled state.

Minding at last the product of the two unitary directed graded elements of motion $T_{f,1} = s_{6,4}s_{2,3}$ and likewise $T_{f,1}^{-1} = s_{2,3}s_{6,4}$. They are both orthogonal matrices, hence unitary, with their matrix representations

$$T_1 = \begin{bmatrix} 1 & 0 & 0 & 0 \\ 0 & 0 & 0 & 1 \\ 0 & 1 & 0 & 0 \\ 0 & 0 & 1 & 0 \end{bmatrix} \text{ and } T_1^{-1} = \begin{bmatrix} 1 & 0 & 0 & 0 \\ 0 & 0 & 1 & 0 \\ 0 & 0 & 0 & 1 \\ 0 & 1 & 0 & 0 \end{bmatrix}$$

Combined with the phase gate ϕ these trigonal permutations of the graded unit monomials bring forth a braiding operation

$$\mathfrak{T} = \phi\, T \tag{246}$$

I have now avoided indices, since it is clear that this holds for $T_{f,1}$ and for the other color spaces. But, suppose there is a graded Lie group that acts like a rotation on the unitary directed graded elements, then those 24 representatives $s_{\chi,\alpha}$ will be 'rotated out' to or 'unfolded into' a sphere of unitary elements that are graded and may have dimensions up to 16, just like the tip of a unit vector in Euclidean 3-space can enfold and scan the surface of the whole sphere S^2 when it is driven by orthogonal rotations of the $SO(3)$ and respectively the $SU(2)$. Those readers who read some of my contributions and watched me attentively, are aware that I am using here a purely fermionic phase gate. I am doing this on purpose because some of us were extremely impatient and claimed that the entanglement had nothing to do with strong interaction. And how! Don't I know it! In my first book on primordial space I have shown that waves of photons are expected to be elements of the even subalgebra of the Minkowski algebra. Therefore we have phase gates in this algebra $Cl_{3,1}^{+}$ and orthogonal operators that permute grades. To generalize the theory towards a full module capable to explain entanglement we have to find out about the role of the motion group $SU(3)$ not only for the Cartan modules of the Minkowski algebra but for the whole universe of quantum motion in this space. So let us go into this.

The Motion Group

Most of the following rigor ultimately clears up the question why the symmetric unitary group $SU(3)$ plays such an important role in particle physics. Though it is based on Clifford algebra and the corresponding mathematical software MAPLE Clifford by Fauser and Ablamowicz, we must henceforth keep in mind that every interaction in this Minkowski space can by carried out by polarity strings which interact by touch. Polarity strings can interact by logic identification in an unspecified domain. A peculiar feature of such domains is their potential for translocal interaction. We know now that the grade 4 iterant algebra is isomorphic with the Minkowski algebra $Cl_{3,1}$. So I will represent the final resulting Clifform of the algebra $su_{Cl}(3)$ not only in Clifford algebra, but also in iterant form.

Theorem 9: In accord with theorem 6 the Minkowski algebra can be conceived as a Lie algebra. The full motion group preserving neutrino fields is a rank 2 space-time group $\mathfrak{L}_2 \subset \mathfrak{L}_3 \subset Cl_{3,1}$. This is a subgroup $SU(3)$ of a rank 3 Lie group $SU(4)$.

Proof: In physics we wish to consider idempotents primitive in the Minkowski algebra as fermion and lepton states. We start with the assumption that there is a lepton, say the f_1, a neutrino that does not partake in strong interaction, that is, it should be invariant under the action of $\mathfrak{L}_2$. This would mean that there is some Eigenform in which the f_1 appears as a fixed point, and the $\mathfrak{L}_2$ would turn out as a stabilizer group for that primitive idempotent. This is a natural assumption. Originally the generators of the *Clifform* of that graded Lie group have been found on a long path of trial and error. Meanwhile it is possible to derive it from comparatively few basic assumptions by a little rigor and empathy. These assumptions read:

1. There is a Group $\mathfrak{L}_2 \subset Cl_{3,1}$ with a fixed point f_1 .
2. Every group element $u \in \mathfrak{L}_2$ as an element of Clifford algebra $Cl_{3,1}$ must have the general form $u = x_2e_1 + x_3e_2 + \ . \ . \ . + x_{15}e_{234} + x_{16}e_{1234}$.
3. Every group element $u \in \mathfrak{L}_2$ transforms elements $\psi \in Cl_{3,1}$ by conjugation just like in a spin group: $\psi \longmapsto u\psi u^{-1}$.
4. Generators of the Lie algebra of the group $\mathfrak{L}_2 \subset Cl_{3,1}$ should be unitary.
5. Group elements $u \in \mathfrak{L}_2$ are calculated from the algebra by an exponential map.
6. Generators of the group algebra should satisfy commutation relations of some form of $SU(3)$.

This (Cliff)form can have different real forms (Magnea, 2002). If it were a *normal real form*, it should be represented by a matrix algebra $SL(3,\mathbb{R})$ within $Cl_{3,1}$, where the latter is known to be isomorphic with the matrix algebra $Mat\ (4,\mathbb{R})$. If it were a compact form, it should be represented by $SU(3,\mathbb{C})$, but now as matrices $Mat\ (4,\mathbb{C})$ in $\mathbb{C}\otimes Cl_{3,1}$. We know that in agreement with point 2., the symmetry $u \in Cl_{3,1}$ has the general (Cliff)form

$$u \stackrel{\text{def}}{=} x_1Id + x_2e_1 + x_3e_2 + x_4e_3 + x_5e_4 + x_6e_{12} + x_7e_{13} + x_8e_{14} + x_9e_{23} + \\ + x_{10}e_{24} + x_{11}e_{34} + x_{12}e_{123} + x_{13}e_{124} + x_{14}e_{134} + x_{15}e_{234} + x_{16}e_{1234} \tag{247}$$

Considering 1., the first point, every group element u should satisfy the equation $f_1u - uf_1 = 0$. This defines a problem of linear algebra that can most conveniently be treated with MAPLE Clifford. Namely it turns out that coefficients $x_1, x_2, \boldsymbol{x_6}, \boldsymbol{x_8}, x_{10}, \boldsymbol{x_{12}}, x_{13}, \boldsymbol{x_{14}}, \boldsymbol{x_{15}}, \boldsymbol{x_{16}}$ can be chosen freely while six of them have to satisfy certain simple linear relations to another six. So they form six pairs according to conditions

$$\text{(i) } x_3 = x_8, \quad \text{(ii) } x_4 = x_{15}, \quad \text{(iii) } x_5 = x_6, \\ \text{(iv) } x_7 = x_{16}, \quad \text{(v) } x_9 = -x_{12}, \quad \text{(vi) } x_{11} = -x_{14} \tag{248}$$

This proposes a pairing of monomials having different grades up to some constant, namely, if we chose six Lie algebra generators to be essentially given by

$$\text{(i) } e_2 + e_{14} \quad \text{(ii) } e_3 + e_{234} \quad \text{(iii) } e_4 + e_{12} \tag{249}$$

(iv) $e_{13}+j$ (v) $-e_{23}+e_{123}$ (vi) $-e_{34}+e_{134}$

Notice, every generator is an element in a 2-dimensional graded subspace, say, $\mathfrak{S}_k$. We can satisfy equations (248) since the exponential mapping enlarges the subspaces of generators over the domain of the chromatic space ch_1 while preserving the correct proportion. Namely, we obtain the *exponential map*:

$$exp\colon \mathfrak{S}_k \longrightarrow \mathfrak{S}_k \oplus ch_1 \tag{250}$$

Example:

$$exp\colon \mathfrak{S}_2 := \{e_3, e_{234}\} \longrightarrow \mathfrak{S}_2 \oplus \{Id, e_{24}\} \tag{251}$$

So, whenever the exponential map adds some further monomial to any $\mathfrak{S}_k$, this is always an element from ch_1, that is, it is either Id or e_1 or e_{24} or a unit space-time volume e_{124} up to a constant. Since the group element is the exponential of any linear combination of generators, the coefficients of Id, e_1, e_{24} and e_{124} must appear as free solutions in the MAPLE run. Those are the four numbers x_1, x_2, x_{10}, x_{13} correlated with chromatic space ch_1 which constitute the isospin and hypercharge. The six Clifford numbers (249) are nothing else than components of shift operators for the real form $SL(3,\mathbb{R})$. In the algebra $Cl_{3,1}$ these can be represented by 4×4-matrices with real entries. We take *const* = ½ to secure unitarity (point 4.) and rearrange terms in accord with a standard representation. So we define at last

$$\begin{aligned}&\Lambda_1 = \tfrac{1}{2}(-e_{34}+e_{134});\;\; \Lambda_2 = \tfrac{1}{2}(-e_{23}+e_{123});\;\; \Lambda_3 = \tfrac{1}{2}(-e_{24}+e_{124});\\ &\Lambda_4 = -\tfrac{1}{2}(e_3+e_{234})\,;\;\; \Lambda_5 = -\tfrac{1}{2}(e_{13}+J);\qquad \Lambda_6 = \tfrac{1}{2}(e_2+e_{14});\\ &\Lambda_7 = \tfrac{1}{2}(e_4+e_{12})\,;\quad \Lambda_8 = \tfrac{1}{2}(-e_1+e_{24})\end{aligned} \tag{252}$$

Then we can calculate the following familiar generators of the special linear group $SL(3,\mathbb{R})$

$$\begin{aligned}&T_{+,1} \overset{\text{def}}{=} \Lambda_1-\Lambda_2;\;\; T_{+,2} \overset{\text{def}}{=} \Lambda_6-\Lambda_7;\;\; T_{+,12} \equiv T_{+,3} \overset{\text{def}}{=} \Lambda_4-\Lambda_5;\\ &T_{-,1} \overset{\text{def}}{=} \Lambda_1+\Lambda_2;\;\; T_{-,2} \overset{\text{def}}{=} \Lambda_6+\Lambda_7;\;\; T_{-,12} \equiv T_{-,3} \overset{\text{def}}{=} \Lambda_4+\Lambda_5;\end{aligned} \tag{253}$$

together with their diagonal operators

$$T_{0,1} \overset{\text{def}}{=} \Lambda_3 \text{ and } T_{0,2} = \Lambda_8 \tag{254}$$

We can obtain the corresponding complex (Cliff)form by performing the Weyl-trick. To generate the compact group by the same Clifform, we have to multiply three generators $\{\Lambda_2, \Lambda_5, \Lambda_7\}$ – for the compact subalgebra $su(2,\mathbb{C})$ – by the imaginary unit. That is, in order to obtain the correct commutation relations, we multiply Λ_2 by $i=\sqrt{-1}$ and obtain the generators of the ancient '*t-spin*':

$$\lambda_1 = \frac{1}{2}(-e_{34}+e_{134});\;\; \lambda_2 = \frac{i}{2}(-e_{23}+e_{123});\;\; \lambda_3 = \frac{1}{2}(-e_{24}+e_{124}) \tag{255}$$

To also have *u-spin* and *w-spin*, we need some more generators (Schmeikal 2009). We also multiply Λ_5, Λ_7 by i and consider a most relevant and rather familiar linear combination for λ_8

$$\begin{aligned}
&\lambda_1 = \tfrac{1}{2}(-e_{34} + e_{134});\ \lambda_2 = \tfrac{i}{2}(-e_{23} + e_{123});\ \lambda_3 = \tfrac{1}{2}(-e_{24} + e_{124});\\
&\lambda_4 = -\tfrac{1}{2}(e_3 + e_{234})\ ;\ \lambda_5 = -\tfrac{i}{2}(e_{13} + J);\ \lambda_6 = \tfrac{1}{2}(e_2 + e_{14});\\
&\lambda_7 = \tfrac{i}{2}(e_4 + e_{12});\ \lambda_8 = \tfrac{1}{\sqrt{3}}(\Lambda_3 + 2\Lambda_8) = \tfrac{1}{2\sqrt{3}}(-2e_1 + e_{24} + e_{124})
\end{aligned} \tag{256}$$

These eight elements, usually taken halved, generate the compact group, and in a peculiar standard representation, in a 4×4-picture, are equal to the Gell-Mann matrices. We just have to omit the first row and column. Note that the non-compact group $SL(3, \mathbb{R})$ provides the same root spaces as the $SU(3, \mathbb{C})$.[3] Those readers who are interested in the non-compact form and its physical theory, are advised to read "*Liouville Type Models in Group Theory Framework*" by Gerasimov, Kharchev, Marshakov, Mironov, Morozov and Olshanetsky (1996).

Matrix- and Iterant Representations

To calculate the appropriate matrices for shift operators and the Gell-Mann matrices, we use a peculiar standard representation of the Clifford algebra, namely the base units

$$e_1 = \begin{bmatrix} 1 & & & \\ & -1 & & \\ & & -1 & \\ & & & 1 \end{bmatrix},\ e_2 = \begin{bmatrix} & 1 & & \\ 1 & & & \\ & & & 1 \\ & & 1 & \end{bmatrix},\ e_3 = \begin{bmatrix} & & 1 & \\ & & & -1 \\ 1 & & & \\ & -1 & & \end{bmatrix},$$
$$e_4 = \begin{bmatrix} & -1 & & \\ 1 & & & \\ & & & -1 \\ & & 1 & \end{bmatrix} \tag{257}$$

Using (254) to (256) we derive from these the familiar shift operators of the $SL(3, \mathbb{R})$

$$T_{+,1} = \begin{bmatrix} 0 & 0 & 0 & 0 \\ 0 & 0 & 1 & 0 \\ 0 & 0 & 0 & 0 \\ 0 & 0 & 0 & 0 \end{bmatrix},\ T_{-,1} = \begin{bmatrix} 0 & 0 & 0 & 0 \\ 0 & 0 & 0 & 0 \\ 0 & 1 & 0 & 0 \\ 0 & 0 & 0 & 0 \end{bmatrix},\ T_{+,2} = \begin{bmatrix} 0 & 0 & 0 & 0 \\ 0 & 0 & 0 & 0 \\ 0 & 0 & 0 & 1 \\ 0 & 0 & 0 & 0 \end{bmatrix},$$
$$T_{-,2} = \begin{bmatrix} 0 & 0 & 0 & 0 \\ 0 & 0 & 0 & 0 \\ 0 & 0 & 0 & 0 \\ 0 & 0 & 1 & 0 \end{bmatrix},\ T_{+,3} = \begin{bmatrix} 0 & 0 & 0 & 0 \\ 0 & 0 & 0 & 1 \\ 0 & 0 & 0 & 0 \\ 0 & 0 & 0 & 0 \end{bmatrix},\ T_{-,3} = \begin{bmatrix} 0 & 0 & 0 & 0 \\ 0 & 0 & 0 & 0 \\ 0 & 0 & 0 & 0 \\ 0 & 1 & 0 & 0 \end{bmatrix}, \tag{258}$$

[3] The *normal real form* of the complex algebra $sl(n; \mathbb{C})$ is the non-compact algebra $sl(n; \mathbb{R})$. This algebra can be decomposed as $K \oplus iP$ where K is the algebra consisting of real, skew-symmetric and traceless $n \times n$-matrices and iP is the algebra consisting of real, symmetric and traceless $n \times n$ matrices. Under the Weyl unitary trick we constructed, this algebra gives us the compact real form of $su(n; \mathbb{C}) = K \oplus P$.

with diagonal operators $T_{0,1} = \begin{bmatrix} 0 & 0 & 0 & 0 \\ 0 & 1 & 0 & 0 \\ 0 & 0 & -1 & 0 \\ 0 & 0 & 0 & 0 \end{bmatrix}$, $T_{0,2} = \begin{bmatrix} 0 & 0 & 0 & 0 \\ 0 & 0 & 0 & 0 \\ 0 & 0 & 1 & 0 \\ 0 & 0 & 0 & -1 \end{bmatrix}$

Carrying out formulas (256), we obtain the familiar Gell-Mann matrices

$$\lambda_1 = \begin{bmatrix} 0 & 0 & 0 & 0 \\ 0 & 0 & 1 & 0 \\ 0 & 1 & 0 & 0 \\ 0 & 0 & 0 & 0 \end{bmatrix}, \lambda_2 = \begin{bmatrix} 0 & 0 & 0 & 0 \\ 0 & 0 & -i & 0 \\ 0 & i & 0 & 0 \\ 0 & 0 & 0 & 0 \end{bmatrix}, \lambda_3 = \begin{bmatrix} 0 & 0 & 0 & 0 \\ 0 & 1 & 0 & 0 \\ 0 & 0 & -1 & 0 \\ 0 & 0 & 0 & 0 \end{bmatrix}, \tag{259}$$
$$\lambda_4 = \begin{bmatrix} 0 & 0 & 0 & 0 \\ 0 & 0 & 0 & 1 \\ 0 & 0 & 0 & 0 \\ 0 & 1 & 0 & 0 \end{bmatrix}, \lambda_5 = \begin{bmatrix} 0 & 0 & 0 & 0 \\ 0 & 0 & 0 & -i \\ 0 & 0 & 0 & 0 \\ 0 & i & 0 & 0 \end{bmatrix}, \lambda_6 = \begin{bmatrix} 0 & 0 & 0 & 0 \\ 0 & 0 & 0 & 0 \\ 0 & 0 & 0 & 1 \\ 0 & 0 & 1 & 0 \end{bmatrix}$$
$$\lambda_7 = \begin{bmatrix} 0 & 0 & 0 & 0 \\ 0 & 0 & 0 & 0 \\ 0 & 0 & 0 & -i \\ 0 & 0 & i & 0 \end{bmatrix}, \lambda_8 = \frac{1}{\sqrt{3}} \begin{bmatrix} 0 & 0 & 0 & 0 \\ 0 & 1 & 0 & 0 \\ 0 & 0 & 1 & 0 \\ 0 & 0 & 0 & -2 \end{bmatrix}$$

We have been brought up with several habits of our teachers who used for example the forms of linear algebra, the matrices and tensor products, and if they were advanced they brought in the graded forms of multivectors. Some saw the advantages of discrete models, location theory or Heyting algebras, but little was known yet of the potential in polarity strings. Most of us feared the idea that there was a primordial chaos of bit strings that interacted in nonlinear ways, and if we would try to investigate such exotic processes with reference to path integrals and weak and strong force, we would open Pandora's box and get lost forever. But that fear is ill-founded. As we know there is a well established correspondence between the Clifford algebra of the Minkowski space and fourfold iterants, we can formulate a surprising theorem.

Theorem 10: The Motion group in the form $SU(3,\mathbb{C})$ has a grade 4 iterant representation

$$\dot{\lambda}_1 = [0,0,-1,-1]\sigma \quad \dot{\lambda}_2 = i[0,0,+1,-1]\sigma \quad \dot{\lambda}_3 = [0,0,+1,-1] \tag{260}$$
$$\dot{\lambda}_4 = [0,+1,+1,0]\tau \quad \dot{\lambda}_5 = i[0,+1,-1,0]\tau \quad \dot{\lambda}_6 = [0-1,0,-1]\varphi$$
$$\dot{\lambda}_7 = i[0,-1,0,+1]\varphi \quad \dot{\lambda}_8 = \frac{1}{\sqrt{3}}[0,-1,0,+1]$$

Proof: carried out by the aid of correspondences (192) using the representation of fourfold iterants e, f, g, permutations $\varphi, \sigma, \tau \in S_4$ and direct calculation. Using the correspondences (192) we obtain

$$\dot{\lambda}_1 = \tfrac{1}{2}(-Id+e)\sigma \quad \dot{\lambda}_2 = \tfrac{i}{2}(g-f)\sigma \quad \dot{\lambda}_3 = \tfrac{1}{2}(g-f) \tag{261}$$
$$\dot{\lambda}_4 = \tfrac{1}{2}(Id-f)\tau \quad \dot{\lambda}_5 = \tfrac{i}{2}(e-g) \quad \dot{\lambda}_6 = \tfrac{1}{2}(-Id+g)\varphi$$
$$\dot{\lambda}_7 = \tfrac{i}{2}(f-e)\varphi \quad \dot{\lambda}_8 = \tfrac{1}{2\sqrt{3}}(-2e+f+g)$$

Hence carrying out, for example the isospin generator

$$\lambda_3 = \tfrac{1}{2}(-f+g) = \tfrac{1}{2}([-1,+1,+1,-1]+[+1,-1,+1,-1]) = [0,0,+1,-1] \tag{262}$$

We obtain beautiful iterants for the Cartan subalgebra $\{\dot{\lambda}_3, \dot{\lambda}_8\}$ of the $SU(3,\mathbb{C})$, the isospin and hypercharge

$$\dot{\lambda}_3 = [0,0,+1,-1] \text{ and } \dot{\lambda}_8 = \tfrac{1}{\sqrt{3}}[0,-1,0,+1] \tag{263}$$

Which represent the early discovery in quantum mechanics that these abelian operators can be diagonalized simultaneously which is due to the fact that they do not contain φ, σ, τ as factors. Calculating one after the other, we have

$$\begin{aligned}
\dot{\lambda}_1 &= \tfrac{1}{2}(-Id+e)\sigma = \tfrac{1}{2}([-1,-1,-1,-1]+[+1,+1,-1,-1])\sigma = [0,0,-1,-1]\sigma \\
\dot{\lambda}_2 &= \tfrac{i}{2}(g-f)\sigma = i[0,0,+1,-1]\sigma \\
\dot{\lambda}_3 &= \tfrac{1}{2}(-f+g) = [0,0,+1,-1] \\
\lambda_4 &= \tfrac{1}{2}(Id-f)\tau = \tfrac{1}{2}([+1,+1,+1,+1]+[-1,+1,+1,-1])\tau = [0,+1,+1,0]\tau \\
\lambda_5 &= \tfrac{i}{2}(e-g)\tau = i[0,+1,-1,0]\tau \\
\lambda_6 &= \tfrac{1}{2}(-Id+g)\varphi = [0-1,0,-1]\varphi \\
\lambda_7 &== \tfrac{i}{2}(f-e)\varphi == i[0,-1,0,+1]\varphi \\
\lambda_8 &= \tfrac{1}{2\sqrt{3}}(-2e+f+g) = \tfrac{1}{2\sqrt{3}}([-2,-2,+2,+2]+[+1,-1,-1,+1])+[+1,+1,-1,-1]) = \\
&= \tfrac{1}{\sqrt{3}}[0,-1,0,+1]
\end{aligned} \tag{264}$$

Now, see how beautifully these operators make use of the design of the iterants. First it turns out as most important that these locations have four characters. They constitute what I called the quad locations. We know that, seeing apart from iterant time and tangle time, that is, from permutations of characters, the three iterants e, f, g together with the identity form a 4-dimensional linear space with a discrete Klein 4 group structure of the corresponding base monomials. Let us consider the traditional partition of isospin in t-spin, u-spin and w-spin. Then take the three quantities that form a Lie subgroup of t-spin:

$$\begin{aligned}
\dot{\lambda}_1 &= [0,0,-1,-1]\sigma \\
\dot{\lambda}_2 &= [0,0,+1,-1]\sigma \\
\dot{\lambda}_3 &= [0,0,+1,-1]
\end{aligned} \tag{265}$$

You see, each of these has a form $[0,0,c,d]$. Further the permutation $\sigma = (1\,2)(3\,4)$ will exchange characters in positions 3 and 4 whereas whatever will be in positions 1 and 2 after the action of σ will be annihilated by the iterant $[0,0,c,d]$. Hence each t-spin operator will have an effect only on positions 3, 4. Similar statements are holding for u-spin essentially given by

$$\begin{aligned}
\dot{\lambda}_4 &= [0,+1,+1,0]\tau \\
\dot{\lambda}_5 &= i[0,+1,-1,0]\tau
\end{aligned} \tag{266}$$

Permutation $\tau = (1\,4)(2\,3)$ exchanges characters 2 with 3 and 1 with 4. So in cooperation with the iterant $[0, b, c, 0]$ it will have an effect on positions 2 and 3 only. For w-spin we have

$$\dot{\lambda}_6 = [0 - 1{,}0, -1]\varphi \tag{267}$$

$$\dot{\lambda}_7 = i[0, -1{,}0, +1]\varphi \text{ with } \varphi = (1\,3)(2\,4).$$

Thus we conclude: t-spin affects characters 3 and 4 in a quaternion location, u-spin affects characters 2 and 3, and w-spin characters 2 and 4. This combinatorial structure, we notice at first without overmuch attention, is the same as the $\{1{,}3\}, \{1{,}2\}, \{2{,}3\}$ featuring the bivectors of the Pauli algebra of the Euclidean 3-space. Now, that we have derived the Lie algebra $su(3)$ in both 'Clifform' and iterant form we shall be surprised how creatively this can act simultaneously on the different subspaces of the Minkowski algebra. We are now in a position that we need not take out a small part and study, say, the motion of a proton, but we investigate the sophisticated interplay of fields with space and time, as those subspaces are nothing else than the mathematical expressions of forces and particles.

Theorem 11: The natural motion group has form $SL(4, \mathbb{R})$ and an iterant representation with no $'i'$.

Passing over from formulas (252) to formulas (256) we applied the Weyl trick. Thus we obtained the corresponding complex (Cliff)form by performing the Weyl-trick. To generate the compact group by the same Clifform, we had to multiply three generators $\{\Lambda_2, \Lambda_5, \Lambda_7\}$ for one compact subalgebra $su(2, \mathbb{C})$ by the imaginary unit. That is, in order to obtain the correct commutation relations, we multiplied Λ_2 by $i = \sqrt{-1}$ and obtain the generators of the familiar '*t-spin* ', another $su(2, \mathbb{C})$ subalgebra. The real motion group does not require this arbitrary complexification. Real motion in iterant space is based on polarity strings and three combinatorial operations only. Thus are the late consequences of Sir Hamilton's mathematics of 'pure time'.

$$\begin{aligned} &\dot{\Lambda}_1 = [0{,}0, -1, -1]\sigma \;\; \dot{\Lambda}_2 = [0{,}0, +1, -1]\sigma \;\; \dot{\Lambda}_3 = [0{,}0, +1, -1] \\ &\dot{\Lambda}_4 = [0, +1, +1{,}0]\tau \;\; \dot{\Lambda}_5 = [0, +1, -1, 0]\tau \;\; \dot{\Lambda}_6 = [0 - 1{,}0, -1]\varphi \\ &\dot{\Lambda}_7 = [0, -1{,}0, +1]\varphi \;\; \dot{\Lambda}_8 = [0, -1{,}0, +1] \end{aligned} \tag{268}$$

The macroscopic iterant group arising from quantum motion of polarity strings does not contain any artificial imaginary unit.

Photons and Fermions Entangled

Fermions need not be decomposed as in a tensor product. They provide the Artin braid group B_3 by their natural appearance in the geometric Clifford and iterant algebras of quantum motion. If we consider photons as articulations in space-time algebra,- we found out in the first volume on »Primordial Space« [4]- we have to consider the subspace of directed unit areas spanned by $\{e_{12}, e_{13}, e_{23}\}$. The important discovery that bosons, that is, photons, W^+, W^-, W^0 are best conceived as commuting and respectively anti-commuting triples in the complex Minkowski algebra or in the even algebra $Cl^+_{4,0}$, satisfying a pseudo-unitary subalgebra $su(2)\oplus su(2)$, goes back to Robert Wallace (2008) who constructed a classification scheme for all elementary particles. I have shown then that the trigonal operators such as $T_{f,1}$ are indeed acting not only in fermion fields, but also on photons where they change the polarization planes. The fundamental photon iterants are respectively the iterants for $\{e_{12}, e_{13}, e_{23}\}$.

φe photon polarizing area e_{12}
$g\tau$ photon polarizing area e_{13}
$f\sigma$ photon polarizing area e_{23}

which, again, looks surprisingly natural. We have the matrix representations in basis (176)

$$e_{12} = \begin{bmatrix} 0 & 1 & 0 & 0 \\ -1 & 0 & 0 & 0 \\ 0 & 0 & 0 & -1 \\ 0 & 0 & 1 & 0 \end{bmatrix}; \ e_{23} = \begin{bmatrix} 0 & 0 & 0 & -1 \\ 0 & 0 & 1 & 0 \\ 0 & -1 & 0 & 0 \\ 1 & 0 & 0 & 0 \end{bmatrix}; \ e_{13} = \begin{bmatrix} 0 & 0 & 1 & 0 \\ 0 & 0 & 0 & 1 \\ -1 & 0 & 0 & 0 \\ 0 & -1 & 0 & 0 \end{bmatrix} \tag{269}$$

Now, we shall not assume a Hilbert space to go on, and we shall not decompose the photons into tensor products of 2×2-matrices, because it is not necessary, but we let them as they are. We reconsider the figure 49 showing the generators of the braid group B_4 which we could verify for fermions. Now let

$$e_{12} = a, \ e_{13} = b, \ e_{23} = c \tag{270}$$

They already satisfy the braiding multiplication table for 'generators and relations' in the 4-strand braid group B_4:

$$\begin{aligned} &aba = b, \ bab = b, \\ &aca = c, cac = a, \\ &bcb = c, cbc = b. \end{aligned} \tag{271}$$

Recall the same equations (219, 220) for fermions. As before we shall use the matrices (269) as phase gates in a quantum computer. We have already noticed that $T := T_{f,1}$ is a unitary combinatorial, monomial swap gate. It acts as a monomial and a grade swap gate. What is it doing? We have

$$\begin{aligned} &T^{-1}aT = b \text{ often written as } aT = Tb \\ &T^{-1}cT = -acT = -Ta \\ &T^{-1}bT = -cbT = -cT \end{aligned} \tag{272}$$

[4] In a chapter on ‚*Topological Evolution of Particles*'

So $T_{f,1}$ changes the polarization plane of the photons. Considering ψ as an element in this even subspace of bivectors as a phase gate for photons, we obtain

$$\psi = RT \text{ hence} \quad (273)$$
$$R = T^{-1}\psi$$

as a braiding operator. It is a matrix that has the potential to entangle photons. This is independent of any Hilbert space construction such as for example GNS construction. Such is possible, but not necessary for a better understanding of the events we seem to encounter in the experiments. So we recapitulate we obtain two triangles, one for fermions and one for photons, both subjected to phenomena of entanglement.

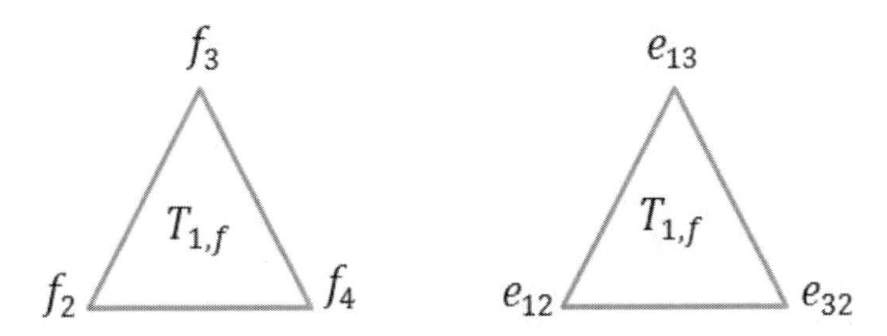

Figure 51. Braid triangles of fermions and photons.

Photons Travel and Entanglement on Primordial Domain

It is interesting to speculate how such polarity strings interact while travelling the void. We know the structures, the forces, the interaction patterns that occur by touch. So what will a photon experience so to say in a permanent statu nascendi? Waves have a capability to walk at random in chaotic conditions of neighborhood.

Primordial light travels on more locations than we were able to consider. The appearances of light are even more variegated than we are aware of.

We do not know anything about the mechanisms that might have the potential to stabilize that shift of a polarity string through the void, a happening that we had decided some time ago to call the path of a photon through the vacuum. There must be some sort of cybernetic process that gives stability to travelling photons. There is some kind of strange attractor. Since deep down there is just a chaotic process of bit strings in motion and interacting with each other.

Now think of the unspecified void as a vivid thing as we showed in the first chapters. It is a self-referent thing interacting with itself, and thereby it creates locally and trans-locally the alternating polarities. We observe a chaotic g-background of alternating polarity.

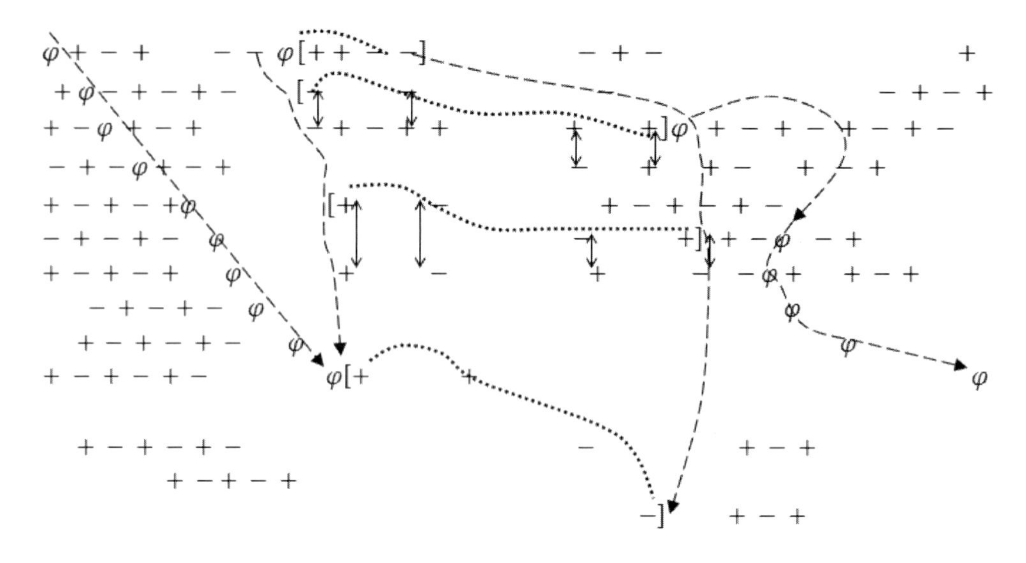

Figure 52. Travel of photon polarised in area e_{12}.

You see what happens in figure 52 (despite that nothing is happening any more as, as you watch this page, since this was printed long ago), you realize the event I expect us to imagine? We begin with an unspecified domain, unmeasured and without orientation, determined by one of three comparatively passive bit strings, the g, 'passive' since it does not contain any of the permutations σ, φ and τ.

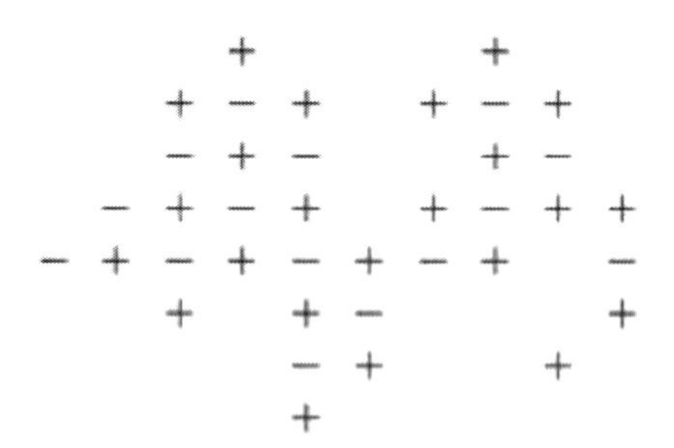

Figure 53. Polarized void in accord with g.

That domain provides a zero energy polarized pattern by *self-reference plus instability*, into which in the upper left region a photon $\varphi[+1,+1,-1,-1]$ enters. As it has the form $\varphi[+1,+1,,-1-1]$ that is, an 'φe' polarized in a directed area we are used to denote as e_{12} it encounters a peculiar interaction with the polarized void. Namely the void in that unspecified domain provides the opposed bit strings we already know from the Schrödinger theory. Here they provide the building blocks of space-time volume.

$$\begin{array}{l} ...+-+-+-+-... \\ ...-+-+-+-+... \end{array}$$

which gain their inequality by mere confrontation; symbolically the 'g'-iterants. So we have a touch between φe and g hence bringing forth, probably, $-e\varphi$ and g or the $\varphi f = \varphi[+ \; - \; - \; +] = [- \; + + \; -]\varphi$ located across the second and third line of figure 52. On its onward movement the φ on the right side decouples and the $[+ \; - \; - \; +]$ encounters again an 'g'-iterant thereby reproducing the $e\varphi$ by a φ coming in from the left. The φ is a potential. It may have got lost in the first step, even. Notice, we have

$$\varphi eg = \varphi f = \varphi[+ \; - \; - \; +] = -f\varphi = [- \; + \; + \; -]\varphi \quad \text{and} \tag{274}$$

this transmutation occurs when we pass from the first to the second line in the figure.

A free f-string beginning in the fifth line then interacts with the g-background and brings forth a free f after a φ has decoupled from the string. We can consider the φ as a free potential capable to create trajectories. It is free because it commutes with the background. We have

$$g\varphi = \varphi g, \quad \text{(table 2)} \tag{275}$$

Along its journey the primordial photon may experience swaps from φe to eφ and back, while the φ decouples and moves around like a free potential ready to work on trajectories-

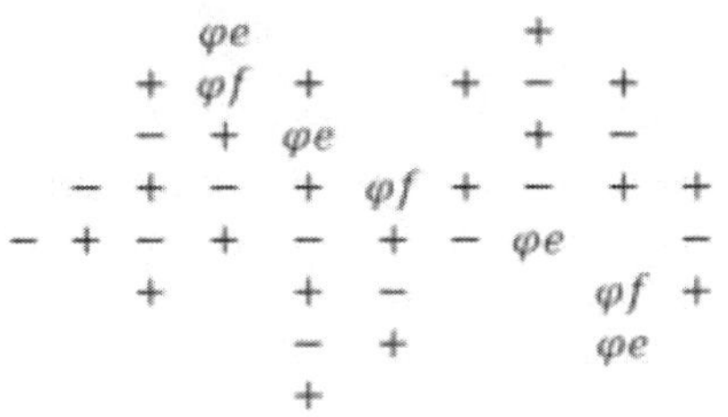

```
         φe              +
      +  φf  +       +   -   +
      -  +   φe          +   -
   -  +  -   +   φf  +   -   +   +
-  +  -  +   -   +   -   φe      -
      +      +   -           φf  +
             -   +           φe
             +
```

Figure 54. Primordial photon, polarizing directed area e_{12}, delivering time $-e_4 \simeq \varphi f$.

As φe is a photon polarized 'horizontally' in the directed plane area e_{12} and g is a unit space-time volume in opposite direction $g \simeq e_{124}$ we are actually witnessing a temporal 'observer' monomial, - the temporal unit vector of the algebra, - supporting re-creation of a polarized photon and reverse, a polarized photon creating a directed time unit. This is not accidental. In the Clifford algebra of the Minkowski space there occurs a 3-partition of the algebra due to the action of the operator $T_{f,1}$ and what is it that changes the polarization plane of the photons? It is the operator $T_{f,1}$. This turns polarization from e_{12} to e_{13} and from there to $-e_{23}$. Thus it generates period 3 rotations in the even subalgebra of $Cl_{3,1}$ and likewise in the subspace $\{\varphi e, g\tau, \sigma f\}$ of iterants. There is a 3-partition between the seal with base unit monomials which squared give the identity $+Id$, quaternion bivectors which squared give $-Id$, and the *time-like observers*, quaternions e_4, e_{123}, e_{1234} which squared also give $-Id$, also called *volume-time* or *timespace*.

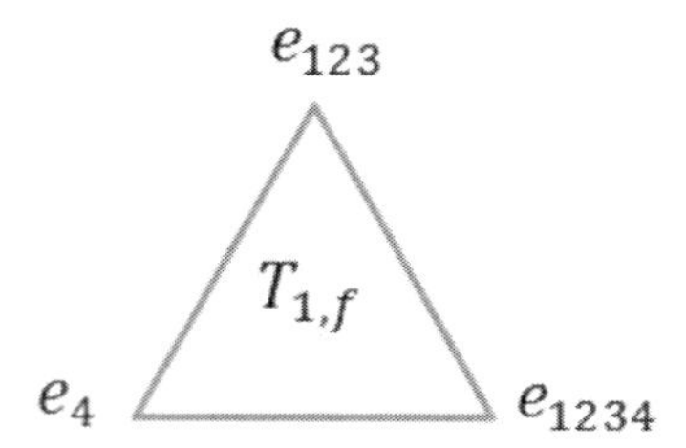

Figure 55. Time-like observer space.

Be aware that what we 'later' observe as units of time $e_4 = ict$, or of spatial volume e_{123} and space-time volume e_{1234}, is deep inside space a mere sample of polarity strings, namely the $f\varphi$, $g\sigma$, τe, strings with the potential to transpose polarities

$$[+1,-1,-1,+1]\varphi,\ [+1,-1,+1,-1]\sigma, \tau[-1,-1,+1,+1].$$

It is formally quite obvious that those strings $\{\varphi e, g\tau, f\sigma\}$ and $\{g, f, e\}$ have a non-obvious phenomenolgical relation which is indicated by this magic travel of the primordial photon which interacts with a zero energy background composed by potential-free $\{g, f, e\}$. It should be assumed that the symmetries are such that primordial photons of the other polarization planes can be linked with the other units of volume time. This is indeed the case. Let us slowly go into it. What does the primordial photon that shall 'later' disclose a polarization plane directed as e_{23} experience in that polarized zero energy pattern stabilized by self-reference? What does the primordial photon $f\sigma$ encounter when it enters the sea of e's? Note that e is another passive bit string, as it does not cause any permutation of characters. It goes through a history as shown below

<table>
<tr><td></td><td></td><td></td><td>+</td><td>+</td><td>−</td><td>−</td><td>$f\sigma$</td><td></td><td></td></tr>
<tr><td></td><td></td><td>+</td><td>+</td><td>−</td><td></td><td>+</td><td>$g\sigma$</td><td>−</td><td></td></tr>
<tr><td></td><td></td><td>−</td><td>−</td><td>+</td><td></td><td></td><td>$f\sigma$</td><td>+</td><td></td></tr>
<tr><td></td><td>−</td><td>+</td><td>+</td><td>−</td><td>−</td><td>+</td><td>$g\sigma$</td><td>−</td><td>−</td></tr>
<tr><td>+</td><td>+</td><td>−</td><td>−</td><td>+</td><td>+</td><td>−</td><td>−</td><td>$f\sigma$</td><td>−</td></tr>
<tr><td></td><td></td><td>+</td><td></td><td>−</td><td>−</td><td></td><td>$g\sigma$</td><td>−</td><td>−</td></tr>
<tr><td></td><td></td><td></td><td></td><td>+</td><td>+</td><td></td><td></td><td>$f\sigma$</td><td></td></tr>
<tr><td></td><td></td><td></td><td></td><td>−</td><td></td><td></td><td></td><td></td><td></td></tr>
</table>

Figure 56. Primordial photon, polarizing directed area e_{23}, delivering volume $e_{123} \simeq g\sigma$.

You see what happens in figure 56 (despite that nothing is happening any more as, as you watch this page, since this was printed long ago), you realize the event I expect us to imagine? We begin with an unspecified domain, unmeasured and without orientation, determined by one of three comparatively passive bit strings, the e representing the space-time volume element of the first Cartan subalgebra

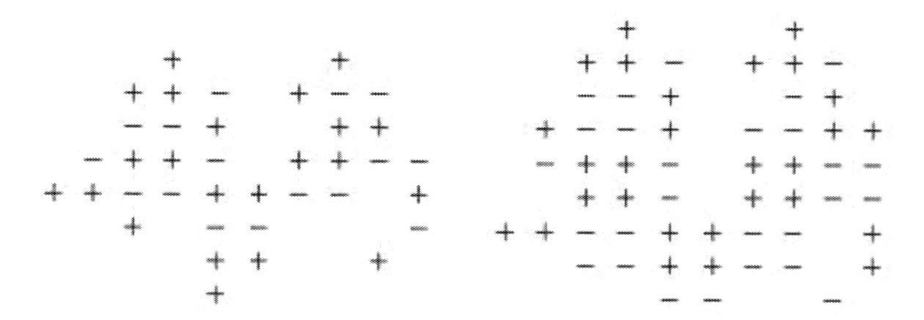

Figure 57. Polarized void arranged by e and g on the left, or e alone, on the right side.

That domain provides a zero energy polarized pattern by *self-reference plus instability*, into which in the upper right hand region a photon $[+1,-1,-1,+1]\sigma$ enters. As it has the form $[+1,-1,-1,+1]\sigma$, that is, an '$f\sigma$' polarized in a directed area we are used to denote as e_{23} it encounters a peculiar interaction with the void polarized by e. Namely the void in that unspecified domain provides the opposed bit strings we already know from the theory of fourfold logic strings

$$\begin{array}{l} ...+ + - - + + - - ... \\ ...- - + + - - + + ... \end{array}$$

which again obtain their inequality by mere confrontation; symbolically the 'e'-iterants. So we have a touch between $f\sigma$ and e hence bringing forth the $[+ \; - \; + \; -]\sigma = g\sigma$ located in the second line of figure 56. So, on its onward movement this $[+ \; - \; - \; +]\sigma$ encounters again a 'e'-iterant thereby reproducing the $f\sigma$. For we have $f\sigma e = g\sigma$ and $g\sigma e = f\sigma$. As $f\sigma$ is a photon polarized 'horizontally' in the directed plane area e_{12} and $g\sigma$ is a directed volume unit we are actually witnessing creation of a temporal volume-observer string, - the spatial time base unit of the algebra, - supporting re-creation of a polarized photon and reverse, a polarized photon creating a directed unit volume. Would the photon not have the possibility to interact repeatedly with the Cartan iterant provided by the void, it would be lost and dwindled away. The same would happen to any photon polarized in any direction. By repeatedly encountering the Cartan strings the propagating photons are reproduced. Last not least let us watch the path in a primordial background of a photon polarized parallel to plane area e_{13}. It is a bit string $g\tau$ interacting a with a background-f, a passive iterant isomorphic to e_{24}. The pattern may look like

<table>
<tr><td></td><td></td><td></td><td>$g\tau$</td><td>+</td><td>−</td><td>−</td><td>+</td><td>+</td><td></td></tr>
<tr><td></td><td></td><td>+</td><td>−</td><td>$e\tau$</td><td>+</td><td>+</td><td>−</td><td>−</td><td></td></tr>
<tr><td></td><td></td><td>−</td><td>+</td><td>+</td><td></td><td></td><td>$g\tau$</td><td>+</td><td></td></tr>
<tr><td></td><td>−</td><td>+</td><td>−</td><td>−</td><td>+</td><td>+</td><td>−</td><td>$e\tau$</td><td>+</td></tr>
<tr><td>+</td><td>+</td><td>−</td><td>+</td><td>+</td><td>−</td><td>−</td><td>+</td><td>$g\tau$</td><td>−</td></tr>
<tr><td></td><td></td><td>+</td><td></td><td>−</td><td>−</td><td></td><td>$e\tau$</td><td>−</td><td>+</td></tr>
<tr><td></td><td></td><td></td><td></td><td>+</td><td>$g\tau$</td><td></td><td></td><td>+</td><td></td></tr>
<tr><td></td><td></td><td></td><td></td><td>$e\tau$</td><td></td><td></td><td></td><td></td><td></td></tr>
</table>

Figure 58. Primordial photon polarizing directed area e_{13}, delivering space-time volume $-e_{1234} \simeq e\tau$.

Here the equations $g\tau f = e\tau$ and $e\tau f = g\tau$ apply, so that we obtain a trace of polarities $g\tau \to e\tau \to g\tau \to e\tau \to \cdots$ on this chaotic f-background. The trace of the photon polarizing directed area e_{13} is accompanied by a trace of strings that represent space-time volume. The trace of a photon polarizing directed area e_{12} was accompanied by a sequence representing reverted time. And the photon polarizing directed area e_{13} is accompanied by a sequence of strings representing the involuted space-time volume, director of the Minkowski algebra.

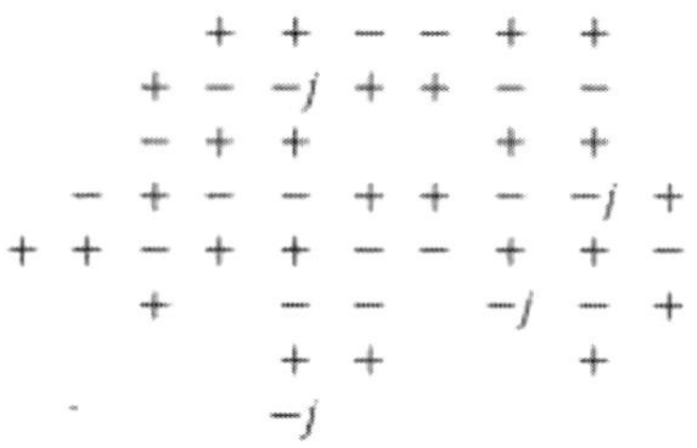

```
        + + − − + +
      + − −j + + − −
      − + +     + +
    − + − − + + − −j +
  + + − + + − − + + −
      +   − −  −j − +
          + +     +
          −j
```

Figure 59. Particle scan of photon e_{13} on f, g-background creating $-e_{1234}$-trace.

Notice, the primordial 'background' we provide for the travelling photons is constituted by fragments – namely, the monomials e, f, g that constitute the Cartan subalgebra $ch_1 \in Cl_{3,1}$ we are attentive for. So this is our peculiar construction in order to make something visible und comprehensible, slowly and step by step. We obtain the following quixotic sketches of particle scans for three types of polarized photons.

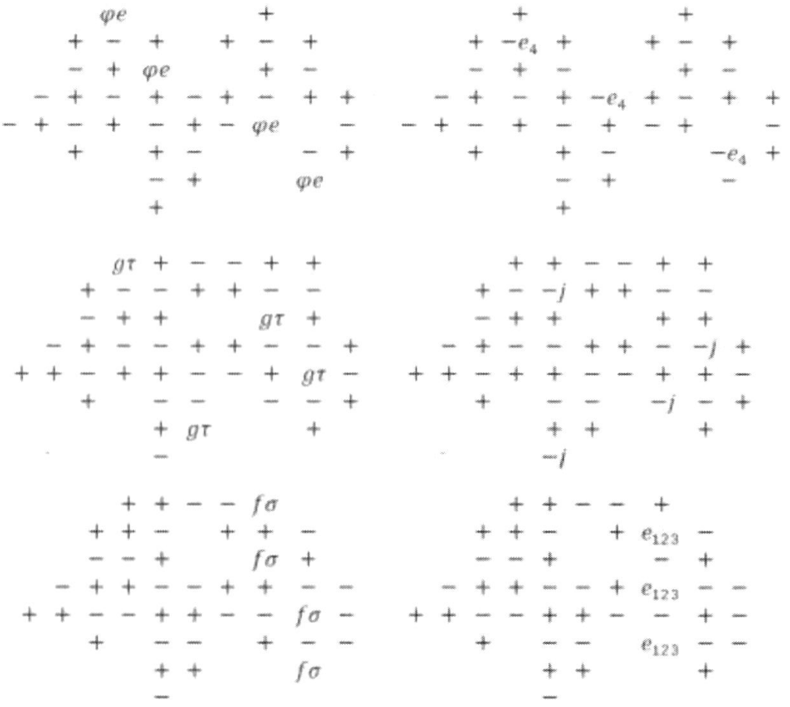

```
φe +
+ − + + − +
− + φe + −
− + − + − + − + +
− + − + − + − φe −
+ + − − +
− + φe
+
```

```
+ +
+ −e4 + + − +
− + − + −
− + − + −e4 + − + +
− + − + − + − + −
+ + − −e4 +
− + −
+
```

```
gτ + − − + +
+ − − + + − −
− + + gτ +
− + − − + + − − +
+ + − + + − − + gτ −
+ − − − − +
+ gτ +
−
```

```
+ + − − + +
+ − −j + + − −
− + + + +
− + − − + + − −j +
+ + − + + − − + + −
+ − − −j − +
+ + +
−j
```

```
+ + − − fσ
+ + − + + −
− − + fσ +
− + + − − + + − −
+ + − − + + − − fσ −
+ − − + − −
+ + fσ
−
```

```
+ + − − +
+ + − + e123 −
− − + − +
− + + − − + e123 − −
+ + − − + + − − + −
+ − − e123 − −
+ + +
−
```

Figure 60. Particle scan of photons e_{12}, e_{13}, e_{23} creating time space.

Once we have formalized and standardized the geometric phenomena to such a degree that we can represent photons and other particles in a standard frame – though we can be sure there are continuously many more possibilities to frame the events by the aid of the motion group – we can study some of those surprising features of quantum motion. One such feature is the link between the polarization of photons and the orientation of time-space. When the system is properly calibrated, a rotation of the polarization planes is correlated with a rotation in time-space, and one peculiar polarization can be adjusted such that the path of that photon which has this oscillation can be associated with a time reverted trace. Under the condition that it is allowed to progressively interact with a Cartan string background, it will have to create sequences of strings that constitute time reverted paths.

If you read carefully, you sure became aware, our formulas have an extension. They are creatures in space, they interact by touch, and at the same time they perform mathematical operations. So they are both material and cognitive. There is no more separation between matter and mind. We give the symbols back their life. They have a life of their own. So, by now we have just seen that the polarization areas go hand in hand with the time-space. This is a thermodynamic space of time-like observers. It sounds all very crazy, I know. But nature is simple and crazy. Her expressions, however, can look complicated.

As the e, f, g are but fragments of a 4-dimensional space spanned by idempotents, we should mind what the photons would experience when they encounter a fermion background of bit strings, and what do the fermions 'see' when a photon enters their domain. You see, we are no longer satisfied with the mere fragments, but we assume that this chaotic process in the basis of polarized strings has come to a certain metastable equilibrium where we can be sure that there are enough fermions with which the photons got to interact. Think of a photon that is absorbed in the electron shell of an atom. Or think about a boson running into a quark. These processes are very similar. Let us for a moment still deal with the primordial, fragmentary case.

The path history of a photon reads like this: photons interacting with primordial fermion strings stabilize their propagation and on the other hand generate primordial strings of the time-space. Thus they are effecting what is otherwise brought forth by a rotation of the motion group. And what kind of rotation can that be? It is the coordinate swap $s_{2,3}$ which we are familiar with since our investigation of the braiding operation. This swap carries every photon to polarized strings of time-space. To clarify circumstances, interaction of a photon with a Cartan iterant e, f, g is similar to the action of a spinor that acts from one side only. Rotation such as that carried out by $s_{2,3}$ act from both sides like a rotor in 3-space.

$$s_{2,3}e_{13}s_{2,3} = -e_4 \text{ is the origin of time reversion at polarization splitting} \qquad (276)$$

What is happening in a quantum eraser? Take any of those devices. A photon source sends off a split photon field circular polarized in directions e_{13} and e_{31}. The two fields will encounter under appropriate chemical conditions reflecting polarity strings $s_{2,3}$. Our photons will be confronted with a comparatively high local density of bit strings $s_{2,3} = [+1,+1,0,0] + [+1,+1,0,0]\sigma$ that carry out the grade swap of an entanglement. This results in one photon e_{31}creating an $+e_4$ environment and a photon oscillating in area e_{13} embedded within $-e_4$ bit-strings. This will not always be the case. But it requires the right chemical environment, in this case the right crystals. The motion group acts not only in high energy physics, but even in solid state physics and biochemistry.

Photon fields create in a fermion environment a *fermion Nirvana*, that is, a self-annihilating field which is peculiar for that fermion environment of polarity strings. The photon is absorbed in this process. But by interaction with the annihilating field it recreates itself. Thus the photon interrupts its own history and reproduces itself. We might say that it polarizes the fermion background, splitting it into fermionic energy density and annihilating background, and it also polarizes the process itself, for it makes a cut in the continuity of its propagation. We know this process very well. A photon is absorbed and re-emitted. We know phenomena of polarization from charge separation in dielectric materials, from dipolar polarization, spin polarization and vacuum polarization, but this kind of polarization is unique as it brings in all types of physical polarity with a permanent and primordially chaotic change between being and not being. The path of photons in fermionic environments is a discontinuous move within the ambiguity of creation and annihilation. Now let us study the paths of photons on a fermion background.

Recall, we had $f\sigma$ photon polarizing area e_{12}, $e\varphi$ photon polarizing area e_{13}, and $g\tau$ photon polarizing area e_{23}. Hence we know the phases of photons have a photon phase template

$$\psi = \alpha\varphi e + \beta g\tau + \gamma f\sigma \tag{277}$$

Or for simplicity formulated in geometric algebra

$$\psi = \alpha e_{12} + \beta e_{13} + \gamma e_{23} \tag{278}$$

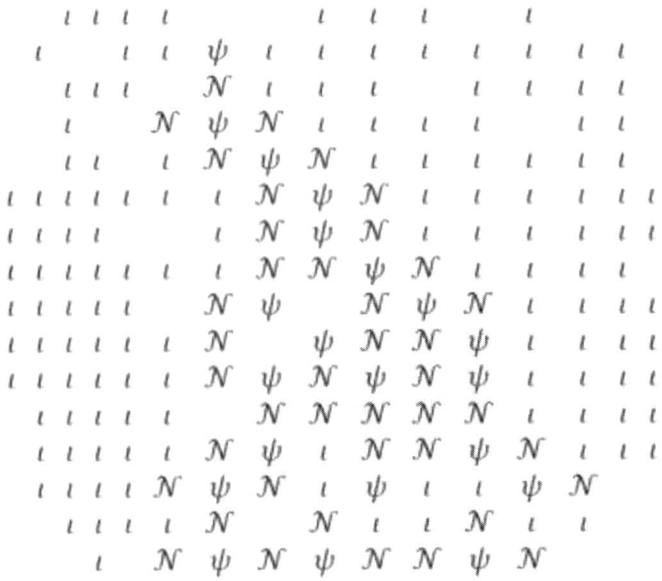

ι ι ι ι ι ι ι ι
ι ι ι ψ ι ι ι ι ι ι ι ι
ι ι ι 𝒩 ι ι ι ι ι ι ι
ι 𝒩 ψ 𝒩 ι ι ι ι ι ι
ι ι ι 𝒩 ψ 𝒩 ι ι ι ι ι ι
ι ι ι ι ι ι ι 𝒩 ψ 𝒩 ι ι ι ι ι ι
ι ι ι ι ι 𝒩 ψ 𝒩 ι ι ι ι ι ι
ι ι ι ι ι ι ι 𝒩 𝒩 ψ 𝒩 ι ι ι ι
ι ι ι ι ι 𝒩 ψ 𝒩 ψ 𝒩 ι ι ι ι
ι ι ι ι ι ι 𝒩 ψ 𝒩 𝒩 ψ ι ι ι ι
ι ι ι ι ι ι 𝒩 ψ 𝒩 ψ 𝒩 ψ ι ι ι ι
ι ι ι ι ι 𝒩 𝒩 𝒩 𝒩 𝒩 ι ι ι ι
ι ι ι ι ι 𝒩 ψ ι 𝒩 𝒩 ψ 𝒩 ι ι ι
ι ι ι ι 𝒩 ψ 𝒩 ι ψ ι ι ψ 𝒩
ι ι ι ι 𝒩 𝒩 ι ι 𝒩 ι ι
ι 𝒩 ψ 𝒩 ψ 𝒩 𝒩 ψ 𝒩

Figure 61. Light travel on fermion background.

Now consider any energy density of a fermion field in the Minkowski algebra. Whether it is one of the 'standard' primitive idempotents in the vertices of the seal or any orthogonal transform that we obtain by the action of the elements of the motion group, that is, let ι be any primitive idempotent, for instance

$$\iota := f_{4,3}$$

a 'minimal' primitive idempotent having a minimal number of dimensions in the unitary motion group. Now we formulate a theorem by the aid of a picture.

This iconic theorem tells us that a photon ψ oscillating in whichever planes of polarization, travelling on a fermion background creates a neighboring field of self-annihilating strings that has the capability to recreate the photon. We use to call this, in its most simple appearance, absorption and emission of photons. But the mechanism applies to gluons as well. In algebraic form. Let

$\iota := f_{4,3}$ without loss of generality, a minimal primitive idempotent and consider the Clifford product

Theorem 12: $\mathcal{N}(\psi,\iota) := \psi\,\iota$
self annihilating field created by photon ψ on its path in a fermionic environment ι

Lemma 5: $\mathcal{N}(\psi,\iota)\mathcal{N}(\psi,\iota) = 0$ Proof:

$$\begin{aligned}\mathcal{N}(\psi,\iota) &= \tfrac{1}{4}\{-\alpha e_1 + \gamma e_3 - \alpha e_4 + \alpha e_{12} + \beta e_{13} + \gamma e_{23} - \\ &-\alpha e_{24} - \beta e_{34} + \beta e_{123} - \gamma e_{134} + \beta e_{234} + \gamma e_{1234}\}\end{aligned} \tag{279}$$

$$\begin{aligned}\mathcal{N}(\psi,\iota)\mathcal{N}(\psi,\iota) &= \tfrac{1}{16}\{\alpha^2 e_1 e_1 + \alpha^2 e_4 e_4 + \alpha^2 e_{12} e_{12} + \alpha^2 e_{24} e_{24} + \\ &+\beta^2 e_{13} e_{13} + \beta^2 e_{34} e_{34} + \beta^2 e_{123} e_{123} + \beta^2 e_{234} e_{234} + \\ &+\gamma^2 e_3 e_3 + \gamma^2 e_{23} e_{23} + \gamma^2 e_{134} e_{134} + \gamma^2 e_{1234} e_{1234} - \\ &-\alpha\gamma e_1 e_3 - \alpha\gamma e_1 e_{23} + \alpha\gamma e_1 e_{134} - \alpha\gamma e_1 e_{1234} + \ldots \\ &-\alpha\gamma e_3 e_1 - \alpha\gamma e_3 e_4 + \alpha\gamma e_3 e_{12} - \alpha\gamma e_3 e_{24} + \ldots = \\ &= \tfrac{1}{16}\{\alpha^2 - \alpha^2 - \alpha^2 + \alpha^2 - \beta^2 + \beta^2 - \beta^2 + \beta^2 + \\ &+\gamma^2 - \gamma^2 + \gamma^2 - \gamma^2 - \alpha\gamma e_{13} + \alpha\gamma e_{13} - \alpha\gamma e_{123} + \alpha\gamma e_{123} + \ldots = 0\end{aligned} \tag{280}$$

The proof continues by carrying out calculations (280) for the 24 standard primitive idempotents of the Minkowski algebra. Finally all these calculations can be made by transforming with any group element t generated by the motion group. We obtain for instance for any such rotation of the Clifford algebra

$$t^{-1}\psi\, t\, t^{-1}\iota\, t = t^{-1}\mathcal{N} t \tag{281}$$

The transformed nilpotent field from the thus transformed photons and primitive idempotents. So we have a continuous manifold of fermions, photons and Nirvanas which correspond with another by the same form of interaction. The photon annihilates itself by any interaction with the fermion environment, but it also recreates itself by letting the specific trace of self-annihilating polarity strings interact with itself. The fermion fields absorb the photon, but they have to give it back. So an absorbed photon is reemitted through the interaction $\psi\,\mathcal{N}(\psi,\iota) = \psi$. Altogether we have the equations

$$\begin{aligned} \psi\,\iota &= \mathcal{N} \\ \mathcal{N}\mathcal{N} &= 0 \end{aligned} \tag{282}$$

But here the story does not halt. But the fermion field has the capability to support creation of the annihilating field while the latter can support re-creation of photons. Photons and fermion spinor-strings in cooperation bring forth the annihilating field. This process involves the totality of photon fields and fermions fields. There is no exception to this interaction which we have depicted by equations (282) and figure 61. Light is not diminished that easily. But the Nirvana is insisting too. Nirvana can't do without fermions. Light can't do without fermions and it can't without Nirvana. Therefore it first creates Nirvana before it goes on. The propagator of light is entirely based on the propagation of a self-annihilating field. But what, if there were no fermion background? As long as there is matter, the photons have the capability to cut themselves off the history and vanish into the annihilating field by which it is recreated. This is the innermost origin of creative destruction.

LIGHT CREATES NIRVANA ON FERMIONS

A photon φe polarized in the directed plane e_{12} penetrates the energy cloud of electrons, represented in the lower corner of the seal by an element of density

$$E^- := \frac{1}{2}(Id + e_2)\frac{1}{2}(Id + e_{14}) \tag{283}$$

The quibbling primordial creatures constituting the electron density iterants have two worms, one of which is rather passive, the component $\frac{1}{2}[0,0,+1,+1]$, it acts by logic identity only, while the other, the $\frac{1}{2}[0,0,-1,-1]\sigma$, commutes actively the first with the second character and the third with the forth in any incoming polarity string

$$E^- := \frac{1}{2}[0,+1,0,+1] + \frac{1}{2}[0,-1,0,-1]\varphi \tag{284}$$

The photon field φe meeting the $E^- := \frac{1}{4}(Id - g - \varphi + g\varphi)$ string transforms this according to

$$\varphi e \frac{1}{4}(Id - g - \varphi + g\varphi) = \frac{1}{4}(\varphi e - \varphi e g - \varphi e \varphi + \varphi e g \varphi) = \frac{1}{4}(e - f + \varphi e - \varphi f) \tag{285}$$

So it creates a nilpotent polarity string which also has two small components

$$N = \frac{1}{4}\left((e-f)+\varphi(e-f)\right) = \frac{1}{2}[0,+1,0,-1] + \frac{1}{2}[0,-1,0,+1]\varphi \tag{286}$$

one of which is passive the other active. This Nirvana which I denoted as N looks very similar like the electron density string, but has the potential to first annihilate itself and then recreate an electron. Let us carry out the 'long rigor':

$$\begin{aligned}
NN &= \frac{1}{4}\left((e-f)+\varphi(e-f)\right)\frac{1}{4}\left((e-f)+\varphi(e-f)\right) = \\
&\frac{1}{4}([0,+1,0,-1]+[0,-1,0,+1]\varphi)([0,+1,0,-1]+[0,-1,0,+1]\varphi) = \\
&\frac{1}{4}([0,+1,0,-1][0,+1,0,-1]+[0,+1,0,-1][0,-1,0,+1]\varphi + \\
&[0,-1,0,+1]\varphi[0,+1,0,-1]+[0,-1,0,+1]\varphi[0,-1,0,+1]\varphi = \\
&\frac{1}{4}([0,+1,0,+1]+[0,-1,0,-1]\varphi+[0,-1,0,+1][0,-1,0,+1]\varphi + \\
&[0,-1,0,+1][0,+1,0,-1]) = \\
&\frac{1}{4}([0,+1,0,+1]+[0,-1,0,-1]\varphi+[0,+1,0,+1]\varphi+[0,-1,0,-1]) = [0,0,0,0]
\end{aligned} \tag{287}$$

hence $NN = \emptyset$

Suppose an electron density string meats a nilpotent string N; then we have an encounter like that

$$\begin{aligned}
E^{-}N &= \left(\frac{1}{2}[0,+1,0,+1]+\frac{1}{2}[0,-1,0,-1]\varphi\right)\left(\frac{1}{2}[0,+1,0,-1]+\frac{1}{2}[0,-1,0,+1]\varphi\right) = \\
&\frac{1}{4}([0,+1,0,+1][0,+1,0,-1]+[0,+1,0,+1][0,-1,0,+1]\varphi + \\
&[0,-1,0,-1]\varphi[0,+1,0,-1]+[0,-1,0,-1]\varphi[0,-1,0,+1]\varphi) = \\
&\frac{1}{4}([0,+1,0,-1]+[0,-1,0,+1]\varphi + \\
&[0,-1,0,-1][0,-1,0,+1]\varphi+[0,-1,0,-1][0,+1,0,-1]\varphi\varphi\,) = \\
&\frac{1}{4}([0,+1,0,-1]+[0,-1,0,+1]\varphi+[0,+1,0,-1]\varphi+[0,-1,0,+1] = [0,0,0,0]
\end{aligned} \tag{288}$$

The electron string is annihilated. But the story does not end here, for we need the spinor's life.

Light on the Spinor

As the electron density is the product of a spinor and its grade involuted, we have to go into details. This is possible, if we are allowed to make a loan from the next but one chapter. The spinor of an electron neutrino is thought to be located on top of the seal. In iterant notation it can be written as

$$|\overline{\nu_e}\rangle = i[+1,0,0,0]\tau \tag{289}$$

Hence the electron spinor should have a form derived from the density

$$E^- := \frac{1}{2}[0,+1,0,+1] + \frac{1}{2}[0,-1,0,-1]\varphi \tag{290}$$

namely as a 'Clifform' which is a Cartan spinor

$$|e^-\rangle = \pm\frac{i}{4}(e_2 + e_{14})(e_3 + e_{23}) \tag{291}$$

that is represented by the iterant

$$|e^-\rangle = \frac{i}{2}\sigma[-1,0,+1,0] + \frac{i}{2}\tau[+1,0,-1,0] \quad \text{up to } \pm \text{ as a minimal form.} \tag{292}$$

We prove this by a short calculation using representation (192)

$$\begin{aligned} &|e^-\rangle = \frac{i}{4}(g\varphi - \varphi)(f\sigma + f\tau) = |e^-\rangle = \frac{i}{4}(g\varphi f\sigma + g\varphi f\tau - \varphi f\sigma - \varphi f\tau) = \\ &= \frac{i}{4}(-gf\varphi\sigma - gf\varphi\tau + f\varphi\sigma + f\varphi\tau) = \frac{i}{4}(-e\tau - e\sigma + f\tau + f\sigma) = \\ &|e^-\rangle = \frac{i}{4}\big((f-e)\sigma + (f-e)\tau\big) = \\ &\frac{i}{4}\big(([+1,-1,-1,+1] + [-1,-1,+1,+1])\sigma + \\ &([+1,-1,-1,+1] + [-1,-1,+1,+1])\tau\big) = \\ &= \frac{i}{2}([0,-1,0,+1]\sigma + [0,-1,0,+1]\tau) \end{aligned} \tag{293}$$

Clearly, being a Cartan spinor, this must turn out as a nilpotent form. We check the product:

$$\begin{aligned} &\big((f-e)\sigma + (f-e)\tau\big)\big((f-e)\sigma + (f-e)\tau\big) = \\ &(f-e)\sigma(f-e)\sigma + (f-e)\sigma(f-e)\tau + (f-e)\tau(f-e)\sigma + (f-e)\tau(f-e)\tau = \\ &(f-e)(-f-e) + (f-e)(-f-e)\varphi + (f-e)(f+e)\varphi + (f-e)(f+e) = \\ &(-Id - g + g + Id) + (-Id - g + g + Id)\varphi + (Id + g - g - Id)\varphi + (Id + g - g - \\ &\quad Id) \\ &= [0,0,0,0] = \emptyset \quad \text{q.e.d.} \end{aligned} \tag{294}$$

This is the last time I carried out the rigor in detail. I hope the reader has seen how it goes. Everything that will be coming for calculation will be abbreviated considerably.

As well shall see, the density $E^- := \frac{1}{2}[0,+1,0,+1] + \frac{1}{2}[0,-1,0,-1]\varphi$ is to be derived from a product of the electron spinor and its grade involuted, that is, we have

$|e^-\rangle\langle e^-| = E^-$ a moderate way to get the particle density from its spinor.

With the electron iterant spinor $|e^-\rangle = \frac{i}{2}([0,-1,0,+1]\sigma + [0,-1,0,+1]\tau)$ up to $\pm$ and orthogonal transformations by $\mathfrak{L}_2$ we are now ready to ask the crucial question: what happens when the photon $f\sigma$ polarized in the directed plane e_{12} penetrates the energy cloud of electrons that is given now by the primordial forms of spinor polarity strings? Let the φe encounter a spinor iterant $|e^-\rangle$, what is the result?

$$|\,\omega(e_{12})\rangle|e^-\rangle = \varphi e\frac{i}{4}\big((f-e)\sigma + (f-e)\tau\big) = \frac{i}{4}(\varphi(g-Id)\sigma + \varphi(g-Id)\tau) = \tag{295}$$
$$= \frac{i}{4}\big((g-Id)\varphi\sigma + (g-Id)\varphi\tau\big) = \frac{i}{4}\big((g-Id)\tau + (g-Id)\sigma\big) =$$
$$= \frac{i}{2}[0,-1,0,-1]\tau + [0,-1,0,-1]\sigma$$

Here I've taken the liberty to denote the photon by a ket which is somewhat ventured. But here and for now it has no effect. This nilpotent creature is a very creative, self-annihilating worm with two parts, let us baptize it, it shall be a $\nu(e_{12}, e^-)$ nirvana, an annihilating field with a special talent. It can recreate an electron by a photon. That this nirvana annihilates itself can quickly be seen

Consider the first term of the Clifford product $\nu(e_{12}, e^-)\nu(e_{12}, e^-)$, this is up to factor $\frac{i}{4}$ equal to

$$(g-Id)\tau(g-Id)\tau = (g-Id)(-g-Id)\tau\tau = -gg - g + g + Id = -Id - g + g + Id = \emptyset \tag{296}$$

and so on with the other terms. The electron field can only stabilize nil, as we can easily see that we have

$$|e^-\rangle\ \nu(e_{12}, e^-) = \nu(e_{12}, e^-)\ |e^-\rangle\ = \emptyset \tag{297}$$

But, surprise, the electron string can be recovered from the electron-photon-nirvana by a photon in the space inverted form; we shall have a model for this, so I don't enumerate those formulas, as they just sketch the kind of process we got to deal with

$$|\,\omega(e_{12})\rangle\ \nu(e_{12}, e^-) = -|e^-\rangle \tag{298}$$

This reversion of sign is typical for what is called a phase flip error in quantum information with 1 qubit, where the third Pauli matrix causes a single state to flip sign which is due to the action of a spinor (Matzke 2002). But here it is triggered by a photon which appears in the garment of a pure quaternion. Thus a polarized photon releases a spinor flip. In Clifford algebraic, non iterant form this is simply

$$e_{12}\,\nu(e_{12}, e^-) = e_{12}\left(\pm\frac{i}{4}(e_2 + e_{14})(-e_{13} + e_{123})\right) = \mp\frac{i}{4}(e_2 + e_{14})(e_3 + e_{23}) \tag{299}$$

So we have a self-annihilating field built up by double-active bit strings

$$\nu(e_{12}, e^-) == \frac{i}{2}[0,-1,0,-1]\tau + [0,-1,0,-1]\sigma \tag{300}$$

an electron spinor field constituted by bit strings

$$|e^-\rangle = \frac{i}{2}([0,-1,0,+1]\sigma + [0,-1,0,+1]\tau) \tag{301}$$

and photons

$$| \, \omega(e_{12}) \rangle = \varphi[+1,+1,-1,-1] \tag{302}$$

which have the power to vanish and reappear and to recreate electrons from a nirvana which is the product of the interaction between photon and electron. But there is a beauty spot. It doesn't seem that the photon can be recovered that easily either from the nilpotent field or from the fermion spinor field. It seems that the emission of photons from fermions requires a bit more consideration than the structure template from bivectors.

Chapter 6

MAJORANA SPACE-TIME SPINORS

Said Feynman "In books it says that science is simple: you make up a theory and compare it to experiment; if the theory doesn't work, you throw it away and make a new theory. Here we have a definite theory and hundreds of experiments, but we can't compare them! It's a situation that has never before existed in the history of physics. We're boxed in, temporarily, unable to come up with a method of calculation. We're snowed under by all the little arrows." (Feynman 1985, p. 139) This situation has not essentially changed. If you look at any of the figures, we are still lost in little arrows. The picture has changed a little. We see more clearly the pattern of interacting locations where we have to put the little arrows. But the theory has remained largely qualitative. Though we can even explain the entanglement and the time reversion involved in these interactions, there is neither a complete theory for the paths of photons, nor a path integral for all the interactions, no exact rigor for the calculation of the strong coupling constant. Yet one has to stay in contact with the ongoing efforts. Life is contact. Part of life has to do with Majorana spinors.

Majorana spinors form a universe in a rather simple structure of geometric algebra. The power of this algebra which consists essentially of 4×4-matrices with real entries is largely underappreciated. It turns out that not only the neutral Majorana fermions, but also the Baryons with $1/3$-charges can be understood in the context of both Majorana algebra and iterant algebra. The study of the mathematics of these particles becomes extremely interesting as soon as we investigate them in terms of processes and discrete iterant algebra. In this lecture the standard model of Hephy was set up by the aid of idempotent eigenforms of Majorana spinors, and some important relations with iterants and models of polarized strings were investigated. The spinor theory thus developed unifies particular features of several traditional approaches to spinors. It is shown that the geometry we perceive as 'macroscopic space-time', the Clifford algebra $Cl_{3,1}$ of the Minkowski space can be understood as a synchronous freeze pattern of an iterant algebra that points beyond that geometry.

"We all think that we know matrix algebra quite well. But it is a recent invention and has some strange wisdom built into its very bones". Such is the beginning of a chapter introducing *matrix algebra via iterants* in his lecture 'Space and time in computation, topology and discrete physics' (Kauffman 1994). And how little is known about the riches of 4 times 4 matrices with real entries! But they represent faithfully the 16-dimensional linear space of the geometric Clifford algebra of the Minkowski space in the Lorentz metric, our space-time. At the same time these matrices, incorporate the imaginary base unit e_4, the unit imaginary and respectively the $'it'$ introduced as iterant by Kauffman, here in its peculiar form as a 4×4 –matrix with real valued entries with the t denoting a real measure of time:

$$i\,t = (\epsilon\eta)\,t = ([-1,+1]\eta_2)\,t \simeq t\,(Id_2 \otimes \eta_2) = \begin{bmatrix} 0 & -t & 0 & 0 \\ +t & 0 & 0 & 0 \\ 0 & 0 & 0 & -t \\ 0 & 0 & +t & 0 \end{bmatrix} \quad (303)$$

This formula does various things: first it introduces the imaginary number i in terms of a product between a unit iterant $[-1,+1]$ and a temporal shift operator η, secondly it ascribes to this time delay shift a degree two, and thirdly it claims a correspondence with a time shift operator of degree four

$$\eta_4 \stackrel{\text{def}}{=} Id_2 \times \eta_2 = \begin{bmatrix} 0 & -1 & 0 & 0 \\ 1 & 0 & 0 & 0 \\ 0 & 0 & 0 & -1 \\ 0 & 0 & 1 & 0 \end{bmatrix} \text{ or shortly } \begin{bmatrix} & -1 & & \\ 1 & & & \\ & & & -1 \\ & & 1 & \end{bmatrix} \quad (304)$$

where Id_2 is the real identity matrix and $\otimes$ denotes the Kronecker product for matrices. To come closer to an understanding of how physical fields constitute the whole of space and time together with its geometric algebra, it is necessary to understand the meaning of time algebra and to lift it to the dimension $n = 4$ of the generating space for its Clifford algebra. The role of the Majorana fermion, in that context, is based on the comprehension of the graded motion group of which this fermion is an eigenform. That is, the motion group is a stabilizer group and the neutral Majorana fermion is its fixed point. The excellent mathematics of this phenomenology has been developed in two comprehensive contributions by Rafal Ablamowicz and Bertfried Fauser (2010). It emerges, in a way, in the center of the HEPhy processes. Let us go into this slowly and speak a little while about developments and about the relevance of the method of iterant algebra. That may be somewhat redundant, but the repetition has an intrinsic value.

The historic construction of the square root of minus one as an algebraic pair in conjugate functions goes back to Sir William Rowan Hamilton's "Essay on Algebra as the Science of Pure Time" (1837).[1] This paper written sometime in 1834 had 3 main parts. In the third part which is titled "The Theory of Conjugate Functions, or Algebraic Couples," Hamilton defines complex numbers by ordered pairs of real numbers. In this way the properties of addition, subtraction, multiplication, and division could be preserved. On page 107 Hamilton introduces the ordered algebraic pair with the allying sentences: "In the theory of single numbers, the symbol $\sqrt{-1}$ is absurd, and denotes an impossible extraction, or a merely imaginary number; but in the theory of couples, the same symbol $\sqrt{-1}$ is significant, and denotes a possible extraction, or a real couple, namely (as we have just now seen) the principal square-root of the couple $(-1,0)$. In the latter theory, therefore, though not in the former, this sign $\sqrt{-1}$ may properly be employed; and we may write, if we choose, for any couple (a_1, a_2) whatever, $(a_1, a_2) = a_1 + a_2\sqrt{-1}$‹interpreting the symbols a_1 and a_2, in the expression $a_1 + a_2\sqrt{-1}$, as denoting the pure primary couples $(a_1, 0)$ $(a_2, 0)$, according to the

[1] Link to D. R. Wilkins on 26th January 2012, http://www.maths.tcd.ie/pub/HistMath/People/ Hamilton/PureTime/

law of mixture". This historic finding about 'the real root of minus one' in the 'algebra of pure time' has been taken up and revived by Louis Kauffman and developed further, so that it could be applied in nowadays quantum mechanics. What was possible for the order two, could also be carried out for higher orders, and so we obtained a new purely combinatorial construction of quaternions and further of the real and complex Clifford algebras $Cl_{p,q}$ having orders $p + q = 4$. Thus we are already dealing with 16-dimensional spaces involving time-evolution in the form of 'diachronic' shift operations. As we saw, the fourfold iterant algebra is nothing else than the real Minkowski algebra. It allows one to look into the inner discrete structure of the process.

For now we leave the algebra of iterants and revive the phenomenology of spinors as far as this is connected with the peculiar structure of the HEPhy standard model of elementary particles. Suppose that the neutral Majorana spinor field does not partake in strong interaction, then this provides a first formal condition. Assume further that we have found a motion group for the strong force action. Then the neutral fermion would stay fixed like the center of a ball, while the charged fermions orbit on that sphere. It would be pleasant if all the spinors in a corresponding minimal supersymmetric model would turn out to be of the same structural type, so that we had similar spinors for neutrinos and quarks. Is this possible, and what would this structural type be like? It is indeed possible to construct such spinors which describe both kinds of particles. Interestingly we are led to such a simple model of consistent representation by just following the various demands that we have developed for spinors and that popped up over the course of the years. The peculiar Majorana fermion, as I can see it, with reference to the standard model, has been represented in a chapter on space-time spinors (2011a, p. 324) in equations (8). Let us now discuss together how this formula is derived. To get the Clifford algebraic spinors and respectively the most simple and relumining matrix representations, various investigations are necessary. We must, so to say, try to grasp the whole and look beyond our own noses. First of all we should be aware of several important differences between the geometric, quadratic Clifford algebras $Cl_{3,1}$, $Cl_{1,3}$, $\mathbb{C}\otimes Cl_{1,3}$ and $\mathbb{C}\otimes Cl_{3,1}$. These differences can be found described in Lounesto's book on *Clifford Algebras and Spinors* (2001, p. 140f., 227 and ch. 13) and in Schmeikal (2001). The first, the real Clifford algebra $Cl_{3,1}$ contains six equivalence classes of isomorphic lattices, each of which can be generated by four mutually annihilating, primitive idempotents. This geometric algebra is isomorphic with the matrix algebra $Mat(4,\mathbb{R})$, sometimes called the 'Majorana algebra'. The second, the real Clifford algebra $Cl_{1,3}$ does not contain these lattices. That makes an important difference. Historically, this geometric feature provided the very first and in my view the most important starting point for a reasonable explanation of the geometric origin of the standard model symmetries of both space-time and particle physics. The four generating fermions form a basic tetrahedral structure which has been realized and investigated by Chisholm and Farwell (1992) in a fundamental paper on the decent of the symmetric unitary group $SU(4)$ within the complex $Cl_{3,1}$. Equally important is our second task, namely to be aware of the historic development of the concept of spinor. It may turn out that the irregular motion of quantum events has a nonlinear and global origin that goes beyond the structure of this metric linear space. Nevertheless, we shall find out that the very appearance of the geometric algebra $Cl_{3,1}$ can tell us something important about the process of this global interaction and the translocal provenance of our so called particles. Let us suppose we regard the historic models as relevant for our understanding. We look at those models, beginning with the Cartan, Pauli, Weyl and Dirac spinors. We usually represent them either as columns with complex entries or as matrix operators where only the first column is different from zero. We never presented them in an iterant algebra. In case we decided for the second option, we are probably aware that 'erzeugende Einheiten', that is, idempotents, base units, spinor spaces as minimal ideals and spin groups are mutually related objects (Schmeikal 2012a,b). To use the most evident example, in Euclidean space the generating primitive idempotent has inhomogenous grade. It is equal to $(Id + e_3)/2$, say, or $(Id - e_3)/2$. From these forms we obtain a basic reflection $r = e_3$ or respectively $-e_3$, which has a definite grade equal one. But in Minkowski algebra this is not the case. The period-2 element $r = (Id - 2f)$ which we deduce from the primitive idempotent and which can serve as a generating reflection in the discrete reorientation group, is a grade-mix. Fortunately it has a wonderful representation in the fourfold real ring $\mathbb{R}\oplus\mathbb{R}\oplus\mathbb{R}\oplus\mathbb{R}$ or likewise in $Mat(4,\mathbb{R})$, and also in one of the six isomorphic Cartan subalgebras of the $Cl_{3,1}$. The difference that is made when a basic reflection is not identical with a base unit vector has been described in Schmeikal (2012a, b). Last not least we shall have to satisfy the predicates provided by Elié Cartan, namely that a spinor is a polarized isotropic vector, in advanced Clifford terminology: a nilpotent with spin. Finally, in some modest parallelism with Crumeyrolle (1987) and Lounesto (2011), we construct the (anti)automorphisms and charge conjugation operators which signify the peculiarity of Majorana fermions. With these equipped we then seek the solution of the existence problem. With Maple Clifford this can be found. It has been pinned down in Schmeikal (2011a, b and 2012a, b), and we shall repeat it here with reference to some new features concerning the idea of eigenform.

As soon as we step out and search for such a set of universal spinors, we encounter a surprising discovery. Namely, since we know that most of the structural demands are already satisfied in the Clifford algebra $Cl_{3,1}$ which is the matrix algebra $Mat(4,\mathbb{R})$, we want to challenge the explanatory power of our thought. At first we search for this spinor, so to say, in an 'utterly real algebra': real Clifford and real entries. Why? Because we want to find out, (i) if the concept exists as a real (pre)structure (a template), and (ii) we want to know what are the consequences of this fact. Now remember how Majorana discovered the real spinors and compare what you know with Todorov's statement: "Sometimes, the Majorana representation is defined to be one with real gamma-matrices. This is easy to realize (albeit not necessary)" and so on. – In the present concept this assignment turns out to be opposite. Indeed, most of the problem can be solved within $Mat(4,\mathbb{R})$ only. This is in perfect resonance with the possibility to represent the Minkowski algebra by 4-fold iterants. But as soon as you construct the charge conjugation and a complete set of anti-fermions, you need to introduce the imaginary unit for the first time. This is inevitable, if you want to factor in all 'known' forces and fields. The Computer program finds the general solution which points you the way to what we will call the minimal solution. For consider the minimal solution for the neutral Majorana fermion, say, the neutrino of the electron. It contains just a single unit entry, while the antineutrino demands an imaginary i. The same holds for the 1/3-charged fermions! What does that mean? That shows us the logical consequence of the statement that the introduction of anti-particles requires the introduction of a temporal shift operator. The procedure that leads to the graded group which provides the fixed point for strong interaction is described in some of my works, for example in Schmeikal (2004). It allows us to carry vectors of any grade to multivectors of another grade in a meaningful way. Let us therefore inquire if in this context the existence of Majorana fermion space-time bundles can be proven.

Let $Cl_{3,1}$ denote the real Clifford algebra in the Lorentz metric. It is often called the Minkowski algebra. It can be represented by the matrix algebra $Mat(4, \mathbb{R})$. Let $\mathbb{C} \otimes Cl_{3,1}$ be the complex $Cl_{3,1}$. Now we look for a polarized isotropic multivector ξ which satisfies

$$\xi \in f_1 Cl_{3,1} \text{ an element in the minimal left ideal generated by } f_1 \tag{305}$$

$$f_1 f_1 = f_1 \text{ a primitive idempotent } f_1 \in Cl_{3,1} \tag{306}$$

The element f_1 is representative for an equivalence class forming a continuous fiber of primitive idempotents in $Cl_{3,1}$ and is – up to permutation coding – represented in a subspace of minimal dimension equal to 4 as

$$f_1 \stackrel{\text{def}}{=} \frac{1}{4}(Id + e_1)(Id + e_{24}) \tag{307}$$

'standard' representation of f_1 primitive and minimal in $Cl_{3,1}$.

$$\xi\, \xi = 0 \text{ nilpotence (sign omitted := Clifford product)} \tag{308}$$

$$\xi \hat{\xi} = f_1 \quad \text{density (} \hat{} \text{ denotes Clifford grade- or main involution)} \tag{309}$$

$$\xi^c \widehat{\xi^c} = -f_1 \quad \text{charge conjugate antifermion} \tag{310}$$

A minimal solution for the charge conjugate spinor ξ^c is found to be given by an element in the real Majorana algebra $Cl_{3,1}$. The corresponding rigor is sketched in Schmeikal (2011a) on page 324:

$$\xi^c = \pm \frac{1}{\sqrt{2}} \phi\, \iota_+ \quad \text{with} \tag{311}$$

$$\phi = \frac{1}{2}(e_1 + e_{24}) \text{ the area extender} \tag{312}$$

$$\iota_+ = \frac{1}{\sqrt{2}}(e_3 + e_{13}) \text{ the torsion momentum, raising shift operator} \tag{313}$$

The area extender annihilates characters 3 and 4 and swops polarity at character 2. As a 4-iterant it takes the form

$$\dot{\phi} = \frac{1}{2}(e + f) = \frac{1}{2}([+1, +1, -1 - 1] + [+1, -1, -1, +1]) = [+1, 0, -1, 0] \tag{314}$$

And the torsion momentum becomes

$$\iota_+ = \frac{1}{\sqrt{2}}(e_3 + e_{13}) = \frac{1}{\sqrt{2}}(\varphi + e\varphi) = \frac{1}{\sqrt{2}}(Id + e)\varphi = \sqrt{2}[+1, -1, 0, 0]\tau \tag{315}$$

Hence $\xi^c = \pm\frac{1}{\sqrt{2}}\phi\,\iota_+ = [+1,0,-1,0][+1,-1,0,0]\tau = [+1,0,0,0]\tau$ (316)

The charge conjugate spinor of a neutral Majorana fermion has a minimal iterant

$\xi^c = [+1,0,0,0]\tau$

Verify that the two components ϕ and ι_+ are part of an entire angular momentum algebra. They bring in the ultimate braid structure I've called a 'surabale'. We just need to realize the reverse torsion element

$$\iota_- \stackrel{\text{def}}{=} \widetilde{\iota_+} = \frac{1}{\sqrt{2}}(e_3 - e_{13}) \text{ is a lowering shift operator} \tag{317}$$

and both the $\iota_\pm$ yield an angular momentum algebra with the commutators

$$[\phi,\iota_+] = \iota_+ \quad [\phi,\iota_-] = -\iota_- \tag{318}$$

Note that this whole drama is performed within the real Clifford algebra $Cl_{3,1}$ in the matrix world of $Mat(4,\mathbb{R})$. For example we obtain a 4×4 –matrix with real entries for the spinor ξ^c

$$\xi^c = \frac{1}{\sqrt{2}}\phi\,\iota_+ = \frac{1}{4}(e_1 + e_{24})(e_3 + e_{13}) = \frac{1}{4}(e_1 + e_{13} - e_{234} - J) \tag{319}$$

The product $\xi^c\widehat{\xi^c}$ is indeed equal to $-\frac{1}{4}(Id + e_1 + e_{24} + e_{124}) \in Mat(4,\mathbb{R})$. But what about the fermion spinor $(\xi^c)^c = \xi$ travelling forward in time? This has a minimal solution

$$\xi = \frac{i}{\sqrt{2}}\phi\,\iota_+ \quad \text{with } i = \sqrt{-1} \quad \text{thus } \xi^c = \mp i\,\xi \tag{320}$$

there is a beautiful logic built into this matrix algebra of space-time spinors. As soon as we want to factor in the tetragonal symmetries of the Hephy model with all its fermion- and antifermion spinors, we automatically have to change the dramaturgy. We have to introduce the imaginary unit. As we have decided to identify the unit imaginary with a second order time-shift operator, this means a definite statement: the representation of Majorana fermions implies the introduction of a time shift operator.

We know that the primitive idempotent $f_1 \in Cl_{3,1}$ can be represented as a matrix in $Mat(4,\mathbb{R})$ with just a single non zero entry of $+1$ in the main diagonal. This reminds us of the density of a pure state as indicated in equation (309). It is due to this peculiar structure of the algebra that we redefine the Dirac bra- and ket-vectors. Namely, we define the fermion spinor $(\xi^c)^c = \xi$ travelling forward in time; this has a minimal solution

$$|\overline{\nu_e}\rangle \stackrel{\text{def}}{=} |\xi\rangle = \frac{i}{\sqrt{2}}\phi\,\iota_+ \quad \text{as a small iterant } |\overline{\nu_e}\rangle = i[+1,0,0,0]\tau \tag{321}$$

With the main involuted $\langle\xi| \stackrel{\text{def}}{=} \hat{\xi}$ (322)

This leads to $\hat{\xi} = \left(\frac{i}{\sqrt{2}}\phi\, \iota_+\right)^\wedge = \frac{i}{\sqrt{2}}\hat{\phi}\hat{\iota_+}$ with the iterants

$$\hat{\phi} = \frac{1}{2}(-e + f) = \frac{1}{2}([-1,-1,+1+1] + [+1,-1,-1,+1]) = [0,-1,0,+1] \quad (323)$$

$$\hat{\iota_+} = \frac{1}{\sqrt{2}}(-e_3 + e_{13}) = \frac{1}{\sqrt{2}}(-f\tau + g\tau) = \frac{2}{\sqrt{2}}[0,0,+1,-1]\tau = \sqrt{2}[0,0,+1,-1]\tau \quad (324)$$

hence

$$\langle\bar{\nu}_e| = \langle\xi| = \frac{i}{\sqrt{2}}\hat{\phi}\hat{\iota_+} = i[0,-1,0,+1][0,0,+1,-1]\tau = i[0,0,0,-1]\tau \quad (325)$$

Using (299) we obtain

$$|\xi\rangle\langle\xi| = \quad (326)$$
$$i[+1,0,0,0]\tau i[0,0,0,-1]\tau = -[+1,0,0,0][-1,0,0,0]\tau\tau = [+1,0,0,0]Id = f_1 \quad \text{q. e. d.}$$

Let's look if those spinors stand the test in the Clifford algebra. Indeed, the ket $|\xi\rangle = \frac{i}{4}(e_1 + e_{13} - e_{234} - J)$ together with the $\langle\xi| = \frac{i}{4}(-e_1 + e_{24})(-e_3 + e_{13})$ yield the same, namely $|\xi\rangle\langle\xi| = f_1$ as requested. This algebraic structure is a surprise. To reconsider, look at Dirac's 'bra' equal to

$\langle\xi| \stackrel{\text{def}}{=} \hat{\xi} = \frac{i}{4}(-e_1 + e_{13} + e_{234} - J)$ and the 'ket'
$|\xi\rangle = \frac{i}{4}(e_1 + e_{13} - e_{234} - J)$.

The ket-multivector has the direction of base unit e_1 reversed, as also the orientation of the space-time unit-volume e_{234}. Now imagine these are to be taken as templates for the phases. In a quite general sense $\langle bra|$ and $|ket\rangle$ of a *standard model Majorana space-time spinor* behave like *advanced* and *retarded waves*. But they do so in a peculiar way. With the components of area extender ϕ and torsion momentum $\iota_\pm$ we can form the skeleton of eight fermions two of which are neutral Majorana fermions while the other six represent quarks and anti-quarks of a definite color. Why? As is known since long, in the Majorana algebra $Cl_{3,1}$ there is a space of minimal dimension spanned by 4 mutually annihilating primitive idempotents. We call this space a color space. It is one of six isomorphic spaces with positive definite signature and isomorphic with the fourfold real ring ${}^4\mathbb{R}$. This space provides a standard minimal representation of the tetrahedral structure of 2×4 Majorana fermions $\{|\nu_e\rangle,|\bar{\nu_e}\rangle,|\bar{s}\rangle,|s\rangle,|\bar{u}\rangle,|u\rangle,|\bar{d}\rangle,|d\rangle,\}$ having electric charges $\{0, 0, 1/3, -1/3, -2/3, 2/3, 1/3, -1/3\}$. These 4 mutually annihilating generators of idempotent lattice are usually written in minimal form as

$$f_1 = \frac{1}{2}(Id + e_1)\frac{1}{2}(Id + e_{24}) \quad (327)$$
$$f_2 = \frac{1}{2}(Id + e_1)\frac{1}{2}(Id - e_{24})$$
$$f_3 = \frac{1}{2}(Id - e_1)\frac{1}{2}(Id + e_{24})$$
$$f_4 = \frac{1}{2}(Id - e_1)\frac{1}{2}(Id - e_{24})$$

Consider the two idempotent constituents not primitive in the algebra together with a third one:

$$a := \frac{1}{2}(Id + e_1),\ b := \frac{1}{2}(Id + e_{24}),\ c := \frac{1}{2}(Id + e_{124}) \tag{328}$$

These quantities mirror the almost trivial braid algebra characteristic for the commutative Klein 4 group. Note, while the a, b, c have no inverses, the e_1, e_{24}, e_{124} are self-inverse and satisfy same algebra.

$$\begin{aligned} aba &= bab \\ aca &= cac \\ bcb &= cbc \end{aligned} \tag{329}$$

They generate a space ch_1 for the red color isomorphic with ${}^4\mathbb{R}$. This space provides the second interface where geometry can be connected with the iterant algebra. The first interface is the time shift operator, the second follows from the structure of the Minkowski algebra. If we introduce a further commuting, imaginary unit i, the four primitive idempotents (327) give us the densities ascribed to the particles

$$\begin{aligned} &f_1 = |\overline{\nu_e}\rangle\langle\overline{\nu_e}| \text{ with } \left|\overline{\nu_e}\right\rangle \stackrel{\text{def}}{=} \frac{i}{\sqrt{2}}\phi\, \iota_+ \quad f_2 = |s\rangle\langle s| \text{ with } \left|s\right\rangle = \frac{i}{\sqrt{2}}\hat{\phi}\, \iota_+ \\ &f_3 = |u\rangle\langle u| \text{ with } \left|u\right\rangle = \frac{i}{\sqrt{2}}\hat{\phi}\, \widehat{\iota_+} \quad f_4 = |d\rangle\langle d| \text{ with } \left|d\right\rangle = \frac{i}{\sqrt{2}}\phi\, \widehat{\iota_+} \end{aligned} \tag{330}$$

And anti-fermions denoted by an overbar not to be confused with Clifford conjugation

$$\begin{aligned} &-f_1 = |\nu_e\rangle\langle\nu_e| \text{ with } \left|\nu_e\right\rangle \stackrel{\text{def}}{=} \frac{-1}{\sqrt{2}}\phi\, \iota_+ \quad -f_2 = |\bar{s}\rangle\langle\bar{s}| \text{ with } \left|\bar{s}\right\rangle = \frac{-1}{\sqrt{2}}\hat{\phi}\, \iota_+ \\ &-f_3 = |\bar{u}\rangle\langle\bar{u}| \text{ with } \left|\bar{u}\right\rangle = \frac{-1}{\sqrt{2}}\hat{\phi}\, \widehat{\iota_+} \quad -f_4 = |\bar{d}\rangle\langle\bar{d}| \text{ with } \left|\bar{d}\right\rangle = \frac{-1}{\sqrt{2}}\phi\, \widehat{\iota_+} \end{aligned}$$

For all fermions $|ket\rangle$ is obtained from $\langle bra|$ by turning from $\langle q|$ to $|\hat{q}\rangle$ by grade involution.

$$\langle q| \xleftrightarrow{\text{main involution}} |\hat{q}\rangle \tag{331}$$

Every space-time spinor has a (Cliff)form $q = \phi\,\eta$ such that reversion is given by the main involution

$\tilde{\phi} = -\hat{\phi}$ as well as $\tilde{\eta} = -\hat{\eta}$ with reversion $\tilde{}$, main involution $\hat{}$ and conjugation $\check{}$

This is a structure which results in a very simple algebra of Clifford conjugation

$$\check{\phi} = \hat{\tilde{\phi}} = -\hat{\hat{\phi}} = -\phi \text{ and } \check{\eta} = \hat{\tilde{\eta}} = -\hat{\hat{\eta}} = -\eta \tag{332}$$

Clifford conjugation $\smile$ of a Cartan space-time spinor in the Minkowski algebra therefore implies a reversion of the components in the product of extension ϕ and torsion. In symbols

$$\text{Clifford conjugation of } q = \phi\,\eta \longrightarrow \widetilde{\phi\,\eta} = \eta\,\phi \tag{333}$$

In this way, in particular by formulas (328), (329) we can profit from advantages characteristic for *-algebras. Particles are structurally separated from anti-particles by a charge conjugation via multiplication with the imaginary unit. Note also that, until now, we have not used any Hilbert space representation. Everything could be derived from the Clifford algebra. We shall next find out, if there are any indicators for the traditional quantum numbers such as hypercharge, strangeness and so on. We shall further show that there are six isomorphic algebras of the above type within the Majorana algebra. Next we find the most natural images of the matrix representations. Finally we show how those fermion states may be represented by higher order iterants so that we have a chance to interpret them as polarity strings in spaces of locations rather then metric space-time algebra. This is useful as we want to learn something about the translocal behavior of particles from the observable space-time structure.

The idempotents f_1, f_2, f_3, f_4 primitive in the algebra $Cl_{3,1}$ are mutually annihilating idempotents

$$f_i f_j = \delta_{ij} f_i \quad \text{with Kronecker Delta } \delta_{ij} \tag{334}$$

If we define $a_i \overset{\text{def}}{=} -f_i$ we are led to formula $a_i a_j = (-f_i)(-f_j) = f_i f_j = \delta_{ij} f_i$ and thus

$$a_i a_j = -\delta_{ij} a_i \quad \text{for the 'anti-idempotents' } a_i \tag{335}$$

These orthogonality relations are preserved by a peculiar graded orthogonal rank 2 Lie group having eight graded generators. These eight elements generate the compact motion group, and in a peculiar standard representation, in a 4×4-picture, are equal to the Gell-Mann matrices. We just have to omit the first row and column. There is a consistent logic in this design of formulas (256). In the appropriate standard basis these are the prominent Gell-Mann matrices. Note that every base unit used in this representation is actually an element in the Clifford algebra $e_{[\dots]} \epsilon Cl_{3,1} \simeq Mat(4, \mathbb{R})$ and therefore a matrix with real entries only. So the complex numbers come in by mere multiplication with the imaginary unit in the three matrices $\lambda_2, \lambda_5, \lambda_7$ generating an image of the $SU(2, \mathbb{C})$. The 'Clifform' of formulas (256) generate the rank 2 Lie group $\mathfrak{L}_2 \simeq SU(3, \mathbb{C})$. Consider the operators

$$\dot{\lambda}_3 = [0,0,+1,-1] \quad \text{and} \quad \text{isospin} \tag{336}$$

$$\dot{\lambda}_8 = \tfrac{1}{\sqrt{3}}[0,-1,0,+1] \quad \text{hypercharge} \tag{337}$$

Referring to the Gell-Mann-Nishijima relation, these Clifford numbers can give us an indicator for the electric charge associated with spinor fields. This charge operator is

$q := \frac{y}{2} + t_3 = \frac{1}{6}(-e_1 + 2e_{24} - e_{124})$ electric charge (338)
$Q = \frac{1}{3}[0, -1, -1, +2]$ charge iterant

The neutral Majorana fermion satisfies

$Qf_1 = 0 = 0f_1$ (anti-)neutrino, (339)

since $\frac{1}{3}[0, -1, -1, +2][+1,0,0,0] = 0$
$Qf_2 = \frac{1}{3}[0, -1, -1, +2][0,0, +1,0] = \frac{1}{3}[0,0, -1,0] = -\frac{1}{3}[0,0, +1,0] = -\frac{1}{3}f_2$ baryons
$Qf_3 = \frac{1}{3}[0, -1, -1, +2][0,0,0, +1] = \frac{1}{3}[0,0,0, +2] = \frac{2}{3}[0,0,0, +1] = \frac{2}{3}f_3$
$Qf_4 = \frac{1}{3}[0, -1, -1, +2][0, +1,0,0] = -\frac{1}{3}[0, +1,0,0] = -\frac{1}{3}f_4$
with representation (192), (193)

We can also construct quantum numbers of strangeness and the baryon number

$s \stackrel{\text{def}}{=} -f_2$ strangeness (340)

$b = y - s = \frac{1}{12}(3Id - e_1 - e_{24} - e_{124})$ baryon number (341)

$S = [0,0, -1,0]$ strangeness iterant
$B = \frac{1}{3}[0, +1, +1, +1]$ baryon iterant

In this way the primitive idempotents turn into eigenforms of geometric operator equations that contain the known quantum numbers as eigenforms. For example, we obtain strangeness $S = -1$ for the 'isospin singlet' f_2 having hypercharge $-2/3$, but strangeness $S = 0$ for the 'isospin dublet' f_3, f_4 and for the 'Majorana fermion' f_1. Carrying out all the Clifford multiplications, these Clifford numbers t_3, y, q, s, b applied to our four fermion densities bring forth a familiar table of 'quantum numbers':

Table 3. Fermion quantum numbers

	$f_1, \|\nu\rangle$	$f_2, \|s\rangle$	$f_3, \|u\rangle$	$f_4, \|d\rangle$	
t_3	0	0	$+1/2$	$-1/2$	*Isospin*
Y	0	$-2/3$	$+1/3$	$+1/3$	*Hypercharge*
Q	0	$-1/3$	$+2/3$	$-1/3$	*Charge*
S	0	-1	0	0	*Strangeness*
B	0	$+1/3$	$+1/3$	$+1/3$	*Baryonnumber*

MAJORANA SPINOR EIGENFORM AND DYNAMIC INVARIANTS

So far we have shown that there exist elements in the Clifford algebra that reproduce the correct quantum numbers for the tetragonal primitive idempotents. But it is not yet clear what that means for the Majorana spinors. Is it possible that they satisfy analogous eigenvalue equations? We shall proof the

Theorem 13: Majorana fermion spinors satisfy the same eigenvalue equations for isospin, hypercharge, electric charge, strangeness and baryon number, as the primitive idempotents they are generating.

We base the proof on the following

Lemma 6: Let f_α a primitive idempotent $f_\alpha \in ch_1$, with ch_1 the color-space generated by the four mutually annihilating primitive idempotents (327), or likewise base units $\{Id, e_1, e_{24}, e_{124}\}$, and $|\alpha\rangle$ a space-time spinor generating this primitive idempotent. Then we get an 'idempotence eigenform'

$$f_\alpha|\alpha\rangle = |\alpha\rangle \tag{342}$$

and in such a case we say that the Majorana spinor $|\alpha\rangle$ is an eigenform of its associated primitive idempotent. This refers to the approach of Kauffman (2003, 2009) and Heinz von Foerster (1981, a, b).

Proof: We consider $f_1 = |\nu_e\rangle\langle\nu_e|$ together with the associated spinor $|\nu_e\rangle \overset{\text{def}}{=} \frac{i}{\sqrt{2}}\phi\, \iota_+$ and carry out the Clifford product

$$\underbrace{\left(\frac{1}{2}(Id+e_1)\frac{1}{2}(Id+e_{24})\left(\frac{i}{\sqrt{2}}\right)\frac{1}{2}(e_1+e_{24})\right)}_{f_1\frac{i}{\sqrt{2}}\varphi}\underbrace{\left(\frac{1}{\sqrt{2}}(e_3+e_{13})\right)}_{\iota_+} = \tag{343}$$

$$\frac{i}{8}(Id+e_1+e_{24}+e_{124})(e_1+e_{24})(\sqrt{2})(\iota_+) =$$

$$\frac{i}{8}(e_1+Id+e_{124}+e_{24}+e_{24}+e_{124}+Id+e_1)(\sqrt{2})(\iota_+) =$$

$$= \frac{i}{8}(Id+e_1+e_{24}+e_{124})(e_3+e_{13}) = \frac{i}{8}(e_3+e_{13}+e_{243}+e_{1243}+e_{13}+e_3+$$

$$+e_{2413}+e_{124}e_{13}) =$$

$$= \frac{i}{4}(e_3+e_{13}-e_{234}-e_{1234}) = |\nu_e\rangle \text{ which is the same as } \frac{i}{\sqrt{2}}\varphi\,\iota_+.$$

To verify this derivation, we only need to know that all units in $ch_1 = span\{Id, e_1, e_{24}, e_{124}\}$ commute while pure spatial units anticommute, and indices can be 'commuted through', that is, we have for example $e_{1243} = -e_{1234}$ and likewise $e_{124}e_{13} = e_{12413} = -e_{12143} = e_{11243} = Id\, e_{243} = -e_{234}$ which explains the negative signs in the last formula. The proof for the other fermions is run quite analogously. Now consider further any element of the Clifford algebra which we regard as an observable operator O having eigenvalues as in formulas (317) to (319), say, $O\, f_\alpha = o_\alpha f_\alpha$ then because of Lemma 6 we have to have

$$f_\alpha|\alpha\rangle = |\alpha\rangle \Longrightarrow O|\alpha\rangle = o_\alpha|\alpha\rangle \quad \text{q.e.d.} \qquad (344)$$

THE STRONG FORCE SEAL RECONSIDERED

We know since long that there are six quarks and six leptons. So there should be some reason observable in the space-time structure itself that could give rise to such a partition. Actually the color space ch_1 forms a Cartan subalgebra of the Clifford algebra $Cl_{3,1}$, and there are six such commutative subalgebras isomorphic with each other. We can draw a picture like that

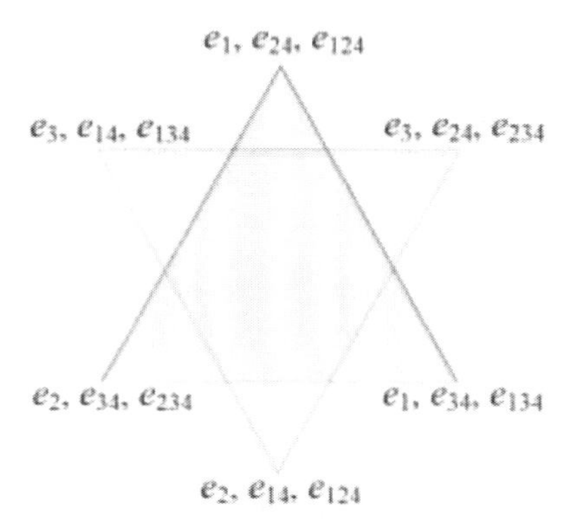

Figure 62. Commutative subalgebras of Clifford algebra $Cl_{3,1}$ with positive signature.

Note, the quadruple $\{Id,\ e_1, e_{24}, e_{124}\}$ provides a peculiar sample of commutative Segré quaternions. In the present form by exterior multiplication the four units generate the discrete Klein 4 group. This follows from their commutativity and norm. Each corner of the seal is constituted by quantities $e_i,\ e_{j4}$ where $e_i^2 = e_j^2 = +Id$ and $i \neq j$. The two generate by exterior multiplication a commutative set of unit monomials of grades $0,1,2,3$ in the form of a quadruple $\{Id, e_i, e_{j4}, e_{ij4}\}$. There are 6 ordered pairs out of the triple of spatial units e_1, e_2, e_3, namely $(1,2), (1,3), (2,3), (2,1), (3,1), (3,1)$, one for each corner. So we can assign a definite color to each corner, and give each triangle either low or high energy fermions, that is $|s\rangle, |u\rangle, |d\rangle$ for the upright and $|t\rangle, |c\rangle, |b\rangle$ for the skewed. So we can fix the electron neutrino and define the u-quark by the aid of area extender φ and torsion momentum ι_+ as in (312, 313).

Red u-Quark

Density $\rho_u = \frac{1}{2}(1 - e_1)\frac{1}{2}(1 + e_{24})$, primitive idempotent in color space ch_1 (345)

spinor $|u\rangle = \pm\frac{i}{\sqrt{2}}\hat{\varphi}_r\hat{\iota}_r = \pm\frac{i}{4}(e_3 - e_{13} - e_{234} + j)$ u-quark, $u \in \mathbb{C}\otimes Cl_{3,1}$

iterant $\pm\frac{i}{4}(f\tau - g\tau + \tau - e\tau) = \pm\frac{i}{4}(Id - e + f - g)\tau =$ (346)

$$\pm\frac{i}{4}([+1,+1,+1,+1]+[-1,-1,+1,+1]+[+1,-1,-1,+1]+[-1,+1,-1,+1])\tau =$$
$$= \pm\frac{i}{4}[0,0,0,+4]\tau = \pm i[0,0,0,+1]\tau$$

If you compare the red u-quark with the electron neutrino iterant

$$|\,u\rangle = \pm i[0,0,0,+1]\tau \qquad (347)$$
$$|\overline{\nu_e}\rangle = i[+1,0,0,0]\tau$$

You realize they involve different primitive idempotents, indicated by different characters, and they act both with a φ permutation on the quad locations of other iterants.

grade involuted $\langle u| = u^{\wedge}$ (348)

$$\langle u| = \pm i[-1,0,0,0]\tau$$

A beautiful blemish of this theory, that $\langle u|$ is equal to $|\overline{\nu_e}\rangle$, or is it a beauty spot? It always reminded me of Feynman's statement: "For example, it appears that the electron, the neutrino, the d-quark, and the u-quark all go together – indeed, the first two couple with the W, as do the last two. At present it is thought that a quark can only change 'colors' or 'flavors'. But perhaps a quark could disintegrate into a neutrino by coupling with an undiscovered particle." (Feynman 1985, p. 150) The iterant picture would suggest more interactions than are seen today. It would even allow for a change of the forces, depending on the 'meta-stable' attractors of the fundamental chaotic process of logic encounters in unspecified domains.

we have $\rho_u = |u\rangle\langle u|$ that is, ρ_u is Clifford product $uu^{\wedge}$
with area extender $\varphi_{red} = \frac{1}{2}(e_1 + e_{24})$, and torsion momentum $\iota_+ = \frac{1}{\sqrt{2}}(e_3 + e_{13})$.
And have the green c-quark in the reverted triangle.

Green c-Quark

Density $\rho_c = \frac{1}{2}(1+e_3)\frac{1}{2}(1-e_{24})$, (349)
spinor $|\,c\rangle = \pm\frac{i}{\sqrt{2}}\phi^{\wedge}_{gr}\,\eta_{gr} = \pm\frac{i}{4}(-e_1 + e_{13} + e_{124} + j)$ c-quark, t∈ $\mathbb{C}\otimes Cl_{3,1}$
iterant $\pm\frac{i}{2}([0,-1,+1,0]+[0,-1,+1,0]\tau)$ (350)
grade involuted $\langle c| = t^{\wedge}$
we have $\rho_t = |c\rangle\langle c|$
with area extender $\phi_{green} = \frac{1}{2}(e_3 + e_{24})$ and torsion momentum $\eta_+ = \frac{1}{\sqrt{2}}(e_1 - e_{13})$

The design of all charged leptons requires some more detail which we do not go into in this chapter. The spinors represented here are just five examples out of a universe of space-time spinors. More can be found in the ICCA9 proceedings (2011a). In order to understand the construction principles of Majorana fermions, it is necessary to understand that a 'wave' of quantum dynamics involves the 'whole space' and the 'whole time', as it is actually constituting this whole 16-dimensional linear space of geometric Clifford algebra. It also involves all possible iterants of grade 4. There is no difference between field and operator.

Matrix Representations

Using a standard representation (176) of Majorana matrices we obtain the Gell-Mann matrices in a 4 × 4-version, and the neutral Majorana fermion can naturally be taken to be the electron-anti-neutrino $\left|\overline{\nu_e}\right\rangle \overset{\text{def}}{=} \pm\frac{i}{\sqrt{2}}\varphi\,\iota_+ = \pm\frac{i}{4}(e_3 + e_{13} - e_{234} - e_{1234})$ which takes the admissible minimal form (351)

$$|\overline{\nu_e}\rangle = \pm\begin{bmatrix}0&0&\mathrm{i}&0\\0&0&0&0\\0&0&0&0\\0&0&0&0\end{bmatrix};\ \langle\overline{\nu_e}| = \mp\frac{i}{4}(-e_3 + e_{13} + e_{234} - e_{1234}) = \mp\begin{bmatrix}0&0&0&0\\0&0&0&0\\i&0&0&0\\0&0&0&0\end{bmatrix}$$

Which according to equations (328) yields the density

$$f_1 = |\overline{\nu_e}\rangle\langle\overline{\nu_e}| = -\begin{bmatrix}0&0&\mathrm{i}&0\\0&0&0&0\\0&0&0&0\\0&0&0&0\end{bmatrix}\begin{bmatrix}0&0&0&0\\0&0&0&0\\i&0&0&0\\0&0&0&0\end{bmatrix} = \begin{bmatrix}1&0&0&0\\0&0&0&0\\0&0&0&0\\0&0&0&0\end{bmatrix} \tag{352}$$

as desired. The neutrino can be represented by the matrix

$$|\nu_{\mathrm{e}}\rangle = \pm\begin{bmatrix}0&0&1&0\\0&0&0&0\\0&0&0&0\\0&0&0&0\end{bmatrix};\ \langle\nu_{\mathrm{e}}| = \mp\begin{bmatrix}0&0&0&0\\0&0&0&0\\1&0&0&0\\0&0&0&0\end{bmatrix};\ \text{with density} \tag{353}$$

$$|\nu_{\mathrm{e}}\rangle\langle\nu_{\mathrm{e}}| = -\begin{bmatrix}1&0&0&0\\0&0&0&0\\0&0&0&0\\0&0&0&0\end{bmatrix} = -f_1$$

The matrix for the electric charge operator (338) in the standard basis is (354)

$$q = \frac{1}{3}\begin{bmatrix}0&0&0&0\\0&-1&0&0\\0&0&2&0\\0&0&0&-1\end{bmatrix}\ \text{giving}\ q|\nu_{\mathrm{e}}\rangle = \pm\frac{1}{3}\begin{bmatrix}0&0&0&0\\0&-1&0&0\\0&0&2&0\\0&0&0&-1\end{bmatrix}\begin{bmatrix}0&0&1&0\\0&0&0&0\\0&0&0&0\\0&0&0&0\end{bmatrix} = 0|\nu_{\mathrm{e}}\rangle$$

So the *typical* Majorana fermion has zero charge. But the other fermions which carry the ugly name 'baryons' have non zero charges. Take for example the s-quark

$$|s\rangle = \begin{bmatrix}0&0&0&0\\0&0&0&0\\0&0&0&0\\0&i&0&0\end{bmatrix}\ \text{we get the eigenvalue equation} \tag{355}$$

$$q\,|s\rangle = -\tfrac{1}{3}\left[\begin{bmatrix} 0 & 0 & 0 & 0 \\ 0 & 0 & 0 & 0 \\ 0 & 0 & 0 & 0 \\ 0 & i & 0 & 0 \end{bmatrix}\right] = -\tfrac{1}{3}|s\rangle \tag{356}$$

Notice, we did not refer to any Hilbert space.

Chapter 7

COLOR BRAIDS

An initial state in quantum motion, as it appears on the space-time surface, is a multivector in the hypercomplex space of Clifford algebra. But this does not say much more than that the transformation of polarities is entirely based on permutations. In this way, by statistical measures, the metric Minkowski algebra shows itself. Below that appearance it might be a chaotic process of creation and annihilation of polarity strings or something similar. We can't be sure yet. A measurement of a macro state is then the result of our observation of a stochastic nonlocal process that we do not yet understand. If we can conceive a state such as $|\, u\rangle = \pm \left(i/\sqrt{2}\right)\varphi_r^{\wedge}\iota_r^{\wedge} = \pm i[0,0,0,+1]\tau$ we think that it occurs in steps that are determined by a unitary linear transformation, so that we can write

$$|\, u\rangle \longrightarrow U|\, u\rangle = |\, Uu\rangle \tag{357}$$

And we can take the symbol $|\, u\rangle$ as a token that really indicates that fermion we call u-quark. The unitary transformations that are relevant for the process of transformations in matter are given by the generators (256) of the motion group and by the photonic quaternion subgroup generated by bivectors which squared give $-Id$. We'll come to that. The unitary elements that we used to denote as "quantum states" of the system are then rotated on some high-dimensional sphere so that their unitary feature is preserved. But the important unit base elements that are to be transformed are derived, as we saw, from the primitive idempotents, and are thus not identical with the unit vectors, say, e_1, e_2, e_3 as in Euclidean space, but with the 'basic reflections' $Id - 2f$ calculated from primitive idempotents. Such an idempotent can represent a particle density, and we can combine this with the square over a little arrow, just as we can assign an amplitude to the associated spinor and a probability to the derived reflection. We can still assume that our macro measurement will return us probabilities in terms of hypercomplex matrices and conjugate transposes, if we like. That situation need not be altered significantly. Besides the t-, u-, and w- (iso)spingroups, the motion group $SU(3)$ contains an important subgroup $SU(2)$ which is responsible for transformations of flavor. There is a second $SU(2)$ for color disjoint from the $SU(3)$. The two $su(2)$ algebras are nicely separated in space-time algebra. Color rotations are generated by the iterant

$$\tau_\chi = \frac{1}{2}(Id + f\sigma - g\tau - e\varphi) \tag{358}$$

This string has period 6. You can verify by hand that $\tau_{col}^3 = -Id$

Probably there is no much difference if you use the elements (203). Color rotations have the purpose to carry us from corner to corner in the seal, that is, from color to color, and the seal has six corners. The rotations bring forth a step in the seal, a step counter- or clockwise by $\pi/3$, and have form

$$\tau_{(123)} := \frac{1}{\sqrt{2}}(e_1 - e_2)\frac{1}{\sqrt{2}}(e_2 - e_3) = \frac{1}{2}(Id + e_{12} + e_{23} - e_{13})$$

an even element of the algebra. Its inverse is given by the reversion

$$\tau_{(123)}^{-1} = \tau_{(132)} = \tilde{\tau}_{(123)} = \frac{1}{2}(-Id - e_{12} - e_{23} + e_{13})$$

Now one can verify by some Clifford calculator that, for example, $f_{1\alpha}$ by rotation $\tau_{(123)}$ is carried to the $f_{2\alpha}$; for example $\tau_{(123)} f_{11} \tau_{(123)}^{-1} = f_{21}$ and so forth for the other lattices of primitive idempotents. We want to demonstrate some exemplary properties of the color rotation. Namely, the $\tau_{(123)}$ is what we usually denote as a quaternion, and we could just as well write in easy quaternion notation

$$\tau_\chi = a + bi + cj + dk \tag{359}$$

And there are uncountable ways to represent this little thing. In our case we would have that $a = b = c = 1/2$ and $d = -1/2$. Most of us have learned to represent this in the Pauli algebra, as the quaternions can be obtained from those matrices by multiplying them with the imaginary unit. So we would write for example

$$\tau = a\,Id + bI + cJ + dK = a\begin{bmatrix}1 & 0\\0 & 1\end{bmatrix} + b\begin{bmatrix}i & 0\\0 & -i\end{bmatrix} + c\begin{bmatrix}0 & 1\\-1 & 0\end{bmatrix} + d\begin{bmatrix}0 & i\\i & 0\end{bmatrix} \text{ with} \tag{360}$$

$I^2 = J^2 = K^2 = IJK = -Id,\ IJ = K, JK = I, KI = j,\ IJ + JI = 0,\ JK + KJ = 0,\ IK + KI = 0$

You see, we are differing here between the imaginary unit written by a small i and a quaternion large I. Here one must take a little care as various computer programs are using the big I for the imaginary unit in the complex numbers. Here in text we are avoiding this. Also we have different and more comfortable matrices in the Majorana representation we are using the identity $Id = diag(+1, +1, +1, +1)$ and the quaternion base units

$$e_{12} = \begin{bmatrix}0 & 1 & 0 & 0\\-1 & 0 & 0 & 0\\0 & 0 & 0 & -1\\0 & 0 & 1 & 0\end{bmatrix},\ e_{23} = \begin{bmatrix}0 & 0 & 0 & -1\\0 & 0 & 1 & 0\\0 & -1 & 0 & 0\\1 & 0 & 1 & 0\end{bmatrix},\ e_{13} = \begin{bmatrix}0 & 0 & 1 & 0\\0 & 0 & 0 & 1\\-1 & 0 & 0 & 0\\0 & -1 & 0 & 0\end{bmatrix} \tag{361}$$

We can easily verify the relationships given in the second line of equations (360). The conjugate transpose of our exemplary quaternion, our color rotation τ is the element

$$\tau_\chi^\dagger = a\, Id - bI - cJ - dK = \frac{1}{2}(Id - e_{12} - e_{23} + e_{13}).$$

We have $\tau^{-1} = \tilde{\tau}$ the Clifford reverted; and the product

$$\tau_\chi\ \tilde{\tau}_\chi := \frac{1}{2}(Id + e_{12} + e_{23} - e_{13})\frac{1}{2}(Id - e_{12} - e_{23} + e_{13}) = Id \text{ is identity.} \quad (362)$$

Verify that the matrix τ satisfies the necessary

$$\tau_\chi \tau_\chi^\dagger = (a^2 + b^2 + c^2 + d^2)Id = \left(\frac{1}{4} + \frac{1}{4} + \frac{1}{4} + \frac{1}{4}\right) Id = Id \quad (363)$$

The length of a quaternion is given by the trace of its product with the conjugate transpose. Mostly we shortly write for this 'length' of the quaternion $\sqrt{|\tau_\chi \tau_\chi^\dagger|} = \sqrt{a^2 + b^2 + c^2 + d^2}$ which is somewhat sloppy, but it works if we take care. The general color rotating quaternion acting on the $SU(3)$ unitary fermion states of the complex Minkowski algebra is given by the exponential map

$$exp\colon\ \{\alpha, \beta, \gamma\} \longrightarrow e^{\alpha e_{12} + \beta e_{23} + \gamma e_{13}} \text{ in the even subalgebra, with } \{\alpha, \beta, \gamma\} \in \mathbb{R} \quad (364)$$

Resulting in a quaternion $aId + be_{12} + ce_{23} + de_{13}$ that has all possible lengths. We call such an element with no scalar component

$$u := \alpha e_{12} + \beta e_{23} + \gamma e_{23} \quad (365)$$

a pure quaternion, for which we have the Clifford product of u with itself

$$uu = (-\alpha^2 - \beta^2 - \gamma^2)Id \quad (366)$$

A pure unit length quaternion has $-\alpha^2 - \beta^2 - \gamma^2 = -1$, hence $u^2 = -Id$. Therefore the set of unit quaternions is often colaterally represented on the unit sphere $S^2 = \{\alpha, \beta, \gamma | \alpha^2 + \beta^2 + \gamma^2 = 1\}$. This can be used to construct a general quaternion of arbitrary length by the form

$$q = a + bu \text{ with any real numbers } a, b \text{ and } u^2 = -Id \quad (367)$$

As unit quaternions are elements of the $SU(2)$ we have the condition $a^2 + b^2 = 1$. Kauffman and Lomonaco (2009 p. 27) realized the braid relation $ghg = hgh$ using quaternions in this reduced form

$$g = a + bu \text{ and } h = c + dv \quad (368)$$

and found the non-trivial solution that the dot product

$$u.v = \frac{a^2 - b^2}{2b^2} \tag{369}$$

which ultimately led to an interesting example of the 'Fibonacci representation' in $SU(2)$. We shall develop this approach further and apply it to the peculiar situation of those polarity strings that constitute the Minkowski algebra. We wish to construct an image of a 3-strand braid group in the $SU(2)$ generated by bivectors e_{12}, e_{23}, e_{13}, that is, pure quaternions, via the homomorphism

$$\rho\colon B_3 \longrightarrow SU(2) \tag{370}$$

and show how those specific braids of the Minkowski algebra entangle the $SU(3)$ spinors and their primitive idempotents, which, I have to confess, are already entangled all by themselves. Hence we search for representative elements $g = \rho(s_1)$ and $h = \rho(s_2)$ which satisfy the generating relations for the braid group B_3, namely the only one $s_1 s_2 s_1 = s_2 s_1 s_2$ which would be $ghg = hgh$ which we multiply by h^{-1} from the left, and by g^{-1} from the right. This gives the very nice form in terms of conjugation

$$h^{-1}gh = ghg^{-1} \tag{371}$$

the less typical braid equation for three strands.

Theorem 14: To every rotation of color $r = e^{\alpha e_{12} + \beta e_{23} + \gamma e_{13}}$ in the quaternion subspace of bi-paravectors there exists a continuous manifold of braid generators s such that the braid relation $rsr = srs$ is satisfied.

Proof: Carried out by Maple Clifford. The general solution which involves all parameters has been found. There are five of them. Two are trivial and the three non-trivial solutions involve 11 occurrences of square roots which contain up to 22 multilinear terms.

So it would not be a towering work to discuss these solutions. But we must work out in that context which is compatible with the standard model motion groups, how the Artin braid theory does apply and where and why it does apply so well. Some fundamental theory on Clifford algebras based on involutive braids and spinors in braided geometry was worked out in the beginning 1990s by Micho Durdewich (1995) and Zbigniew Oziewicz (1996). Durdevich used the denotation of 'braided Clifford algebras' which he understood as deformations of braided exterior algebras in the tradition of Chevalley and Kähler. He constructed peculiar elements of those groups, among others the secondary flip-over operator which as he says 'naturally enters the game', (Durdewich 1995, p. 4). Let V be a $\mathbb{C}$-space, begins Oziewicz (1996), $\sigma \in End(V^{\otimes 2})$ be a pre-braid operator and let $F \in lin(V^{\otimes 2}, \mathbb{C})$; and derives a sufficient condition on (σ, F) that there exists a Clifford algebra $Cl(V, \sigma, F)$ which is the Chevalley F-dependent deformation of an exterior algebra $Cl(V, \sigma, 0) \equiv V^{\wedge}(\sigma)$. These articles, especially the one's by Oziewicz are not easily read to the end by non-mathematicians. But they are of outstanding importance. Unfortunately most physicists do not see how those concepts can or should be applied to their experimental data. This is partly so, because there is a) not much mathematical practice, and b) no whole theory of matter and space-time. It is clear that what is said by those trailblazing papers is actually put into realization in this chapter. And it is realized such that it explains the real physic events. But for physicists it is necessary to see the phenomena that run parallel to this mathematics.

So, please, let us understand the role of the color rotation operator which does not only apply to the colored phenomena, but also to those events where the colors sum up to uncolored. Consider

$$\tau_\chi := \frac{1}{2}(Id + e_{12} + e_{23} - e_{13}) \tag{372}$$

With known quaternion 'coordinates' $\left\{\frac{1}{2}, \frac{1}{2}, \frac{1}{2}, -\frac{1}{2}\right\}$ together with its Clifford inverse

$$\tau_\chi^{-1} = \widetilde{\tau_\chi} := \frac{1}{2}(Id - e_{12} - e_{23} + e_{13}) = Id \tag{373}$$

We put $h = \tau$ and let g be unknown, that is,

$$g := x_1 Id + x_2 e_{12} + x_3 e_{23} + x_4 e_{13} \tag{374}$$

Recall, g is a color operator that flips space ch_1 to space ch_2 as we have

$$\tilde{h} f_{1,1} h = f_{2,1} \tag{375}$$

We obtain six manifolds of solutions.

1. Solution set (376)
   $x_1 = \frac{1}{2}$; $x_2 = \sqrt{4z^2 + (2 - 4x_4)z + 4x_4^2 - 2x_4 - 1}$
   $x_3 = -\frac{1}{2} + x_4 - \sqrt{4z^2 + (2 - 4x_4)z + 4x_4^2 - 2x_4 - 1}$
   $x_4 = x_4$ ... can be selected freely
2. Solution set
   $x_1 = x_2 = x_3 = x_4 = 0$
3. Solution set
   $x_1 = x_2 = x_3 = \frac{1}{2}$; $x_4 = -\frac{1}{2}$
4. Solution set
   $x_1 = +\frac{3}{2}\sqrt{3z^2 + 3z + 1}$; $x_2 = -\frac{1}{2}\sqrt{3z^2 + 3z + 1}$;
   $x_3 = -\frac{1}{2}\sqrt{3z^2 + 3z + 1}$; $x_4 = +\frac{1}{2}\sqrt{3z^2 + 3z + 1}$
5. Solution set
   $x_1 = \frac{1}{2}$; $x_2 - \frac{1}{2}$; $x_3 = x_4 = \frac{1}{2}$
6. Solution set
   $x_1 = \frac{1}{2}$; $x_2 = \frac{1}{6}$; $x_3 = \frac{1}{6}$; $x_4 = \frac{5}{6}$

Consider the sixth solution. It gives us a

$$g = \frac{1}{2}Id + \frac{1}{6}e_{12} + \frac{1}{6}e_{23} + \frac{5}{6}e_{13} \tag{377}$$

Note, the term $u = g - \frac{Id}{2} = \frac{1}{6}e_{12} + \frac{1}{6}e_{23} + \frac{5}{6}e_{13}$ in g is not a pure quaternion, since we have $uu = -\frac{3}{4}Id$

We know that g together with h make the braid. In Clifford algebra notation we form

$$h^{-1}f_{1,1}h = \tilde{h}f_{1,1}h = f_{2,1} = \frac{1}{2}(Id + e_2)\frac{1}{2}(Id + e_{34}) = \frac{1}{4}(Id + e_2 + e_{34} + e_{234}) \tag{378}$$

and applying to the $f_{2,1}$ operation g

$$f_{braid} = g^{-1}f_{2,1}g = \frac{1}{4}Id - \frac{1}{9}e_1 + \frac{2}{9}e_2 - \frac{1}{36}e_3 - \frac{7}{36}e_{14} - \frac{1}{9}e_{24} - \frac{1}{9}e_{34} + \\ + \frac{2}{9}e_{124} + \frac{1}{36}e_{134} - \frac{1}{9}e_{234} \tag{379}$$

gives this wonderful braid containing base monomials from the whole seal with the magic number 7 which may indicate the invisible presence of the Fibonacci numbers. Now we can ask, if the f_{braid} is a primitive idempotent at all. It should be, because all the orthogonal groups relevant for unitary transformations of spinors in that system of $Cl_{3,1}$ preserve the property of orthogonality, idempotence and mutual annihilation. We can carry out the rigor by Clifford Maple as before, and obtain

$$f_{braid}f_{braid} = f_{braid} \tag{380}$$

We want to transform the electron spinor

$$|e^-\rangle = \pm\frac{i}{4}(e_2 + e_{14})(e_3 + e_{23}) \tag{381}$$

by a unitary braid element h. Notice, the gradeinverted spinor is

$$\langle e^-| = \pm\frac{i}{4}(e_2 - e_{14})(e_3 - e_{23}) \tag{382}$$

With the familiar identity

$$|e^-\rangle\langle e^-| = f_{4,1} \tag{383}$$

So we carry out the conjugate transformation

$$\psi(h) = h^{-1}\,|e^-\rangle\, h = \frac{i}{4}(e_1 + e_{13} + e_{124} + e_{1234}) = \frac{i}{4}(e_3 + e_{24})(e_1 - e_{13})$$

which is the spinor ascribed to a lepton in the upper right of the seal belonging to the primitive idempotent

$$f_{5,1} = \frac{1}{2}(Id + e_3)\frac{1}{2}(Id + e_{24})$$

So this is indeed a counter clockwise $\pi/3$-turn. This will now be followed by the second generator of B_3 which results in the braided element

$$\text{(i)}\quad \Psi(h,g) = g^{-1}\psi(h)\, g = i\Big(-\tfrac{1}{9}e_1 - \tfrac{1}{36}e_2 + \tfrac{2}{9}e_3 - \tfrac{1}{36}e_{12} - \tfrac{2}{9}e_{13} - \tfrac{1}{9}e_{23} - \quad (384)$$

$$-\tfrac{1}{9}e_{124} + \tfrac{1}{9}e_{134} - \tfrac{7}{36}e_{234} + \tfrac{1}{4}e_{1234}\Big)$$

This spinor has a grade involuted

$$\widehat{\Psi}(h,g) = g^{-1}\psi(h)\, g = i\Big(+\tfrac{1}{9}e_1 + \tfrac{1}{36}e_2 - \tfrac{2}{9}e_3 - \tfrac{1}{36}e_{12} - \tfrac{2}{9}e_{13} - \tfrac{1}{9}e_{23} -$$
$$+\tfrac{1}{9}e_{124} - \tfrac{1}{9}e_{134} + \tfrac{7}{36}e_{234} + \tfrac{1}{4}e_{1234}\Big)$$

and the product

$$F(h,g) = \Psi(h,g)\widehat{\Psi}(h,g) = \tfrac{1}{4}Id - \tfrac{7}{36}e_1 - \tfrac{1}{9}e_2 - \tfrac{1}{9}e_3 - \tfrac{1}{9}e_{14} + \tfrac{2}{9}e_{24} - \tfrac{1}{36}e_{34} - \quad (385)$$
$$-\tfrac{2}{9}e_{124} - \tfrac{1}{36}e_{134} + \tfrac{1}{9}e_{234}$$

has the property of primitive idempotence. Note, that it stays within the boundaries of the Cartan algebras given by the seal.

Lemma 7: The element $F(h,g)$ is an idempotent element in pure positive signature $\{+,+,+,+,+,+,+,+,+,+\}$ primitive in the Minkowski algebra $Cl_{3,1}$

Proof: of primitive idempotence was given by Maple Clifford.

Notice the strict separation between actor and acted on. The braid operators are all in the quaternion algebra of purely spatial bivectors. They provide the 'photonic actor' in color rotations. The acted on spinors are all provided by the seal, that is, the 10-dimensional subspace with positive definite signature $\{+,+,+,+,+,+,+,+,+,+\}$ in the standard representation.

UNIVERSAL GATES OF MINKOWSKI ALGEBRA

It is well known that the spinor of a spin ½ particle can be thought to represent one single qubit of so called quantum information. Therefore a fermion represented in the Lie algebra $su(2,\mathbb{C})$ within the Pauli algebra of Euclidean 3-space could be associated with quantum information. It has variously been noticed that a pair of spin ½ particles would provide an appropriate means to describe entangled states and the logic gates ascribed to them. Strange enough, as the entangled photons are not spin ½ particles. Therefore the role of $su(4)$ was recognized as relevant (Rau 2009). Various subgroups such like $su(2) \times su(2) \times u(1)$, the

Fano plane of projective geometry, the subalgebra $so(5)$ with ten generators were investigated. Planat (2009) realized the importance of the octahedral symmetry in connection with what he called Clifford group dipoles. He gave a most competent, explosive recital of the finite Coxeter groups until to the enactment of the one and only 'Weyl/Coxeter group $W(E_8)$' with the largest cardianality of 696729600. But the fundamental superstructure of the geometric algebra of the Minkowski space was not yet realized. The theory of graded reorientation was not seen as forming an emergent whole that can be derived by comprehending quantum motion. One of the best pieces on quantum gates and 'quantum computation using geometric algebra' is the dissertation of Douglas J. Matzke (2002) who investigated very carefully the fundamental gates and the computational basis in relation to 2-qubits. This work made me aware of the importance of what he called the 'even grade plane' in the reorientation of the single qubit Hilbert space to the Hadamard- or Fourier basis of superimposed states. For 2-qubits in Minkowski algebra this can be the Clifford number $(Id - e_{23})/\sqrt{2}$ as we shall see next. In their refreshing paper and slides Kauffman and Lomonaco (2009, p. 14) explained what a two-qubit gate is, and when a two-qubit gate is called 'universal'. Namely, a *two-qubit gate* is a unitary linear mapping

$$G: V \otimes V \longrightarrow V \otimes V \tag{386}$$

where V is a complex vector space of dimension 2.

The gate G is said to be universal for quantum computation[1] if G together with local unitary transformations from V to V generates all unitary transformations of the complex vector space of dimension 2^n to itself. A gate G is said to be entangling if there is a vector

$$|\alpha\beta\rangle = \alpha \otimes \beta \in V \otimes V \tag{387}$$

Such that the element $G\,|\alpha\beta\rangle$ is not decomposable as a tensor product of two qubits. Under these circumstances the element $G\,|\alpha\beta\rangle$ is called entangled. The Brylinskis (2002) gave a general criterion for a quantum gate to be universal.

Theorem 15: A gate G is universal if and only if it is entangling.

Proof: Brylinski and Brylinski 2002

As a matter of fact, the Minkowski algebra is not a complex vector space of dimension 2. And quantum entanglement is not the same as topological entanglement. Then what?

WHEREABOUTS OF TWO-QUBITS IN MINKOWSKI ALGEBRA

It was at about the same time when I realized the meaning of a transposition map for the graded Clifford algebra of the Minkowski space that Mermin (1993) discussed some paradoxes we are confronted with in the theory of quantum mechanical measurement. Michel Planat (2009) encapsulated this follow up of the discourse around the Kochen-Specker theorem by exhibiting a pair of two qubit entangling gates which are braiding matrices that

[1] The authors made the annotation "or just universal".

also generate an octahedral symmetry. Briefly, the Kochen-Specker theorem discloses possible situations where the structure of the eigenvalues of measurement operators cannot be brought into agreement with the eigenstates. These peculiarities which are also hidden in the structure of the idempotent manifolds of the Minkowski algebra could, in the beginning of that discourse, be dealt with by the aid of the marvel of tensor products of the Pauli spin matrices. Clearly these matrices bring in the whole cockroach problem of the '2 qubit braiding gates'. The basic building blocks of an algebraic proof of the theorem by Kochen and Specker are two triples of commuting tensor products of the Pauli matrices which constitute 2-qubit states in the matrix algebra $Mat(4,\mathbb{C})$, namely $\sigma_1 \otimes \sigma_1, \sigma_2 \otimes \sigma_2, \sigma_3 \otimes \sigma_3$ and $\sigma_1 \otimes \sigma_3, \sigma_3 \otimes \sigma_1, \sigma_2 \otimes \sigma_2$. One of the most prominent long runners among those braiding gates is the matrix

$$R = \frac{1}{\sqrt{2}} \begin{bmatrix} 1 & 0 & 0 & 1 \\ 0 & 1 & -1 & 0 \\ 0 & 1 & 1 & 0 \\ -1 & 0 & 0 & 1 \end{bmatrix} \tag{388}$$

It is one of the unitary solutions of the Young-Baxter equation. In the standard basis (176) of $Cl_{3,1}$ this is the matrix that represents

$$R = \frac{1}{\sqrt{2}}(Id - e_{23}) \tag{389}$$

With reference to quantum computing – and I confess this is a rather cryptic formulation – matrices like the R should be imagined to act on the standard basis $\{|0\,0\rangle, |0\,1\rangle, |1\,0\rangle, |1\,1\rangle\}$ of $H = V \otimes V$. May I request the reader to recall the chapter about 'freedom of entanglement' where the Young-Baxter equation is discussed. - Letting the R act on such a computational basis, we get

$$\begin{aligned} R\left|0\,0\right\rangle &= +\tfrac{1}{\sqrt{2}}|0\,0\rangle - \tfrac{1}{\sqrt{2}}|1\,1\rangle \\ R\left|0\,1\right\rangle &= +\tfrac{1}{\sqrt{2}}|0\,1\rangle + \tfrac{1}{\sqrt{2}}|1\,0\rangle \\ R\left|1\,0\right\rangle &= -\tfrac{1}{\sqrt{2}}|0\,1\rangle + \tfrac{1}{\sqrt{2}}|1\,0\rangle \\ R\left|1\,1\right\rangle &= +\tfrac{1}{\sqrt{2}}|0\,0\rangle + \tfrac{1}{\sqrt{2}}|1\,1\rangle \end{aligned} \tag{390}$$

which is the Bell basis of entangled states.

Paradoxes Brought Forth a Transposition Map

Another such matrix is the one we used in (225)

$$R' := \begin{bmatrix} a_1 & 0 & 0 & 0 \\ 0 & 0 & a_2 & 0 \\ 0 & a_3 & 0 & 0 \\ 0 & 0 & 0 & a_4 \end{bmatrix} \tag{391}$$

Kauffman and Lomonaco pointed out that this braiding matrix effectuates the following change of the computational basis

$$\begin{aligned} R'|0\,0\rangle &= a_1|0\,0\rangle \\ R'|0\,1\rangle &= a_3|0\,0\rangle \\ R'|1\,0\rangle &= a_2|0\,1\rangle \\ R'|1\,1\rangle &= a_4|1\,1\rangle \end{aligned} \tag{392}$$

Look at this very attentively! We saw that the matrix R' can be seen as the product of a phase gate and a swap gate. Those swap gates point towards the discrete multivector symmetries of a geometric algebra which determine its structural features. Planat brings in a second matrix

$$S = \frac{1}{2}\begin{bmatrix} 1 & -1 & 1 & 1 \\ 1 & 1 & -1 & 1 \\ 1 & -1 & -1 & -1 \\ 1 & 1 & 1 & -1 \end{bmatrix} \tag{393}$$

both R and S are entangling while only R satisfies the Young Baxter equation. Their matrix product

$$RS = \frac{1}{\sqrt{2}}\begin{bmatrix} 1 & 0 & 1 & 0 \\ 0 & 1 & 0 & 1 \\ 1 & 0 & -1 & 0 \\ 0 & 1 & 0 & -1 \end{bmatrix} \tag{394}$$

is equal to the tensor product $H \otimes I_2$ with the Hadamard matrix

$$H = \frac{1}{\sqrt{2}}\begin{bmatrix} 1 & 1 \\ 1 & -1 \end{bmatrix}. \tag{395}$$

Kauffman and Lomonaco denote the matrix Q

$$Q = \frac{1}{\sqrt{2}}\begin{bmatrix} 1 & 1 & 0 & 0 \\ 1 & -1 & 0 & 0 \\ 0 & 0 & 1 & 1 \\ 0 & 0 & 1 & -1 \end{bmatrix} \tag{396}$$

by the same tensor product $H \otimes I_2$ which shows that we have not yet come to a clear convention where the interpretation (as Kronecker product) and the preferred order of the factors of a tensor product is concerned. But it is obvious that Q is a universal gate and it satisfies the Young-Baxter equation. Anyway, it appears that we have to respect a set of predicates which characterize such quantum gates. Namely, a matrix can be entangling or not, it can be a braiding matrix or not, it can satisfy the Young-Baxter equation or not, it can represent a universal gate and so on. We have to bring some more order into this mathematical 'Panoptikum', so that physicists can use it. In the above quoted paper Planat stresses that matrices R and S are *distinguished members of the two-qubit Clifford group* C_2. He mentions that R and S generate a finite group with 96 elements isomorphic with $\mathbb{Z}_4 S_4$.

Clearly, this group coming up in the context of the Majorana algebra, in the form of two generators in $Mat(4,\mathbb{R})$, are a subgroup of the complete reorientation group of the basic not grade preserving multivector group of $Cl_{3,1}$ (Schmeikal 2004). If we shrivel the geometry and reduce it to the habitual use of Hilbert space and qubit-terminology we will immediately loose the beautiful phenomenology of Minkowski space-time algebra. The structure of the $Cl_{3,1}$ is given by the seal of Cartan subalgebras and by two quaternion algebras. Therefore the basic swaps which produce all the reorientations as discrete automorphisms are generated by the 6 times 4 minimal primitive idempotents of the form

$$f_{\chi,\alpha} = \frac{1}{2}(Id \pm e_i)(Id \pm e_{j4}) . \tag{397}$$

Here the index χ denotes the number of the color-space, that is, the corners of the seal, and it runs from 1 to 6. The index α means the 4 sign combinations $(\pm,\pm)$ in the factors. The e_i and e_j must not be equal and must be unequal to e_4. So we have two spatial base units $e_i \neq e_j$ with $e_i^2 = e_j^2 = Id$. The generating 'Coxeter reflections' as elements with period 2 are given by the 24 Clifford numbers

$$s_{\chi,\alpha} = Id - 2f_{\chi,\alpha} \tag{398}$$

These 24 numbers embody the whole mystery of spinor construction in the space $Cl_{3,1}$. For that construction is not trivial, but it must satisfy many requirements that have been discovered by great thinkers like Cartan, Chevalley, Crumeyrolle, Lounesto; and may the others forgive me for not having mentioned them. These sets of numbers are fundamental for 2-qubits in the Clifford algebra $Cl_{3,1}$.

$$\begin{aligned} f_{\chi,1} &= \frac{1}{2}(Id + e_i)\frac{1}{2}(Id + e_{j4}) \\ f_{\chi,2} &= \frac{1}{2}(Id + e_i)\frac{1}{2}(Id - e_{j4}) \\ f_{\chi,3} &= \frac{1}{2}(Id - e_i)\frac{1}{2}(Id + e_{j4}) \\ f_{\chi,4} &= \frac{1}{2}(Id - e_i)\frac{1}{2}(Id - e_{j4}) \end{aligned} \tag{399}$$

These are the mutually annihilating primitive idempotents in the standard basis of the Clifford algebra $Cl_{3,1}$. These give rise to the combination of terms in the spinors which parallel them, up to constants given by such products as

$$\begin{aligned} |\ldots\rangle &= \phi_\chi \eta_\chi \\ |\ldots\rangle &= \phi_\chi \hat{\eta}_\chi \\ |\ldots\rangle &= \hat{\phi}_\chi \eta_\chi \\ |\ldots\rangle &= \hat{\phi}_\chi \hat{\eta}_\chi \end{aligned} \tag{400}$$

Where the empty bras $|\ldots\rangle$ denote leptons and fermions. These represent the $\{|0\ 0\rangle, |0\ 1\rangle, |1\ 0\rangle, |1\ 1\rangle\}$, the mysterious 'computational basis' within the entanglement of the Minkowski space.

Two-qubits Where and Wherefrom?

What 'complex two-dimensional vector space' are we talking about? Where does it originate? Which considerations lead to the assumption that in the Minkowski algebra there are spinors which represent leptons, fermions and hence the standard model partition of particles? This can only be understood if you read the book including the sections on Majorana space-time spinors and Majorana spinor eigenforms. As we said, such a spinor must satisfy many demands that have been discovered by Cartan, Chevalley, Crumeyrolle and Lounesto. What is a spinor in the geometric algebra of Minkowski space? How does it look like. We gave several examples, such as a red u-quark. Such a particle has to reproduce the primitive idempotent

$$f_{1,3} = \frac{1}{2}(Id - e_1)\frac{1}{2}(Id + e_{24}) \tag{401}$$

Why? Because the matter is complicated, and we have to begin somewhere. May be, we should rather begin with the lepton and then let the quarks follow. Then we could link the first primitive idempotent with the $|\overline{\nu_e}\rangle$ neutrino or alternatively with the electron $|e^-\rangle$. Such could be the structural logic in the game. Therefore we assign densities to the following particles

$f_{1,1} = \rho_{\overline{\nu_e}} = \frac{1}{2}(Id + e_1)\frac{1}{2}(Id + e_{24})$ ... primitive idempotent linked to the $|\overline{\nu_e}\rangle$ or to $|e^-\rangle$ (402)
$f_{1,2} = \rho_s = \frac{1}{2}(1 + e_1)\frac{1}{2}(1 - e_{24})$ ... primitive idempotent linked to the red s-quark
$f_{1,3} = \rho_u = \frac{1}{2}(1 - e_1)\frac{1}{2}(1 + e_{24})$ ... primitive idempotent linked to the red u-quark
$f_{1,4} = \rho_d = \frac{1}{2}(1 - e_1)\frac{1}{2}(1 - e_{24})$ ... primitive idempotent linked to the red d-quark

Now it is essentially unimportant which way we wish to represent these Clifford numbers which replace elementary particles. We can choose the terms of Clifford algebra as they are, or the 4×4-matrices or the polarity strings. Every analysis must lead to the same resulting statements. The primitive idempotents already seem to show us the polarities that will enter the computational basis

$f_{1,1}$ (+)(+) (403)
$f_{1,2}$ (+)(−)
$f_{1,3}$ (−)(+)
$f_{1,4}$ (−)(−)

But the primitive idempotents are not yet the spinors. But we have for the spinors

$$|\overline{\nu_e}\rangle \stackrel{\text{def}}{=} \pm\frac{i}{\sqrt{2}}\varphi_r\,\iota_r = \left(\pm\frac{i}{\sqrt{2}}\right)\frac{1}{2}(e_1 + e_{24})\frac{1}{\sqrt{2}}(e_3 + e_{13}) = \pm\frac{i}{4}(e_3 + e_{13} - e_{234} - e_{1234}) \tag{404}$$
$$|\,d\rangle = \pm\frac{i}{\sqrt{2}}\hat{\varphi_r}\,\iota_r = \left(\pm\frac{i}{\sqrt{2}}\right)\frac{1}{2}(-e_1 + e_{24})\frac{1}{\sqrt{2}}(e_3 + e_{13}) = \pm\frac{i}{4}(-e_3 - e_{13} - e_{234} - j)$$
$$|\,u\rangle = \pm\frac{i}{\sqrt{2}}\hat{\varphi_r}\,\hat{\iota_r} = \left(\pm\frac{i}{\sqrt{2}}\right)\frac{1}{2}(-e_1 + e_{24})\frac{1}{\sqrt{2}}(-e_3 + e_{13}) = \pm\frac{i}{4}(e_3 - e_{13} - e_{234} + j)$$

$$|\, s\rangle = \pm\frac{i}{\sqrt{2}}\varphi_r\, \hat{\iota_r} = \left(\pm\frac{i}{\sqrt{2}}\right)\frac{1}{2}(e_1 + e_{24})\frac{1}{\sqrt{2}}(-e_3 + e_{13}) = \pm\frac{i}{4}(e_3 - e_{13} + e_{234} - j)$$

So we can summarize the algebraic situation by a table in which we can correlate the '2-qubit signature' of the primitive idempotents in the 2nd column with the logic combination of not involuted and grade involuted components φ, ι. We try the following, being aware that might be a mistake:

$$\begin{array}{llll} f_{1,1} & (+)(+) & |\overline{\nu_e}\rangle & \varphi_r\, \iota_r \\ f_{1,2} & (+)(-) & |\, d\rangle & \hat{\varphi_r}\iota_r \\ f_{1,3} & (-)(+) & |\, u\rangle & \hat{\varphi_r}\hat{\iota_r} \\ f_{1,4} & (-)(-) & |\, s\rangle & \varphi_r\, \hat{\iota_r} \end{array} \tag{405}$$

You see, there is a muddle coming in, for we cannot enforce a correlation between $(-)(-)$ and $\hat{\varphi_r}\hat{\iota_r}$. But the $\hat{\varphi_r}\hat{\iota_r}$ migrates from position 4 to 3 which in turn results in an exchange of positions $|\, d\rangle$ with $|\, s\rangle$. Notice that the first concord of $(+)(+)$ with $|\overline{\nu_e}\rangle$ makes us expect that any grade involuted *hat constituent* correlates with a $(-)$. But this is precisely what is not the case. Here we are already confronted with the fundamental braiding operation in the Clifford algebra of the Minkowski space. Therefore it is one of the mind-boggling exercises for students if we ask them to construct a representation of the Clifford algebra $Cl_{3,1}$ where the primitive idempotents $f_{1,1}$, $f_{1,2}$, $f_{1,3}$, $f_{1,4}$ appear as diagonal matrices with the entries $diag[1,0,0,0]$, $diag[0,1,0,0]$, $diag[0,0,1,0]$, $diag[0,0,0,1]$. Take me at my word! Throughout this book I have used the following iterant or vector representation of the first Cartan subalgebra together with the identity Id. Now calculate according to their definition the primitive idempotents $f_{1,1}, f_{1,2}, f_{1,3}, f_{1,4}$. You do not get $[1,0,0,0]$, $[0,1,0,0]$, $[0,0,1,0]$, $[0,0,0,1]$, but instead

$$\begin{array}{ll} f_{1,1} = diag[1,0,0,0] & f_{1,2} = diag[0,0,0,1] \\ f_{1,3} = diag[0,0,1,0] & f_{1,4} = diag[0,1,0,0] \text{ in the matrices} \end{array} \tag{406}$$

Let us go on with the u-quark, in order to see the wonderful entanglement in isospin and angular spin. See, apart from the fact, we may ask why the fermions do not have spin $\pm 3/2$ and $\pm 5/2$ supernumerary to the ±½: It has a spinor composed by the area extender $\varphi_r = \frac{1}{2}(e_1 + e_{24})$, and torsion momentum $\iota_r = \frac{1}{\sqrt{2}}(e_3 + e_{13})$. Now investigate the commutation relations brought in by these little actors. We have

$$\text{spinor } |\, u\rangle = \pm\frac{i}{\sqrt{2}}\hat{\varphi_r}\hat{\iota_r} = \pm\frac{i}{4}(e_3 - e_{13} - e_{234} + j) \text{ u-quark, } u \in \mathbb{C} \otimes Cl_{3,1}$$

with area extender $\varphi_r = \frac{1}{2}(e_1 + e_{24})$, and torsion momentum $\iota_r = \frac{1}{\sqrt{2}}(e_3 + e_{13})$. Now consider the Clifford reverted torsion momentum $\tilde{\iota}_r = \frac{1}{\sqrt{2}}(e_3 - e_{13})$. It is easy to see that the three quantities $\varphi_r = \frac{1}{2}(e_1 + e_{24})$, $\iota_r = \frac{1}{\sqrt{2}}(e_3 + e_{13})$, $\tilde{\iota}_r = \frac{1}{\sqrt{2}}(e_3 - e_{13})$ satisfy the commutation relations of Euclidean 3-space. This analogy is not far-fetched! But it has a very direct meaning. We first want to baptize these Clifford numbers by combining them with the $\pm$sign typical for shift operators.

$$\iota_+ \stackrel{\text{def}}{=} \iota_r = \frac{1}{\sqrt{2}}(e_3 + e_{13}),\ [\varphi_r, \iota_+] = +\iota_+ \left[\tfrac{1}{2}e_1,\ \iota_+\right] = +\iota_+ \tag{407}$$
$$\iota_- \stackrel{\text{def}}{=} \tilde{\iota}_r = \frac{1}{\sqrt{2}}(e_3 - e_{13}),\ [\varphi_r, \iota_-] = -\iota_+ \left[\tfrac{1}{2}e_1,\ \iota_-\right] = -\iota_+$$

It would be enough, if we took $\frac{1}{2}e_1$ as the first component of 'torsion momentum' in order to obtain the same commutation relation, as you see

$$\left[\tfrac{1}{2}e_1,\ \iota_+\right] = +\iota_+\ ,\ \left[\tfrac{1}{2}e_1,\ \iota_-\right] = -\iota_+ \tag{408}$$

The $\frac{1}{2}e_1$ plays the role of the angular momentum operator l_x in quantum mechanics of older times. In Vienna in the 60s there were several specialists who taught us how to handle the mathematics of gamma matrices. The angular momentum algebra, as is shown for instance in Messiah part I, pins down as usual

$$[l_z, l_+] = +l_+ \text{ and} \tag{409}$$
$$[l_z, l_-] = -l_- \text{ with the shift operators } l_+ = \frac{1}{\sqrt{2}}(l_x + il_y),\ l_- = \frac{1}{\sqrt{2}}(l_x - il_y)$$

If you replace l_x by the first Pauli matrix σ_x and l_y by the spin matrix σ_y you have in the terminology of the Pauli algebra

$$\sigma_+ = \frac{1}{\sqrt{2}}(\sigma_x + i\sigma_y) \text{ and} \tag{410}$$
$$\sigma_- = \frac{1}{\sqrt{2}}(\sigma_x - i\sigma_y)$$

Together with a measurement of the third spin component we get $[\sigma_z, \sigma_+] = +\sigma_+$ and $[\sigma_z, \sigma_-] = -\sigma_-$, two relations we are still quite familiar with. The more advanced comes in as soon as we realize that we can replace the imaginary unit by the 'pseudo scalar' $\sigma_{xyz} = \sigma_x\sigma_y\sigma_z$ and since we know that the sigma matrices represent the unit vectors of the Euclidean 3-space, we end up with the equations $\sigma_+ = \frac{1}{\sqrt{2}}(\sigma_x + \sigma_{xyz}\sigma_y), \sigma_- = \frac{1}{\sqrt{2}}(\sigma_x - \sigma_{xyz}\sigma_y)$ which are the same as $\sigma_+ = \frac{1}{\sqrt{2}}(\sigma_x - \sigma_{xz}),\ \sigma_- = \frac{1}{\sqrt{2}}(\sigma_x + \sigma_{xz})$. This is exactly the form of the torsion momentum algebra in the basis of Clifford algebra $Cl_{3,0}$.

$$e_+ = \frac{1}{\sqrt{2}}(e_1 + e_{31})\ \ e_- = \frac{1}{\sqrt{2}}(e_1 - e_{31}) \tag{411}$$
$$[e_3, e_+] = +e_+ \text{ and } [e_3, e_-] = -e_-$$

These are essentially the same equations as in (407), just that the torsion momentum is measured on the first component. Clearly this algebra is preserved under any Weyl rotation acting on the basis. The angular momentum algebra works well, even if we wish to measure the first angular momentum component instead of the third, as is convention. But now comes the clue! This determines the appropriate order of terms. (Note that what follows, the four lines, is still in the wrong order!)

$$
\begin{aligned}
&|\overline{\nu_e}\rangle \overset{\text{def}}{=} \pm\frac{i}{\sqrt{2}}\varphi_r\,\iota_r = \left(\pm\frac{i}{\sqrt{2}}\right)\frac{1}{2}(e_1+e_{24})\frac{1}{\sqrt{2}}(e_3+e_{13}) = \pm\frac{i}{\sqrt{2}}\varphi_r\,\iota_+ \\
&|\,d\rangle = \pm\frac{i}{\sqrt{2}}\hat{\varphi_r}\iota_r = \left(\pm\frac{i}{\sqrt{2}}\right)\frac{1}{2}(-e_1+e_{24})\frac{1}{\sqrt{2}}(e_3+e_{13}) = \pm\frac{i}{\sqrt{2}}\hat{\varphi_r}\iota_+ \\
&|\,u\rangle = \pm\frac{i}{\sqrt{2}}\hat{\varphi_r}\hat{\iota_r} = \left(\pm\frac{i}{\sqrt{2}}\right)\frac{1}{2}(e_1-e_{24})\frac{1}{\sqrt{2}}(e_3-e_{13}) = \pm\frac{i}{\sqrt{2}}\tilde{\varphi}_r\,\iota_- \\
&|\,s\rangle = \pm\frac{i}{\sqrt{2}}\varphi_r\,\hat{\iota_r} == \left(\pm\frac{i}{\sqrt{2}}\right)\frac{1}{2}(e_1+e_{24})\frac{1}{\sqrt{2}}(-e_3+e_{13}) = \mp\frac{i}{\sqrt{2}}\varphi_r\,\iota_-
\end{aligned}
\tag{412}
$$

Every fermion field linked with one of the six Cartan subalgebras in the Minkowski algebra and respectively its $SU(4)$ provides one such 2-qubit Hilbert space. For one color, say, red, the Hermitian elements of the computational basis are

$$
|\overline{\nu_e}\rangle = \begin{bmatrix} 0 & 0 & i & 0 \\ 0 & 0 & 0 & 0 \\ 0 & 0 & 0 & 0 \\ 0 & 0 & 0 & 0 \end{bmatrix}, |d\rangle = \begin{bmatrix} 0 & 0 & 0 & 0 \\ 0 & 0 & 0 & 0 \\ 0 & 0 & 0 & 0 \\ 0 & i & 0 & 0 \end{bmatrix}, |s\rangle = \begin{bmatrix} 0 & 0 & 0 & 0 \\ 0 & 0 & 0 & -i \\ 0 & 0 & 0 & 0 \\ 0 & 0 & 0 & 0 \end{bmatrix},
\quad
|u\rangle = \begin{bmatrix} 0 & 0 & 0 & 0 \\ 0 & 0 & 0 & 0 \\ -i & 0 & 0 & 0 \\ 0 & 0 & 0 & 0 \end{bmatrix}
\tag{413}
$$

Obviously, this pleasant design is resulting from the fact that those spinors reproduce the primitive idempotents of the first color space which we have chosen diagonal in the standard representation.

Theorem 15 claims that a two-qubit state $|\psi\rangle = a|0\ 0\rangle + b|0\ 1\rangle + c|1\ 0\rangle + d|1\ 1\rangle$ turns out to be entangled in case that $a\,d - b\ c \neq 0$. This is in accordance with what we found out in the section 'Freedom of entanglement', Lemma 4. But now we are a little bit more clever as we found out about the 'wherefrom' of the 2-qubit representations. We can use this as an indicator for entanglement if we have the matrix of a state.

$$
\begin{array}{lll}
|\overline{\nu_e}\rangle = 1\,1 & +\ + & \varphi_r\,\iota_r \\
|d\rangle = 0\,1 & +\ - & \hat{\varphi_r}\iota_r \\
|s\rangle = 1\,0 & -\ - & \varphi_r\,\hat{\iota_r} \\
|u\rangle = 0\,0 & -\ + & \hat{\varphi_r}\hat{\iota_r}
\end{array}
\tag{414}
$$

This is now the correct order of terms. Not the sign combination of the constituent idempotents, but the facts of Clifford grade inversion determine about the 2-qubit states.

A non-entangled state in this 2-qubit basis would have the form of a linear superposition of states $|\psi\rangle = a|0\ 0\rangle + b|0\ 1\rangle + c|1\ 0\rangle + d|1\ 1\rangle = a|u\rangle + b|d\rangle + c|s\rangle + d|\overline{\nu_e}\rangle$ provided that $a\,d - b\ c \neq 0$

$$\psi = \begin{bmatrix} 0 & 0 & dI & 0 \\ 0 & 0 & 0 & -cI \\ -aI & 0 & 0 & 0 \\ 0 & bI & 0 & 0 \end{bmatrix}. \tag{415}$$

As we let the 2-qubit entangling element act on this linear superposition, we obtain $R\,\psi$. We have

$R = \frac{1}{\sqrt{2}}(Id - e_{23})$, hence

$$R\,\psi = \frac{1}{\sqrt{2}}\begin{bmatrix} 1 & 0 & 0 & 1 \\ 0 & 1 & -1 & 0 \\ 0 & 1 & 1 & 0 \\ -1 & 0 & 0 & 1 \end{bmatrix}\begin{bmatrix} 0 & 0 & dI & 0 \\ 0 & 0 & 0 & -cI \\ -aI & 0 & 0 & 0 \\ 0 & bI & 0 & 0 \end{bmatrix} = \frac{1}{\sqrt{2}}\begin{bmatrix} 0 & bI & dI & 0 \\ aI & 0 & 0 & -cI \\ -aI & 0 & 0 & -cI \\ 0 & bI & -dI & 0 \end{bmatrix} \tag{416}$$

This signifies the matrix of an entangled operator, as it cannot be represented in the linear basis (413). So, - under the assumption that this proof is reliable in the other Cartan subalgebras too, we have shown that Bell's 'change of basis matrix' is a universal gate for all two-qubit spinor spaces in the whole Clifford algebra of the Minkowski space. The power of this statement is not to be underestimated. After all there are three isomorphic operators of this form that contain exactly one spatial bivector with square $-Id$. One of these is badly needed. For nature, as I often said, it makes no difference. She performs every transposition possible in the geometric basis, or in other words, she allows for every conceivable path and flip when she makes the things we are confronted with. To say it otherwise, we need the electron for the anti-neutrino that we already have. In order to have this field present, a swap is needed. And this swap can be written down without hesitation. If the electron is located opposite the neutrino in the seal, it should belong to the corner having a primitive idempotent

$$f_{4,1} = \rho_{\overline{\nu_e}} = \frac{1}{2}(Id + e_2)\frac{1}{2}(Id + e_{14})$$

This is comparatively simple a situation where a mere exchange of e_1 with e_2 is required. Such an exchange is indeed brought forth by the Clifford number

$$R' = \frac{1}{\sqrt{2}}(-e_{13} + e_{23}) \tag{417}$$

We have $(R')^{-1} = \widetilde{R}'$ and

$$\widetilde{R}' = \frac{1}{\sqrt{2}}(+e_{13} - e_{23}) \tag{418}$$

These two bivectors perform the swaps of locations between $|\overline{\nu_e}\rangle$ and the $|e^-\rangle$

$$\widetilde{R}'\, f_{1,1}\, R' = f_{4,1} \tag{419}$$

$$R' = \frac{1}{\sqrt{2}}\begin{bmatrix} 0 & 0 & -1 & -1 \\ 0 & 0 & +1 & -1 \\ +1 & -1 & 0 & 0 \\ +1 & +1 & 0 & 0 \end{bmatrix}, \; \widetilde{R'} = -R' = \frac{1}{\sqrt{2}}\begin{bmatrix} 0 & 0 & +1 & +1 \\ 0 & 0 & -1 & +1 \\ -1 & +1 & 0 & 0 \\ -1 & -1 & 0 & 0 \end{bmatrix} \tag{420}$$

R' is so to say close friend with braids. It is also a matrix that entangles the basis (413). Applying the swap, we land in the seal at the high energy fermions' corner with possibly these 'particles':

$$\begin{aligned}
|e^-\rangle &\stackrel{\text{def}}{=} \pm\frac{i}{\sqrt{2}}\varphi_r\,\iota_r = \left(\pm\frac{i}{\sqrt{2}}\right)\frac{1}{2}(e_2+e_{14})\frac{1}{\sqrt{2}}(e_3+e_{23}) = \pm\frac{i}{\sqrt{2}}\varphi_r\,\iota_+ \\
|\,c\rangle &= \pm\frac{i}{\sqrt{2}}\hat{\varphi}_r\iota_r = \left(\pm\frac{i}{\sqrt{2}}\right)\frac{1}{2}(-e_2+e_{14})\frac{1}{\sqrt{2}}(e_3+e_{23}) = \pm\frac{i}{\sqrt{2}}\hat{\varphi}_r\iota_+ \\
|\,b\rangle &= \pm\frac{i}{\sqrt{2}}\hat{\varphi}_r\hat{\iota}_r = \left(\pm\frac{i}{\sqrt{2}}\right)\frac{1}{2}(e_2-e_{14})\frac{1}{\sqrt{2}}(e_3-e_{23}) = \pm\frac{i}{\sqrt{2}}\tilde{\varphi}_r\,\iota_- \\
|\,t\rangle &= \pm\frac{i}{\sqrt{2}}\varphi_r\,\hat{\iota}_r == \left(\pm\frac{i}{\sqrt{2}}\right)\frac{1}{2}(e_2+e_{14})\frac{1}{\sqrt{2}}(-e_3+e_{23}) = \mp\frac{i}{\sqrt{2}}\varphi_r\,\iota_-
\end{aligned} \tag{421}$$

Chapter 8

MOTION AND METHOD

The swaps and rotations that entangle two-qubit subspaces of Minkowskian space-time algebra are systematic, are uncounted and innumerable. Entangled pairs are not exceptions that we have created in our laboratories. But entanglement is the rule of nature. The pattern of mutual dependence is yet so complex and sensible to distortion that we usually overlook what was and what is entangled and what is not. It is a great achievement of some very persevering physicists to produce entangled quantum states that follow their nature and go beyond space and time. It requires a lot of intuition and philosophy to sustain the obstinacy that allows us to find and to realize what we are after. As for my own work, it focuses on the mathematics of those phenomena which are so difficult to be understood: the standard model of HEPhy and the entanglement of fields. The young ones among us, the ones who are busy with categories and who are avoiding frames of any sort, may reproach me for being a late follower of Paul Dirac who tried to legitimate, - or would we say "to rescue" – the Hamiltonian method, as I am demonstrating my conviction throughout all my books that we can learn something from the Minkowski algebra. I say firmly that geometric algebra of space and time teaches us a very important lesson, I say so almost resolutely, and I keep a distance to all those fantastic friends of mine who are almost convinced to have refuted Einstein and disproved relativity. But I do so in a strong and emphatic relation to those young among us who have turned their attention towards quantum informatics, to those who say frankly to me: Bernd, what is a Clifford algebra good for in physics? I love their protest, but I pity their rapidity.

Nature has an unbelievable nature. For cognition she is hard to grasp. Nature is both extension and thought. Her intelligence reaches far beyond everything human thought can ever conceive. This is not a philosophic statement. It is a fact. The smallest parts of nature are in permanent motion, they jump from here to there, exchange forces beyond space and time, communicating from location to location, while no one of us knows where these locations actually are, since nature makes the metric, but she does not have it! When the writer of this book designs king Solomon's seal, he does so because of an insight into the facts of life. He is not just following a Jewish belief, - although he is sure a friend of those old beliefs. The abelian subalgebras of the Clifford algebra of the Minkowski space in the opposite Lorentz metric – and here I do not feel ashamed or browbeaten by those great names, which we since long denote after Elié Cartan, form a seal. That seal is a mathematical matter. It belongs to the symmetric unitary $SU(4)$ and its uncountable appearances in the Clifford algebra $Cl_{3,1}$. The seal is a template for motion. It can be ignored and it can be circumvented by quantum motion, but only with the effect of reproducing it. We spoke about quantum jumps and we still do (Wiseman and Gambetta 2012) – and, what indeed – the flickering quanta reproduce the 'template', if we like it or not. We are living in a material world. Time is constantly turned into space. "So all those procedures with the aid of which we erase time deserve our attention" wrote Erhart Kästner "not frighten away, indeed, not ignore, but really annihilate. [..] Fall into the landscape: pure delight, time goes out".

Realizing the template without realizing it, by just following the laws of touch and interaction – oh, what a word sequence Paul Dirac had coined then: 'infinitesimal contact transformation with a generating function' – the polarity strings build up a seal and a space and a time. What a curious enterprise! Where is the motion? Everywhere! Everywhere in the space and time algebra! You cannot pick out a single bit string. Does it follow Lagrange? "We have a time-variable t occurring already as soon as we introduce the Lagrangian." (Dirac 1964, p. 7) So, how can I introduce a Lagrangian if I want to explain the emergence of time? I know there is time. But it is not just 'there'. It can be reverted any time. It has the power to vanish. Time can be annihilated. So my approach cannot be based on the introduction of Lagrangian dynamics. Dirac must have suffered when he realized the proportions of 'the quite serious difficulties' in setting up an accurate quantum theory, 'difficulties of a fundamental character which people have been worrying over for quite a number of years'. And he begins with that gentle reproach on page 3: "Some people are so much impressed by the difficulties of passing over from Hamiltonian classical mechanics to quantum mechanics that they think that maybe the whole method of working from Hamiltonian classical theory is a bad method". But Dirac appreciates that those people had made "quite considerable progress" by setting up alternative methods for getting more comprehensive quantum field theories. He himself remains very determined as to how he wants to proceed. "To get a general formalism which will be applicable, for example, to the nonlinear electrodynamics which I mentioned previously, I don't think one can in any way shortcut the route of starting with an action integral, getting a Lagrangian, passing from the Lagrangian to the Hamiltonian, and then passing from the Hamiltonian to the quantum theory. That is the route which I want to discuss in this course of lectures". He confirms his intent many a time. "Let us now develop this Lagrangian dynamics" [...] But he is aware of the difficulties. He is not secure at all, and he captures the means which every radical constructivist would take hold of. He introduces '*primary constraints*' of the Hamiltonian formalism. He decides, it is convenient to introduce the Poisson bracket formalism, to obtain symmetric and linear terms, and the Jacobi identity that connects three terms. He realizes the Hamiltonian cannot be uniquely determined. He coins the denotation of *weak equations*; and the *total Hamiltonian* to compensate the ambiguities. But it follows from all the order he brings into the formalism that the "Lagrangian equations of motion lead to the inconsistency 1 = 0". Dirac proves an innocent spirit. "We cannot take the Lagrangian to be completely arbitrary. We must impose on it the condition that the Lagrangian equations of motion do not involve an inconsistency". Then he introduces '*secondary constraints*' on the Hamiltonian variables [...] "together with a number of conditions", as he says. He cannot halt here, but he defines "any variable, R, a function of the q's and the p's, to be *first-class* if it has zero Poisson brackets with all the ϕ's: $[R, \phi_j] \approx 0,\ j = 1, \ldots, J$." Dirac cares a lot about every unwanted negligibility which enters his formalism. ... "Otherwise R is *second-class*. If R is first-class, then $[R, \phi_j]$ has to be *strongly equal* to some linear function ..." [...] and then "The whole thing vanishes weakly." Those are the words of a radical constructivist. "Thus (1-33) gives the *total Hamiltonian* in terms of a *first-class Hamiltonian* H' together with some *first-class* ϕ's" (pp. 2ff to 19). People like statements such as these, even today. Dirac convinced me, he is a human being and a brilliant thinker in mathematical physics. But he himself was not convinced of the penetrating power of his work. In his last pages he says once more: "The difficulties are quite serious, and they have led some physicists to challenge the whole Hamiltonian method. A good many physicists are now working on the problem of trying to set up a quantum field theory independently of any Hamiltonian. [...] their hope is that ultimately they will get enough conditions imposed on these quantities of physical importance to be able to calculate them. They are still very far from achieving that end, and my own belief is that it will not be possible to dispense entirely with the Hamiltonian method. The Hamiltonian method dominates mechanics from the classical point of view". His last words in these lectures: "When we go over to quantization we have the difficulties arising which I have discussed". Yet I remember, when I was young I had a beloved teacher, Walter Thirring. Thirring established,- we scholars were sure, the quantum field theory without ever neglecting the Hamiltonian. He always first considered the Lagrangian. This potential of our teacher to factor in the wisdom of many a school, escorted me from the first of Thirring's lectures I have heard, until to the last letter he wrote me. Consider a Lagrangian ... and then he showed that a mix between Minkowskian and Euclidean space-time is possible. Nature has more ways than we are able to conceive.

Dirac defined "any dynamical variable, R, a function of the q's and p's, to be first class" under special conditions, and he considers four different kinds of constraints, namely "first class and second class, which is quite independent of the division into primary and secondary". "The initial variables which we need are the q's and p's [...] These initial conditions describe what physicists would call the initial physical state of the system [...]; but "that means that the state does not uniquely determine a set of q's and p's, even though a set of q's and p's uniquely determines a state. So we have the problem of looking for all the sets of q's and p's that correspond to one particular physical state" (p. 20). That leads to the 'infinitesimal contact transformations'. Like Feynman and Schrödinger, so Dirac is attentive to the entirety of the degrees of freedom of motion.

Nature takes every possible freedom to move. When a step is made, locations of any degree may be carried to other locations. What moves is the information stored in a polarity string. The features of locations are trans-locally communicated. In her innermost nature does not differ between line elements and areas, areas and volumes. But while information is exchanged, such volumes, such areas, such lines are being constituted. Nature makes orientation and metric. Such is the process. She does not have it from the beginning. But she is free to make it for us. The making of a step ultimately leads us back to Hamilton's mathematics of 'Pure time', and leads us forward to the iterant algebras. Louis Kauffman and me, at many occasions, took polarity strings as basic elements of interaction and motion. Thus what we call 'physical states' is constituted by the motion of such iterants which have the power to move not only themselves, but also the others by their potential to interact in terms of what we may call 'moving logic'. This approach is promising, as it allows for a discrete simulation of the path integral for any field, and no particle field that we know of today is excluded from that approach. We need at most two imaginary units and the real Clifford algebra $Cl_{3,1}$ to represent all particles we know of nowadays, and it is possible to establish a correspondence between such a geometric algebra and the iterant algebra.

One of the great surprises is the appearance of entanglement as the natural endowment of the geometric algebra of Minkowski space. Though our quantum erasers will carry on reversing the course of history, elementary particles in the large colliders will not soon be able to move faster than light. But light itself does not care much about it, as it has the potential to link material events beyond the limits of time. Photons are everywhere before their location becomes apparent. Also they do not have structural limitations. They penetrate, it seems, every subnuclear domain without hesitation. Probably the Higgs boson does not care much about the fact if we were able to observe it in a state of quantization or not quantized. The photons let us quarrel.

Consider at last the spinor of an electron $|e^-\rangle = \pm\frac{i}{4}(e_2 + e_{14})(e_3 + e_{23})$ as it travels through some braid of the space-time. This was in equations (381 to 384). It experienced a counter clockwise $\pi/3$-turn that was followed by a second braid generator of B_3. So we obtained the 'braided electron spinor' (384)

$$\text{(i)}\ \ \Psi(h,g) = g^{-1}\psi(h)\,g = i\Big(-\tfrac{1}{9}e_1 - \tfrac{1}{36}e_2 + \tfrac{2}{9}e_3 - \tfrac{1}{36}e_{12} - \tfrac{2}{9}e_{13} - \tfrac{1}{9}e_{23} - \tag{422}$$

$$-\tfrac{1}{9}e_{124} + \tfrac{1}{9}e_{134} - \tfrac{7}{36}e_{234} + \tfrac{1}{4}e_{1234}\Big)$$

This spinor has a grade involuted

$$\text{(ii)}\ \ \widehat{\Psi}(h,g) = g^{-1}\psi(h)\,g = i\Big(+\tfrac{1}{9}e_1 + \tfrac{1}{36}e_2 - \tfrac{2}{9}e_3 - \tfrac{1}{36}e_{12} - \tfrac{2}{9}e_{13} - \tfrac{1}{9}e_{23} -$$

$$+\tfrac{1}{9}e_{124} - \tfrac{1}{9}e_{134} + \tfrac{7}{36}e_{234} + \tfrac{1}{4}e_{1234}\Big)$$

We know since long that these spinors can engross the whole 16-dimensional Clifford space, but their product, the idempotent (362) stays within the boundaries of the 10-dimensional space of the seal as You can see from the expression for $\Psi(h,g)\widehat{\Psi}(h,g)$

$$F(h,g) = \Psi(h,g)\widehat{\Psi}(h,g) = \tfrac{1}{4}Id - \tfrac{7}{36}e_1 - \tfrac{1}{9}e_2 - \tfrac{1}{9}e_3 - \tfrac{1}{9}e_{14} + \tfrac{2}{9}e_{24} - \tfrac{1}{36}e_{34} - \tag{423}$$
$$-\tfrac{2}{9}e_{124} - \tfrac{1}{36}e_{134} + \tfrac{1}{9}e_{234}$$

By now we can interpret $\Psi(h,g)$ and $\widehat{\Psi}(h,g)$ as waves and as phase gates. As waves it is obvious that they represent advanced and retarded waves. That is, one part of the electron's field comes exactly from the opposite direction than the other part. But both together bring forth the density of a single electron.

$$\Psi(h,g) = g^{-1}\psi(h)\,g = i\Big(-\tfrac{1}{9}e_1 - \tfrac{1}{36}e_2 + \tfrac{2}{9}e_3 - \ .\ .\ .$$
$$\widehat{\Psi}(h,g) = g^{-1}\psi(h)\,g = i\Big(+\tfrac{1}{9}e_1 + \tfrac{1}{36}e_2 - \tfrac{2}{9}e_3 - \ .\ .\ .$$

We should acknowledge that any slight asymmetry will determine about if and where the electron 'quantizes', and shows us its presence. This interpretation of an electron field as a product of advanced and retarded field from opposite locations is by no means far-fetched or artificial. But it is a natural consequence of our assumption that iterating fields are real active elements in the process that constitutes space, time and particles. Now, recall Dirac saying "the Hamiltonian method dominates mechanics from the classical point of view" (p. 86). A very similar statement is holding for the present approach. I would say: as long as I am convinced that the laws of relativity in the Minkowski algebra can tell us something about quantum motion, the method of geometric algebra dominates mechanics from a viewpoint of 'classical relativity'. This is actually the case, no doubt. But what we learn about quantum motion is much more than what we lose by the domination. After all, the method is not only one of mathematical selectedness, but it has deep sociological roots, as it comes up to expectations of Cartan, Chevalley, Crumeyrolle, Hamilton, Lounesto, Chisholm and the whole field of Clifford algebra. As sociologists say, all the knowledge can only be found in the whole of society.

It is an obvious advantage that such extension of Dirac's spinor theory, in correspondence with path integrals involving advanced and retarded waves, correctly plays back the annihilation properties. For consider the electron spinor in its minimal standard form of representation

$$|e^-\rangle = \pm\frac{i}{4}(e_2 + e_{14})(e_3 + e_{23}) \text{ and } \langle e^-| \text{ and } |e^-\rangle^\wedge \text{ its grade involuted}$$

and the positron, - here I have avoided the introduction of a further imaginary unit, although this could be a way out of formal difficulties, if there were any.

$$|e^+\rangle = \pm\frac{1}{4}(e_2 + e_{14})(e_3 + e_{23}) \text{ and } \langle e^+| = |e^+\rangle^\wedge$$

Each spinor reproduces the density of the particle, either $+f_{4,1}$ for the $|e^-\rangle$ and $-f_{4,1}$ for the $|e^+\rangle$. Further we have mutual annihilation for electron-positron encounters. We can draw this diagram:

$|e^-\rangle\,|e^+\rangle$

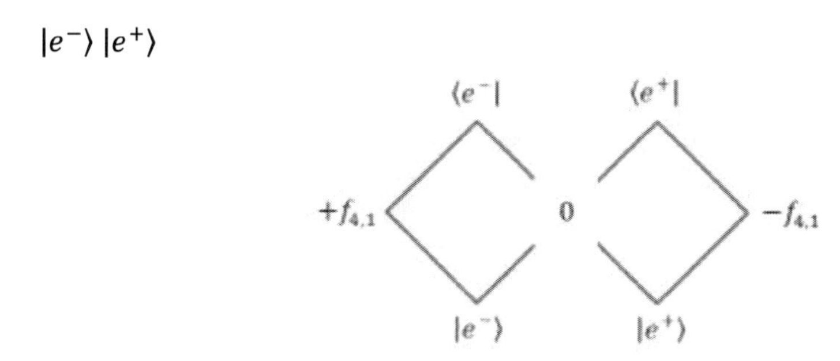

Figure 63. Pair annihilation diagram for electrons.

For the s-quark, e. g., the story is the same. We chose the red one. See, just the symbols differ.

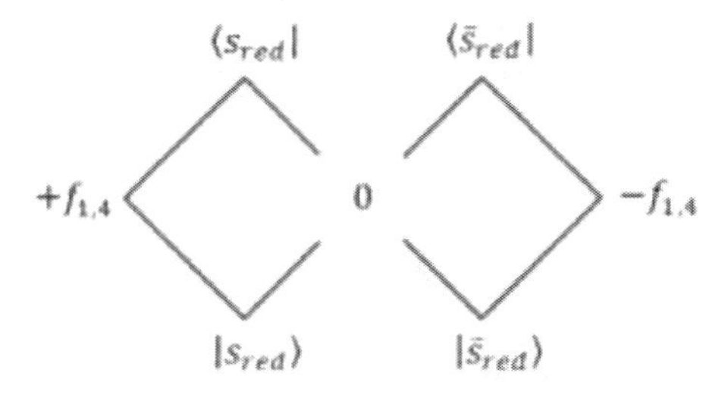

Figure 64. Pair annihilation diagram for red/antired s-(anti)quarks.

$$| s\rangle = \pm\frac{i}{\sqrt{2}}\varphi_r \hat{\iota_r} == \left(\pm\frac{i}{\sqrt{2}}\right)\frac{1}{2}(e_1 + e_{24})\frac{1}{\sqrt{2}}(-e_3 + e_{13}) = \mp\frac{i}{\sqrt{2}}\varphi_r \iota_-$$
$$| \bar{s}\rangle = \pm\frac{1}{\sqrt{2}}\varphi_r \hat{\iota_r} == \left(\pm\frac{1}{\sqrt{2}}\right)\frac{1}{2}(e_1 + e_{24})\frac{1}{\sqrt{2}}(-e_3 + e_{13}) = \mp\frac{1}{\sqrt{2}}\varphi_r \iota_-$$

Recapitulating a 'green t-quark'

$$| t\rangle = \pm\frac{i}{\sqrt{2}}\phi_{gr} \hat{\eta_{gr}} = \pm\frac{i}{4}(-e_1 + e_{13} + e_{124} + j) \tag{424}$$

with area extender $\phi_{green} = \frac{1}{2}(e_3 + e_{24})$ and torsion momentum $\eta_+ = \frac{1}{\sqrt{2}}(e_1 - e_{13})$. It has a diagram

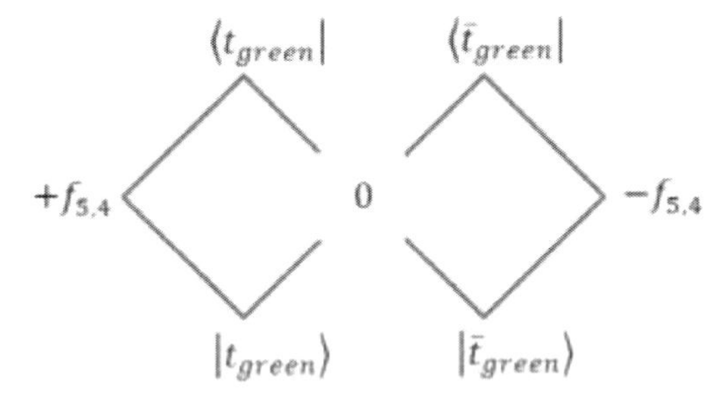

Figure 65. Pair annihilation diagram for green/antigreen t-(anti)quarks.

As all the space-time spinors have the same algebraic form, their diagrams are all the same. They bring about the positive and negative energy densities of fermions and anti-fermions, in accordance with the structure of the seal. The large differences of amplitudes and masses we are observing are the result of a most vigilant symmetry breaking the dynamics of which is still largely unknown to physicists. There is however a showy difference of phenomenology – we are already familiar which since long – a most conspicuous difference in behavior of fermions and photons and respectively bosons with respect to annihilation. That is a peculiar difference that creates a partition in the whole geometric algebra and hence in the dynamics of the system.

Chapter 9

ENVISIONED MEMORY

Fermions, as we know them, have a peculiar annihilation property. When an electron collides with a positron the two create two photons by annihilating each other. Structurally this fact is expressed by the diagram in figure 63. Formally it means in the first instance that in fermion space

$$|e^{-}\rangle|e^{+}\rangle = 0 \tag{425}$$

So, besides the capability of these advanced and retarded spinors to create a positive or negative energy density

$$\begin{aligned} |e^{-}\rangle\langle e^{-}| &= \rho(e^{-}) = +f_{4,1} \\ |e^{+}\rangle\langle e^{+}| &= \rho(e^{+}) = -f_{4,1} \end{aligned} \tag{426}$$

they have the additional potential to annihilate their fermion energy. But the peculiar division between fermion and boson energy in the geometric algebra leaves us alone with another mystery. Namely, bosons and photons can so to say hide themselves in a fermion field. We observe this as absorption. As we know, photons cannot annihilate themselves in the same way fermions are doing that. But they have the capability to extinguish their amplitudes through opposite polarization. Let us inquire how this goes. Consider a photon travelling in direction of, say e_1. It has some dynamic polarization, and a wave function in the even subalgebra:

$$\langle\approx\rangle := ae_{12} + be_{13} + ce_{23} \quad \text{for example realized by the function} \tag{427}$$

$$\langle\approx\rangle = e^{ikx}e_{12} + e^{i\omega t}e_{13} + e^{i\omega t}e_{23} \tag{428}$$

I've used here the three wave symbol to indicate that the photon field is a wave in space and time and at the same instant one of two generators of a 3-strand braid group B_3. The field experiences some circular polarization in space with circle frequency ω, and the oscillation involves all directed spatial areas. By fortune this photon is a pure quaternion which, for the present, has no special significance. What matters is that we are here iterating polarized space areas. The oscillation involves not only polarities in a string at some unspecified domain but also, and we should make that explicit, a rather complicated oscillation of polarized directed areas in some location, though the element takes the simple form of a linear iterant or polarized string. The string is a linear arrangement, but represents areas. The wave $\langle\approx\rangle$ possesses an inverse which is given by the Clifford reversion and some normalization factor, namely, we have

$$\langle\approx\rangle^{\sim} := -nae_{12} - nbe_{13} - nce_{23} \quad \text{with } n = \frac{1}{a^2+b^2+c^2} \tag{429}$$

In the above special case we have the photon field

$$\langle\approx\rangle := (\cos kx)e_{12} + (\cos\omega t)(e_{13}+e_{23}) + i(\sin kx)e_{12} + i(\sin\omega t)(e_{13}+e_{23}) \tag{430}$$
$$\langle\approx\rangle^{\sim} := -n(\cos kx)e_{12} - n(\cos\omega t)(e_{13}+e_{23}) - n\,i(\sin kx)e_{12} - n\,i(\sin\omega t)(e_{13}+e_{23})$$

We have in the Clifford algebra $Cl_{3,1}$ that $\langle\approx\rangle^{-1} = \langle\approx\rangle^{\sim}$ hence the identity

$$\langle\approx\rangle\langle\approx\rangle^{\sim} = Id \tag{431}$$

This identity parallels the fermion annihilation form (425). Fortunately the oppositely polarized photon fields do not leave a zero, but an identity. The annihilated photons do not touch the space-time structure as is brought forth by the fermion interactions. We must understand that a photon that is absorbed in a fermion field creates a new field.

$$|\Psi\rangle = |e^-\rangle\langle\approx\rangle^{-1} \tag{432}$$

Saying that a photon is absorbed by an electron turning this into a new field denoted by $|\Psi\rangle$. Because of equation (431) this is the same as

$$|e^-\rangle = |\Psi\rangle\langle\approx\rangle \tag{433}$$

saying that the field $|\Psi\rangle$ emits a photon $\langle\approx\rangle$ thereby reproducing the electron. This is the whole trick, and it is the only formal possibility to describe this event unless one constructs a design for all possible paths and factors in the coupling constants. The fermion partition of the Minkowski algebra is so strangely divided from the even subspaces of bosons that one sometimes gets the feeling that our whole hocus-pocus about paths integrals and coupling constants, charges, hypercharges and all the rest of it is but an indication that we have to take great pains over the elimination of that interval. There is a broad hint that photons and bosons interact with the spinor world of fermions in a way which we cannot fully understand so as if there was an invisible actor that has to carry out decisions and realize activities that are beyond our cognitive formalism. But it is possible to comprehend a little of what is going on.

We assume that $a^2+b^2+c^2=1$, hence $n=1$: The electron that has absorbed that photon $\langle\approx\rangle$ turns into a field

$$\Psi = \frac{i}{4}(b\,(Id + e_2 + e_{14} - e_{124}) + c(e_1 - e_4 - e_{12} - e_{24}) + \tag{434}$$

$+a(e_{13} + e_{34} - e_{123} + e_{234}))$

This field which is calculated by formula (432) is not identical with an electron space-time spinor, but interestingly it has a component proportional to the density of the electron. To a certain extent the photon can substitute the retarded spinor-component of the electron. Just recall, we have $|e^-\rangle\langle e^-| = \rho(e^-) = +f_{4,1} = (Id + e_2 + e_{14} - e_{124})/4$. The second and the third term having factors c (not to be confused with light speed) and respectively a from the photon amplitudes, are nilpotents. They too are not fermion spinors since multiplied with their gradeinverses remain nil. The Ψ has the potential to recover the electron spinor by emitting a photon. It is due to this strange transformation that the wave-function Ψ discloses a property very close to idempotence, namely we have

$$|\Psi\rangle|\Psi\rangle = ib|\Psi\rangle \text{ and indeed} \tag{435}$$
$$|\Psi\rangle\langle\approx\rangle = |e^-\rangle$$

This mechanism seems to be behind every absorption of a photon or boson by a fermion-field whatever may be the precise form of that photon or boson and fermion. The polarization properties determine the transformation process experienced by the fermion. Now it is interesting to go a little deeper into this phenomenon of pair creation, the Laser induced Breit-Wheeler process with which we began the discourse. Suppose a photon penetrates both the advanced and the retarded polarity string of an electron. What is going to happen?

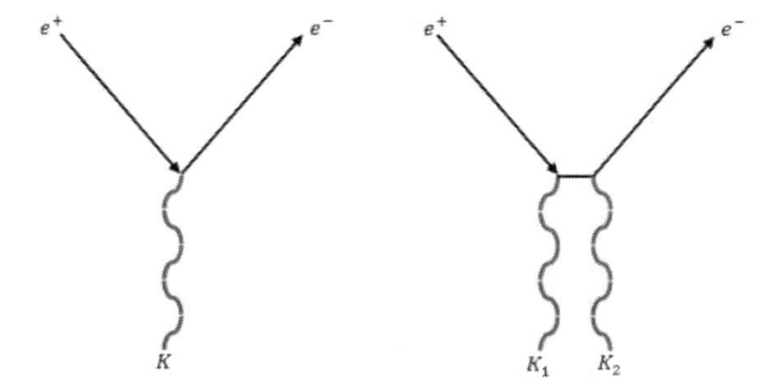

Such penetration of the advanced and retarded electron spinor components leads to an encounter of the following kind

$\langle\approx\rangle\ \langle\approx\rangle$

$|e^+\rangle \qquad \langle e^+|$

$|e^-\rangle\langle e^-|$

$+A^2 f_{4,1}$

$A^2 = a^2 + b^2 + c^2$

Photons, e. g. $\langle\approx\rangle = e^{ikx}e_{12} + e^{i\omega t}e_{13} + e^{i\omega t}e_{23}$ interact with both the advanced and the retarded component of the electron and create a positron environment with the structural positron density of the negative primitive idempotent $-f_{4,1}$ but with the amplitude $a^2 + b^2 + c^2$ borrowed from the photon field. These photons interact in the same way with the positron thereby creating an electron environment with the typical positive density $+f_{4,1}$ and the analogous amplitude brought forth by photons. We saw photons have the power to vanish and reappear and to recreate electrons from a nirvana which is the product of the interaction between photon and electron. This belongs to the bad character of light that it is able to cut itself off the flow of time. It gives matter a shower of eternity.

In theorem 14 we found to every rotation of color $r = e^{\alpha e_{12}+\beta e_{23}+\gamma e_{13}}$ in the quaternion subspace of bi-paravectors there is a continuous manifold of braid generators s such that the braid relation $rsr = srs$ is satisfied. The general solution which involves all free parameters has been found. There are three non-trivial solutions involving many terms. Among these there is a special solution we should scrutinize a little while, for it discloses to us a bit of the mystery of time. Kauffman and Lomonaco realized the braid relation $ghg = hgh$ using pure quaternions (365), that is,

$$g = a + bu \text{ and } h = c + dv$$

and found the non-trivial solution (369) where the dot product satisfies

$$u.v = \frac{a^2-b^2}{2b^2}$$

Let

$$g = e^{i\theta} = a + bi \quad \text{with } a = \cos\theta \text{ and } b = \sin\theta$$

further

$$h = a + b\big((c^2 - s^2)i + 2csk\big) \text{ with}$$
$$c^2 + s^2 = 1,$$
$$c^2 - s^2 = \frac{a^2-b^2}{2b^2}.$$

Then the elements g and h can be rewritten in matrix form as G and H with a phase gate

$$G = \begin{bmatrix} e^{i\theta} & 0 \\ 0 & e^{-i\theta} \end{bmatrix}$$

and a braiding operator $H = FGF^{\dagger}$ derived by conjugation with the unitary matrix

$$F = \begin{bmatrix} ic & is \\ is & -ic \end{bmatrix} \in SU(2)$$

This formalism offers the opportunity for a generalization to four strands and the motion group $SU(3)$, since 3 is a Fibonacci number. Obviously, the generalization seems to work for $n =$ Fibonacci. The inventors say, this most simple example of a topological theory of quantum entanglement is given by numbers

$$\begin{aligned} g &:= e^{7\pi i/10} \\ f &:= i\tau + k\sqrt{\tau} \\ h &:= fgf^{-1} \end{aligned} \tag{436}$$

where $\tau^2 + \tau = 1$. Then g, h satisfy the generating relations for the 3-strand braid group B_3 'dense in the $SU(2)$' as Kauffman accentuates. He denoted this as the '*Fibonacci* representation of B_3 to $SU(2)$'. (Kauffman and Lomonaco 2009, ch. 6). Those numbers g, f, h can and should be conceived as Clifford numbers in the Minkowski algebra as we have done. Yet, the following I am doing in the coming pages is quite speculative. We are asking: how do bit strings remember what they experienced in the course of their interaction? Of course a single string cannot store its whole history if it has the simple structure of iterants. But perhaps there are other ways for polarity strings which make it reasonable to accept that the history of interaction is somehow stored in the present arrangements. To put it simple, matter might be a quantum computer. An intuitive impulse of mine would offer some outlook to the fractal integrative power of the golden rabbit series, if I may say so. I do not want to make things too complicated, but I would like us to open our minds towards this perspective, as the golden series has strong ties with iteration and, without doubt, has deep topological meaning. Now I would like you to let me loosen my speech even more and tell you a little bit about my impression as to what is going on in mathematical physics at times. Sure many of us are suffering a little because of the dogmatic character of the mainstream science and toughness of the big mind where fundamental concepts are concerned. So, some have compiled new mathematical ideas. They brought these into resonance with basic processes in physics and tried to configure a new whole. I must confess that some of us do this with great devotion, so one cannot say their work would be superficial. But sometimes it turns out really difficult to bring the right bricks together, and we are using too many great names for rather elementary things. Proceeding thus, we are creating impressions that bring about feelings of hope, of power and sometimes even leverage or breakthrough, which are just as illusory as the dogmas we have created during the last hundred years. We are using catch words like Majorana fermions, universal quantum gates, Temperley-Lieb recoupling theory, quantum universal Fibonacci model, the colored Jones polynomials, the Fibonacci recoupling model, some appealingly, some legitimately, so to say legitimated by history, like the 'Majorana fermion', some not yet utterly socialized, others for mere promotion purposes. It's okay. But let me have a more accurate left eye on the 'Majorana fermion', a denotation I am using, and a right eye on the 'quantum universal Fibonacci model' which indeed leaves the impression of a great invention, and I will actually reflect on it in this section in some purposive way. We shall ponder over energy and entropy. Let us speculate!

Dear Readers, let a baby rabbit be zero, 0. It stays 0 for some weeks, and then becomes a unit rabbit, a 1, say, a mature rabbit.[1] Therefore we have a rule

[1] The author is not 'only' a mathematical physicist, but he holds his venia docendi in sociology. You can ponder over the various aspects and viewpoints that come to his mind when he writes those sentences.

$$0 \longrightarrow 1$$

for this one rabbit that we identified, - and the rule has to be satisfied within some statistical period of time,- otherwise the rabbit will be too old to be part of the game, but that's another story. Therefore, suppose the mature rabbit also 'generates' a new rabbit, then we have the rule

$$1 \longrightarrow 10$$

which enlarges the bit string a little. You must forget that mathematicians have forgotten to ask if the 1 was a father rabbit or a mother rabbit. So, seeing apart from this fact, we have two rules altogether

$$\begin{aligned} 0 &\longrightarrow 1 \\ 1 &\longrightarrow 10 \end{aligned} \tag{437}$$

Thus, the following lines I am writing down are generations of rabbits, and each generation (line) produces the following bit string (line)

1
10
101
10110
10110101
1011010110110
101101011011010110101
...

May be that this wonderful pattern would look quite different, if mathematicians would have had the nerve to factor in the mother rabbits

11
0110
101101
011010110
1010110110110101
01101101011011011010110110
10101101011011010110110110101101101011010110101
......

But we must be satisfied with what we get. What we get is what we have. The first rabbit seems to be deathless, and all the others seem to follow him undyingly into eternal mathematical life. Strange, but not so bad. One can cut off some generations. That wouldn't alter the golden string. Clearly, there must be some survival rules. There must be a ban that rabbits eat rabbits.

At about the same time when I 'reconstructed' the standard model of High Energy Physics, Chyi-Lung Lin from the Department of Physics at Soochow University in Taipei, Taiwan, published a small masterpiece on Generalized Fibonacci Sequences (and the Triangular Map). From that work one can learn how things go. May be, we should reflect about the '$\tau^2 + \tau = 1$' from above. "The Fibonacci numbers correspond to the choices $F(1) = 1$ and $F(2) = 1$". And I was delighted that Chyi-Lung Lin was aware of the choice.

May be, most of us would prefer $F(2) = 2$. But Lin is right. We can go on with the series by calculating $F(n) = F(n-1) + F(n-2)$. So we get $1, 1, 2, 3, 5, 8, 13, \ldots$, to which we can assign a limit ratio X. Lin stresses the identity $\equiv$. That is something great, he expresses a human feeling: this limit ratio is 'really' equal to this number X. What a surprise! This number is $(1+\sqrt{5})/2$, "a number well known in the golden section as the golden mean or golden ratio". This number "satisfies the equation $X = 1 + X^{-1}$". Yes, it satisfies $X^2 - X - 1 = 0$ which has two roots, and if the positive one is chosen, we have the golden mean or golden ratio. I am using quotation marks not to make a distance to Lin, but to dwell and to annotate that I am thinking in the same way Chyi-Lung Lin is or was thinking. What does all this have to do with Clifford? I am not sure, William Kingdon Clifford might have known nothing about it, though I doubt that, but Clifford algebra as we have worked out till today, after the poor guy had died much too soon, has to do with the Fibonacci. Namely, Lin has shown that we do not just have a Fibonacci that gains its actual value from the two previous ones, but he said, - I am using some of our process-oriented words - there are martingales with more terms than two. We can calculate the sums of the last three numbers, or of the previous four, or even five, and so on. In the end we shall have an incredibly long polynomial for the limit ratio, and a series F_∞ with $X_\infty \equiv (sic!) \lim_{n\to\infty} F_\infty(n)/F_\infty(n-1) = 2$. What a washout! From the golden ratio to a mere 2. But what!? The population doubles under iteration and equals up to a factor P

$$1,1,2,4,8,16, \ldots, 2^n$$

the dimension of a Clifford algebra generated by a space $\mathbb{R}^n$, how beautiful! Let us speculate!

You must know, I adore the creations of Louis Kauffman who gave me 'quite some impulse', and I appreciate his innocence. He really doesn't care about what people think about the new paths he is pursuing, being present everywhere like a photon, he says that a fermion can be marked or unmarked, interact with itself in order to reproduce itself, or interact with itself without leaving any mark, like the nothingness outside the brackets. Kauffman seems to have surveyed the whole district of algebra, shortly before categories rose to their mind blowing significance, trefoil knots, 'Reidemeister One, Two Three', topological logic, he didn't miss a thing, until, it seems to me, he decided to say: There is another brand of fermion algebra where we have generators $c_1, c_2, \ldots, c_k$ and $c_i^2 = 1$ while $c_i c_j = -c_j c_i$ for all $i \neq j$. These are the *Majorana fermions*. But this idea that a fictitious particle, say, an anyon, follows the ambiguous interaction rule

$$PP = P \qquad \text{or } PP = * \tag{438}$$

is really strange. For someone in HEPhy, this would be a particle that can either reproduce itself or produce an unmarked state * and the star can interact with P to reproduce P. Something like this:

$$\begin{aligned} &(PP)P \to PP \to P \ \text{ or} \\ &(PP)P \to * P \to P \end{aligned} \tag{439}$$

I have not seen a particle doing this, though the idea seems appealing. Let us put it this way: The density of a Majorana neutrino can be both an idempotent (marked state) and capable not to partake in strong interaction (unmarked state as fixed point of the motion group). Yet, these Fibonacci trees seem to me to provide a rather restricted interpretation. It could even be understood as a refined re-interpretation Louis Kauffman has worked out to avoid the maturity story of the rabbits. Because the whole consideration leads into some sort of binary Fibonacci sequence, no doubt, but it is not exactly the golden sequence which, of course has no PPP. Is a marked particle a mature particle? How can we imagine a fermion to grow up and reach sexual maturity? But I fear I am wrong. I am speaking like a journalist and I see those things through the eyes of the young one. I was a scanner then, and I was young, indeed. Or should I reflect that statement in self-reference and revert the propositions: I was young, and I was a scanner. My profession was scanning particles. I had no name, was entirely unimportant. Yet, the marked and the unmarked fermions, then, I have not seen them. But perhaps, if our construction, our imagination would have been different, then, if it would have been closer to Kauffman's idea, I would have seen them, the marked and the unmarked states. Indeed, I asked myself if the fermions had to go through some process of physical evolution until they gain the potential to set free radiation. Then the $|\Psi\rangle$ would perhaps have to wait a while, until a certain threshold is reached or the thermodynamic environment is appropriate. Only then it would be ready to re-emit the $\langle\approx\rangle$ and disclose an electron. I am not sure. Which would be the mature fermion, the $|\Psi\rangle$ or the $|e^-\rangle$. I would speculate the $|\Psi\rangle$ is pregnant. But apart from these gender mainstreaming speculations, I would be glad enough you have realized that the generational progress of the golden sequence does not say much about the distribution of 0s and 1s at a certain time, because we do not know when which mature rabbit has died, and every died rabbit cannot contribute to the sequence, that's obvious. So there comes in some fuzziness. But in case that (438), (439) are somehow satisfied by the Majorana fermions, then, wrote Kauffman: "This is the bare minimum that we shall need. The fusion rule is

$$P^2 = 1 + P\,. \tag{440}$$

This represents the fact that P can interact with itself to produce the neutral particle (represented as 1 in the fusion rule) or itself (represented by P in the fusion rule). We shall come back to the combinatorics related to this fusion equation" (Kauffman on knot logic, 2013, p. 28). There is no doubt, that the braid groups in the Minkowski algebra play a fundamental role in particle physics. But it is also important to understand how a spinor of Majorana fermions can be constructed so that it satisfies some minimal conditions in geometric algebra, in the sense of, at least, Cartan, Chevalley, Crumeyrolle and Chisholm, to factor in the big Cs once again. Let us have an idea. Let me work it out slowly. May be we should replace the 0 and 1, the small and the mature rabbit not by 'anyons', artificial particles, but by real iterants and real fermions.

I think we should assume that there is a natural iteration process. This process possesses material features of extension and intelligence. So it has the potential to constitute in a nonlinear chaotic happening the morphogenetic root-structures of orientation. This can be held in high regard if we understand the role of the dihedral and quaternion groups for the

Minkowski algebra. Suppose we replace the zero baby rabbit by a unit space-time iterant $e_1 = e \simeq 0$. We would make the rules as follows:

The first rule $0 \rightarrow 1$: $e \simeq 0 \rightarrow f \simeq 1$ (441)
The second rule $1 \rightarrow 1\,0$: $f \simeq 1 \rightarrow fe = g \simeq 1\,0$

This really refers to the most simple case. We begin with an unspecified self-referent domain that gives rise to the most simple passive iterants. A passive iterant is not entirely passive, since it interacts by touch. But it processes the bit strings by logic identification and nothing else. It does not permute the order of characters in other iterants. In this sense it is passive. Strictly, it is less active than those iterants that permute the characters in touching strings. Suppose first the young rabbit 0 is replaced by a polarity string f the 'Schrödinger iterant'. Then we would just exchange terms and set the rules like this:

The first rule $0 \rightarrow 1$: $f \simeq 0 \rightarrow e \simeq 1$ (442)
The second rule $1 \rightarrow 1\,0$: $e \simeq 1 \rightarrow ef = g \simeq 1\,0$

That would bring on the iterated sequences of iterants:

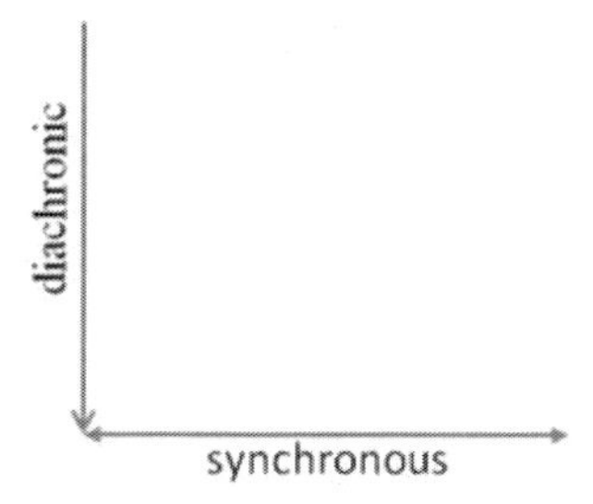

Figure 66. Synchronous and diachronic dimension.

Notice that each generation contains an exact image of the previous one. The result of the 'present step' that led to the present generation always contains a copy of the past generation. So nothing gets lost. That is the mystery of the rabbit explosion, and that's the reason why this idealistic model does not care about mortality rates. It wants to compile just this extraordinary property of the iteration. Now, be aware, this pattern has a linguistic design. Each generation can be seen as a line of rather disjoint synchronous events on the one hand, where the members are coexisting in some unspecified domain, or it can be seen as one element – a single step – in the vertical diachronic procession of generations. So we have a synchronous (horizontal) and a diachronic (vertical) dimension. Let us repeat: an exact correspondence between generation and time is not possible. Time in this model means generation time: steps in an iteration of iterants.

The iterants have the rabbit-habit to recreate ever more of their species from generation to generation. We must somehow assume that they do not eat each other up, as we had to with the rabbits. They must not interact in an unbounded manner, otherwise the whole pyramid would collapse into a single g. Note, we are still operating within the pleasant limits of the commutative Cartan subalgebra of the motion group. Every iterant is passive. Next we introduce an important survival rule for iterants: Elements from different generations must not interact! Now let us be attentive because this is exactly the point where Parmenides argument comes in: the past is gone, hence it cannot interact with the present. But in as much as the past is contained in the present – as an intact copy – it will have its chance to affect the restructuring of the synchronous presence anyway! So we have a chance to let the strings interact. We introduce the rule: Interactions in the present are possible and are determined by the physical algebra. In our case we let *neighboring* iterants e and f interact and produce a g. That is, we allow for a substitution $e\,f \to g$. Then we are operating quasi like an outside actor who makes the next step in the iteration, and transmute the whole pyramid:

e
g
ge
geg
gegge
geggegeg ... generation $n-2$
geggegeggegg ... generation $n-1$
geggegeggeggegeggegeg ... generation n

And see! this does not alter the situation. The golden series is still present. You see, the braid algebra of the color space is perfectly compatible with the Fibonacci iteration. Let us repeat some of the properties of the binary string

$$\mathcal{F} = 1\,0\,1\,1\,0\,1\,0\,1\,1\,0\,1\,1\,0\,1\,0\,1\,1\,0\,1\,0\,1\,1\,0\,1\,1\,0\,1\,0\,1\,1\,0\,1\ldots$$

This can be denoted as the Fibonacci word of, say, generation n. It can be recursively determined by the equation

$$\mathcal{F}(n) := \mathcal{F}(n-1)\mathcal{F}(n-2) \tag{443}$$

a mere concatenation of two Fibonacci words. Each generation's sequence is the start for the next generation's sequence. The imagined result of this iteration is one infinitely long sequence, called the *golden sequence* or the *golden string*. Since the bit-sequence of any generation is called a finite Fibonacci word, the whole infinite string is called the Infinite Fibonacci Word or just *the Fibonacci Word*. The length of the word $\mathcal{F}(n)$ is equal to the Fibonacci number $F_n = |\mathcal{F}(n)|$. We take the Fibonacci numbers by the series $1, 1, 2, 3, 5, 8, 13, \ldots$ which is in exact correspondence with the above transmuted pyramid over the alphabet $\{e, g\}$. For example, if you convince yourself, the 6th line has eight letters. The Fibonacci word is the purest and most simple articulation of the Eleatic idea of a moving presence which has integrated past events into its eternal present structure. The past determines the present by becoming a distinguished part of the present.

What is a distinguished part of the present? That depends on the consciousness that is aware of the present. It might be a Peano curve imagined as a dynamic propagation of location, watching itself,

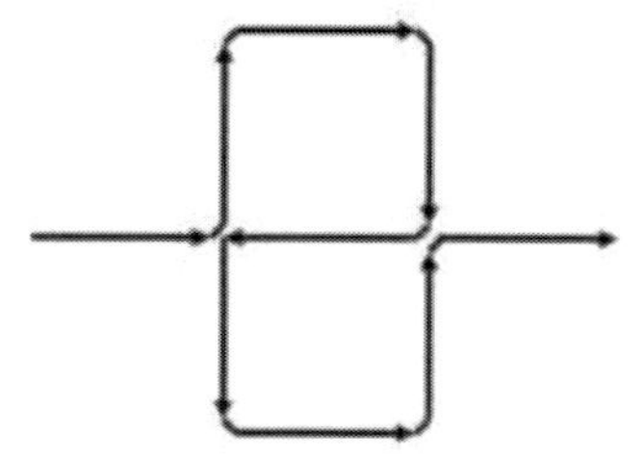

which appears to us in the form of an element that constitutes the orientation of a larger location. Or it might be the whole pattern of flowing energy thus brought about in a plane we imagine. But we do not have an actual plane, and no metric either. We just have that net of contra-oriented little whirls.

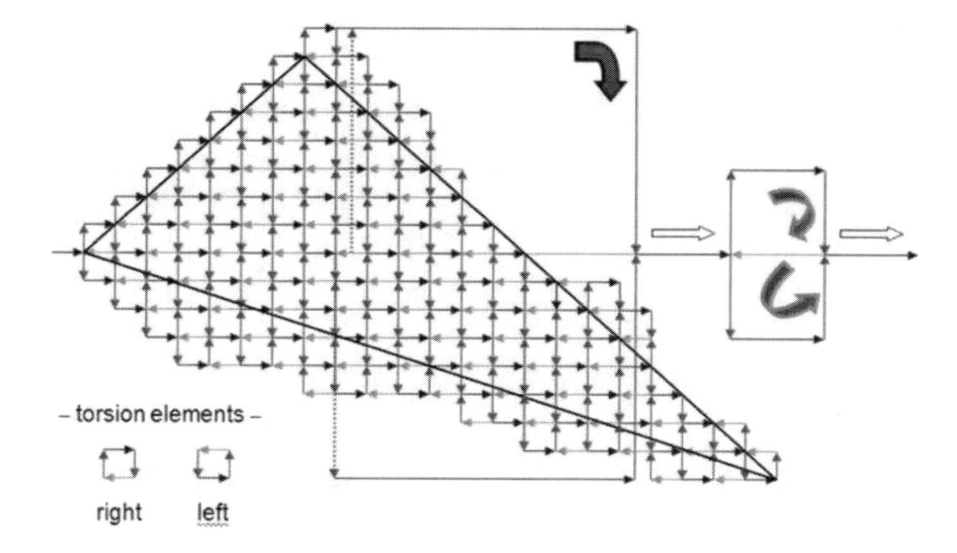

We can think about a local spin veil having the capability to flip from ½ to –½ as a whole net.

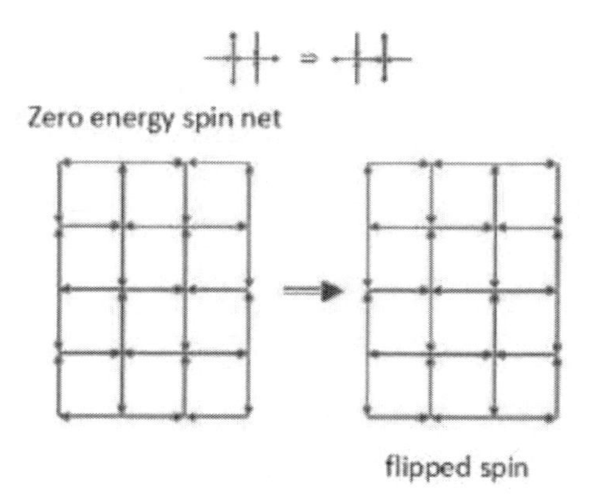

This figure may remind us of a turbulent flow, of some convections cells with spin orientation. But we need only this network of coupled spinors, no plane. Their essential feature is that the thing, the veil, can disappear into the void. The sum of all spins is zero. The void as an 'unmarked state' in the sense of George Spencer-Brown's "Laws of Form" provides the entry point for mathematical physicists. How does a location remember what it experienced recently, and what is 'recent'? Let us recall observation in a very fundamental case – the awareness of an iterant that there is a 'mirror image' of 'it' – as related to a local step in time. We repeat this. Let us be sure there are some unspecified locations somehow identifiable by polarity strings that pop off before our void background. These are thought to come about in a process of self-referent observation. We pose the question who or what observes any entity in a primordial domain? Is it possible that the iterants themselves are both observer and observed? You can regard the iterant $e = [+1, -1]$ as representing a local sequence of isospins, as a spin-chain ... ↑↓↑↓↑↓ ⋯ or anything similar. Who or what observes? We realized that observation, the cognitive act of comparison and creation of time are one. We articulated this by constructing the time shift operator η. Observation introduces a time-shift, a separation between observer and observed, a comparison between two elements. We can translate the following statement into a simple mathematical formula: »The iterant $e = [+1, -1]$ in observation separates from the other iterant $e' = [-1, +1]$ thereby reproducing unity (equation 59).«

$$[+1, -1]\eta[-1, +1]\eta = [+1, -1][+1, -1]\eta\eta = Id$$

A polarity string $[+1, -1]$ observing its dual iterant view $[-1, +1]$ seems to see unity. There is some mystery of self-reference hidden in this algebraic conception of observation. If we go into this more deeply, we find out it is connected with Spencer-Brown's 'laws of form'. First, consider a free Schrödinger field $S := \cos\alpha + i\sin\alpha = [\cos\alpha, \cos\alpha] + [\sin\alpha, -\sin\alpha]\eta$. We investigated what it means, if the free field S in a relation of self-reference observes itself as observer on the one hand and as observed on the other. We just followed the recipe "first observation of X, then observation of Y" as introduced in the equation (36). Consider S observing itself as an iterant view. As an iterant S has the abstract form $S = [a, a] + [b, -b]\eta = A + B\eta$, with $a = \cos\alpha$, $b = \sin\alpha$. Therefore the rule, 'First observe S then observe S' yields

$$\begin{aligned}
&S\eta S\eta = (A + B\eta)\eta(A + B\eta)\eta = (A + B\eta)(A'\eta + B')\eta = \\
&(AA'\eta + AB' + BA + BB\eta)\eta = (AA + BB) + (AB + AB')\eta = \text{since } AA' = AA \\
&= [a^2 + b^2, a^2 + b^2] + ([ab, -ab] + [-ab, ab])\eta = \\
&[a^2 + b^2, a^2 + b^2] = [\sin^2\alpha + \cos^2\alpha, \sin^2\alpha + \cos^2\alpha] = [+1, +1] = Id
\end{aligned}$$

So we find out, the free field observing itself as observer and observed creates unity while the observer η that brings forth time, is diminished. Observer and observed are one. The field reproduces unity. So the free Schrödinger field in self-referent observation does the same as the iterant view $[+1, -1]$ in observing its dual. It annihilates time-shift and recreates unity. In a Kauffman context or in a Spencer-Brown view of form, the two 'dual iterants' represent the 'primordial waveform' of two alternating series of two signs or states called by Spencer-Brown the marked and the unmarked state. The sentences

$I \stackrel{\text{def}}{=}$ *marked, unmarked, marked, unmarked, marked, unmarked, marked, unmarked, ...*
$J \stackrel{\text{def}}{=}$ *unmarked, marked, unmarked, marked, unmarked, marked, unmarked, marked, ...*

reveal the iterant views I, J of an alternating series (Kauffman 2013, p. 14) of marked and unmarked states as representatives of the reentering mark as in $i = [-1/*]$. Denoting by ⊐ the reentering mark, we have its two fundamental primordial waves being I, J. The mark signifies a 'distinction'. We make a distinction by cleaving a domain. We strip the locus off that which it is not. The marked location, if we think in the context of locations, is distinguished from the unmarked, which may be reached once we cross the boundary that points to the outer which it is not. The mathematics that was advised by the 'laws of form' begins with a distinction, with a cleaving. Thus we identify the markedness of something we have in mind. Spencer-Brown has said "we take the form of distinction for the form". Hence the geometric distinction between inner and outer could best be represented by the distinction, the mark of a circle or rectangle make in the plane

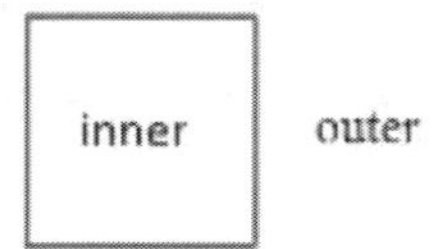

The act of drawing a distinction, writes L. Kauffman, involves a circulation as in drawing a circle, or moving back and forth between the two states. Self-reference and reference are intimately intertwined (Kauffman 2013, p. 13). This can immediately be realized by juxtaposing I and J or the iterants $[+1, -1]$ and $[-1, +1]$. Each one refers to itself by refering to the other. Now there are two Laws of Form. These arise from the difference of the fundamental relationship of two marks with reference to 'inner' and 'outer'. Namely, either one mark is inside the other or neither is inside the other, that is, both are in the outer of the other:

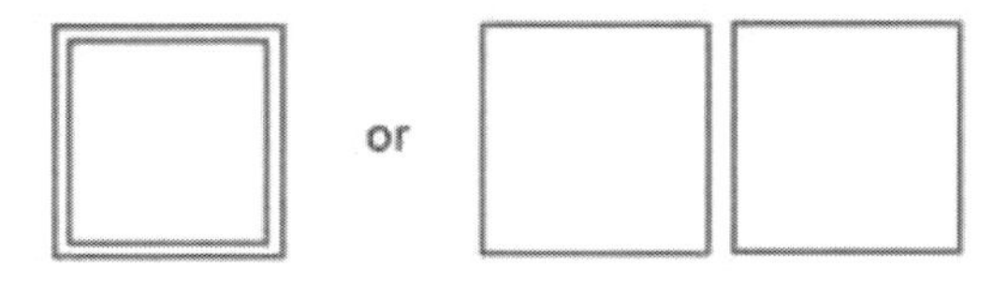

The mathematics in Laws of Form begins with two transitional rules for these two basic expressions

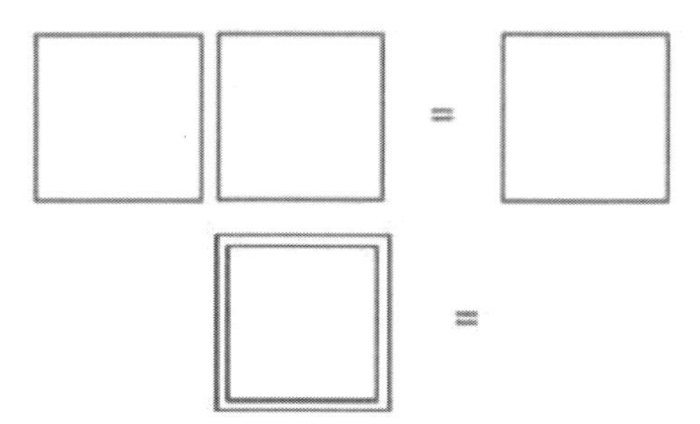

The first rule is called 'the law of calling'. Two adjacent marks fuse to a single one: a call made twice is a call. The second rule is called 'the law of crossing'. It says: going from the unmarked inner to the marked outer and overstepping the boundary twice leads to the unmarked, or two marks, one inside the other annihilate each other to form the unmarked, indicated by nothing at all. Likewise out of the void (right hand side of equation) a pair of nested marks may emerge. From these two equations, from these symmetric transitional rules, the calculus of indications is derived. It is useful to understand the principle of distinction that delineates the use and explains the meaning of the mark:

The mark indicates a state at the inside, the inner state. The mark indicates that the inner state is not the outer state which is obtained by crossing from the inner to the outer state. We write

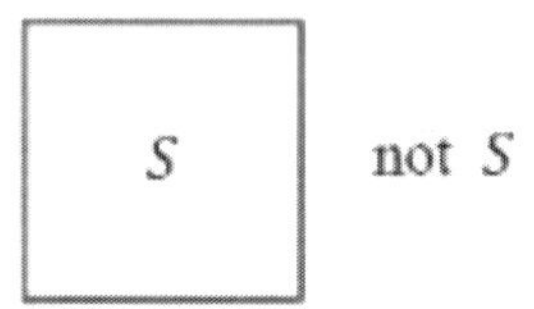

From this *principle of distinction* it follows that the outer of an 'empty mark' indicates the marked state as its inner is unmarked. Further the outer of a mark which has a marked, non-empty inner, signifies the unmarked state. Kauffman stresses that the form produced by a description may not have the properties of the form being described. Investigate carefully what follows:

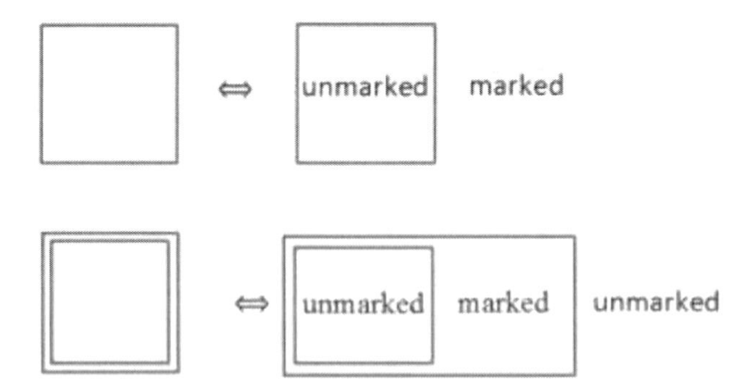

Expressions in the calculus of indications (CI) can disappear. Markedness acting on itself becomes unmarked. This 'contradiction' is *resolved in time* by the iterant views I, J. Similar like the polarity strings the forms in CI behave and act like material entities in space and time, and they can disappear in the void background. These expressions that represent actual forms of physics may behave like linear flow or turbulent flow, they can store information like any memory and have many measures of entropy and metric space. Let us speculate further! They know about natural numbers:

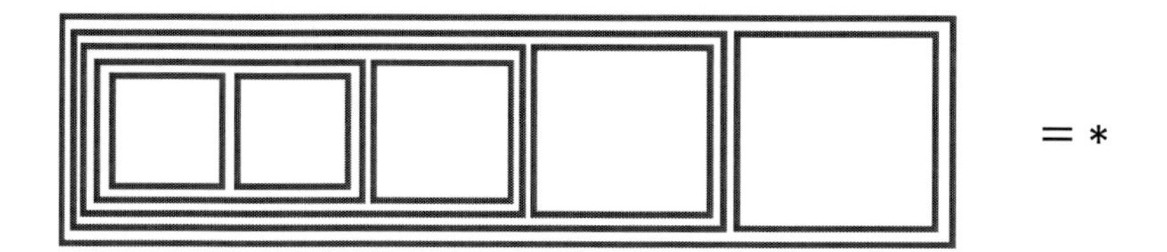

"Let us go slowly!" I proposed. "Please, consider page unnumbered (no blemish!) 'rough draft paper' »Laws of Form – an Exploration in Mathematics and Foundations«, link taken from Wikipedia, where on page 17 there is carried out a reduction to the marked state, and a reduction to the unmarked state. These reductions are what I like to call *diachronic procedures*. They bring in time and the memories of the design engineer. Consider the above figure. It represents the "synchronous form", the "score" for the procedure to be carried out (the symphony!) Please reduce it! Honestly, by doing step after step, and tell me what you are doing or what you have done! Tell me what you have done diachronically when you "reduced" the synchronous well-formed term, the indicated expression. Tell me!

May I do it for you, for the reader? Yes? Thank you! So: You have in mind the marked and the unmarked and the two rules: *calling* and *crossing*. You (me) look at the figure. I (You) see it. You (me) realize it has inner and outer. You (..) are aware, it is recommendable to begin with the reduction in the inner. You scan from left to right, or you see at once: there are two marked states next to each other in the innermost on the left, and you recall you can 'apply calling'. So you 'con-fuse' them by applying calling. You apply and count one, for you have applied a rule for the first time. Next you can apply at the same location the rule of crossing. You count 2. You have now obtained at that very location two nested marks. Hence you recall that you can apply 'crossing', and you count 3. Again you get two nested marks. So you apply crossing to get the unmarked and you count 4. In the end of the diachronic procedure there remains the expression

So you recall 'crossing', apply it, you obtain the UNMARKED STATE, and count 5. Next you send me an email saying: "Right, the diachronic procedure of reduction of your expression has 5 steps (at least!) … You may be dissatisfied with your answer and with my drawing, so you go ahead telling me that … a. s. o.

Kauffman replied: "You are right. And we can make expressions that take 6, 7, ... steps. So in a way these expressions "know" the numbers. This is missed if we jump too soon to algebra. It is only in algebra that we can say that *A* < > = < > for any *A*. One has to learn to play the music before the algebra makes sense. [...] Getting back to the Calculus of Indications. <<>> ⟶ removes two marks. <><> ⟶ <> removes one mark. So one can design various examples that require different numbers of steps.

<<<<<<>>>>>> has 2x3 marks and requires 3 steps for reduction. There may be some number theory here. [...] We could call

1 = <<>>
2 = <<><>>
3 = <<><><>>
4 = <<><><><>>
...

since n takes n steps to reduce to void. In fact, one way to make integer arith from LOF is deny the law of calling and keep the law of crossing with some attention to context. We allow that $<< A >> = A$ for any A.

1 is the additive void in this system.

1 = <<>> =

Then we can use the above expressions for natural numbers to define $a + b \; = << a >< b >>$. Note that

$1 \, + \, 1 \; = < <<<>>> <<<>>> > = < <> <> > = \, 2.$

What about

$Void + <<><>> = < <> <<><>> > = < <> <> <> > = \, 3$

So you see, it works! This re-invention of the natural numbers is due to Spencer-Brown and it can be found in some of the more recent editions of LOF. In order to include multiplication, we need a multiplicative void and a definition of multiplication. This goes as follows:

Define $a \times b$ so that $a \times (b + c) = a \times b + a \times c$.

This means that $a \times << b >< c >> = << a \times b >< a \times c >>$.

I will let you play with this and say more next time. The key is that one must have a multiplicative void (0) distinct from the additive void. It turns out that

$0 \; = < >.$

For example $a + 0 \; = << a >< 0 >> = << a > <<>>> = << a > \; > = << a >> \; = a$."

"This is very interesting and practicable, indeed." said I. "You know, I wonder about the following: there seems to be a state in cognition where time fades and observer becomes observed. This seems to me like a link between 'It, me, ego, observer' - how ever we call it - and the 'local world'. Let me put it this way, ... when this connection is there, the whole complicated expression that represents or may represent the 'event + observer', is reduced to void, and the arrow of time turns indefinite. This 'connection' I guess is not bound to the human observer, the you or me, but it takes place in nuclear domains too, - like an ambiguity between observer and observed, or, say, inner and outer, or, say, an ambiguity between reference and self-reference (... iterants ...).

There is something important coming up when we try to operationalize motion by a discrete calculus CI, namely we verify a typical second order cybernetic feature of observation. Consider some quantum performing some discrete steps of motion, Brownian motion if we wish. These steps are represented by

<<>>, <<><>>, <<><><>>, <<><><><>>, ...

We may claim that the particle performs some objective discrete motion indicated by 1, 2, 3, 4, ... , but how does it perform this? As we are in CI, we actually count the number of steps that we remember we have to make in order to reduce the expressions to void. Namely: 1.) we recall the law of crossing and apply it to the first term and count 'one'. 2.) We recall the law of calling in the second term and reduce it to get <<>>, we count 1 and we apply crossing and count 2, we proceed by going to the third term, 3.) apply 'calling' twice and crossing once and count three, we consider the fourth term, 4.) apply 'calling' three times and 'crossing' one time, we count 4, and so on. If we would not remember the context and its rules we would not count anything, hence we had neither scale nor measure for such discrete motion. So we ask, what is that particle doing? No one would be ready to neglect lightheadedly that this quantum is in some objective state of motion. The particle moves objectively, after all, but the perceiving subject does not recognize the order of that motion. What does this mean? This means that independent of any observation the particle is just as it is, and it is beyond the order the observer tends to project onto it. Then, is there motion, if nobody recognizes it? It seems that there is motion, both objectively and subjectively, even if nobody observes, constitutes order, does a scaling, measurer location and time, and so forth. Even without the observers can it be said that this particle moves, while it 'exists' in self reference and refers to the unmarked which it is not, and it is this interval between the marked particle and the unmarked where the property of change is anchored. There is awareness in change".

Meanwhile we have learned, I hope, that the alternating forms, the iterants, the iterants of marked and unmarked states, the spinors of space and time, form an algebraic whole. What is important with these forms is that they can disappear in the void. The advanced and retarded iterants of space-time spinors annihilate each other in a peculiar pattern of reduction. "The reentering mark has a value that is either marked or unmarked at any given time. But as soon as it is marked, the markedness acts upon itself and becomes unmarked. "It" disappears itself. However, as soon as the value is unmarked, then the unmarkedness "acts" to produce a mark" (Kauffman, "draft of book in preparation", Laws of Form – An Exploration in Mathematics and Foundations p. 13f.). [..] You might well ask how unmarkedness can 'act' to produce markedness. How can we get something from nothing? The answer in Laws of Form is subtle. It is an answer that destroys itself. The answer is that

Any given "thing" is identical with what it is not.

Thus markedness is identical to unmarkedness. Light is identical to darkness. Everything is identical to nothing. Comprehension is identical to incomprehension. Any duality is identical to its confusion into union. There is no way to understand this "law of identity" in a rational frame of mind. In Tibetan Buddhist logic there is existence, nonexistence and that which neither exists nor does not exist [BL]. Here is the realm of imaginary value" (ibidem, p. 14). It is only in this irrational frame of mind that the dilemma of thought Parmenides points at, can be circumvented. Then Louis Kauffman responded:[2]

I shall try to go slowly. I imagine a series of observed or observable events $X \longrightarrow X' \longrightarrow X'' \longrightarrow X''' \longrightarrow \cdots$ The $', '', ''', \ldots$ are the primitive Peano markings of the counts. They could just as well be $<<>>, <<><>>, <<><><>>, \ldots$

We could have

$$X, X <<>>,\ X <<><>>, X <<><><>>, \ldots$$

They are the marks the observer makes on his wooden stick, his memory device, OR they are the ticks of his clock. Each $X <<><>\ldots<>>$ is not necessarily a number. It can be a 'whole world at some time'.

And a difference $X - X'$ is the observer's distinction between a world at one time and a world at 'next' time. Next is our construction as well. And the allowing for a delay in observing $DX = X' - X$ versus the absence of delay in observing X is our distinction between time and no-time. X alone is in fused no-time suspended in eternity. Something happens and an X' occurs, memory occurs, difference DX occurs.

To go to the next level and compare the sequence of 'measurements' $X\,DX$ versus $DX\,X$ is to go to a next level of remembered paired worlds and higher descriptions. Physics starts coming recognizably out at the level of $[X, DX] = X\,DX - DX\,X$, but this description has gone too quickly.

Each X is timeless in its own presence.

Each X is void.

Nothing gives birth all too quickly to apparent something by giving birth to opposites that can just as well collapse back to void. In the beginning there is no directionality to time. It could be

$$X \rightarrow X'$$

or

$$X' \rightarrow X$$

and the simplest process is

$$X \longrightarrow X' \longrightarrow X \longrightarrow X' \longrightarrow \ldots$$

As we remember this, the differences and the commutators evolve , $sqrt(-1)$ appears and non-commutative worlds appear and start on in the world of descriptions that describe

[2] E-mail transaction on Sunday 21st July 2013.

themselves to form the appearance of the continuum, the differential geometry and all that we use so quickly and all too easily in our 'modeling' that is a reflection of our creating.

The advantage of using indices <<><>...<>> is that they all can collapse back to void in a finite number of steps.

Image of an expert shuffler of cards opening and closing the deck.
Expanding and contracting arrays of cards.
Stacks and volleys, orders arising and falling.
All happening nowhere in nothing for nothing.
And if there is no intent, then an intent can arise.
If there is no distinction, then a distinction can arise.
If there is nothing then there can be something falling into nothing.
If there is something falling into nothing there can be something.
If there can be something, we can have something different.
The difference devil is born.
Pandora's box is open.
Let it go once again."

Kauffman said, he will relax and try again in about 24 hours. He would travel downstate to visit friends tomorrow. But throughout his travel we were discussing about these forms of void that bring on a memory. The world is its own memory. It is good to be alone and think about it.

For now, let us recall the pyramid of the iterants e, g and go a little deeper into this distinction between 'synchronous' and 'diachronic', because it hides an illusion which has already been realized by anthropologists. Namely, - in the case of the Fibonacci words, - the transverse section of the movement at any given generation n looks as if it were a synchronous, quasi detemporalized image of the process which freezes out at a certain point of the iteration. But we know, observing this whole construction, so to say, as observers from the outside, that the left hand side of the section represents the beginning of the iteration, and the further we move to the right in that section, the closer we come to the present. Thus the diachronic is perfectly embedded into the synchronous cross section. So, if we allow for a substitution $e\, f \to g$ only for the instant neighbors, we really try to fixate the presence at a peculiar time point. But the fixation turns out as an illusion, as we cannot prevent the interpretation that a pair $e\, f$ interacted translocally, if that transaction was compensated by another pair. One has the feeling that something like that may be going on all the time in reality. There are many mathematical clues to this golden series iteration, and there are two which I want to pick out and highlight a little.[3] The first is that the golden string – understood as *the Fibonacci Word* – contains infinitely many copies of itself. The proof is given by reverting the rules of generation. The sequence contains a copy of itself since we can apply the above process backwards:

Start by pointing at the left hand end of the infinite Fibonacci sequence with your left hand and with the right get ready to start writing another series. If you are 'in touch' with 10, then write down 1. Otherwise you will be pointing at a 1. Then write down a 0. Move your

[3] Visit the beautiful page: file:///F:/Fibo/The%20Fibonacci%20Rabbit%20sequence%20-%20the%20Golden%20String.htm.

left hand past the symbols you have just read and repeat the previous step as often as you like. You will find that your right hand is copying the original sequence, but at 0·618034... of the speed of the left hand. We are merely using the substitution rules $1 \to 10$ and $0 \to 1$ in reverted order. Now we can use the backward substitution twice by considering $1 \to 10 \to 101$ and $0 \to 1 \to 10$ which gives us the guidline: if your left hand touches a location 101 then write with the right hand a 1, otherwise you will be in touch with a pair 10, in which case your right hand sets a 0. Again, your right hand will pin down a copy of the original string. This can be continued by considering the words 10110 and 101 which are transposed to 1 and 0 respectively, and so forth. There is no end to the sequence of instructions how the right hand can derive a copy from words 'read' by the left hand. In any case there arises a copy of the golden string. It would be interesting to consider the bits in the Fibonacci word as placeholders for algebraic terms and allow for transpositions in accordance with observations. May be that under given constraints large local fragments of the word are preserved, so that we develop the imagination that there is a memory which allows us to reconstruct the past. My feeling is that something like that is going on in reality and that the past we are constructing is not the past. But we cannot show or prove what the past really was or is, since it is not. It is not a part of the presently observed strings. That may sound pessimistic, but it can let us stay more humble.

The second viewpoint I would like to mention, is that of an 'enforcing neighborhood' or what is called 'forced pattern'. As we have the rule that 0 is followed by 1, we have the smallest pattern that enforces its follower. The pair 11 must be followed by a 0. So the triple 111 does not occur. There is only one pattern of each length where there is a choice. For the rabbit sequence matters are quite clear, but in physics the case must really be investigated with an unbounded, creative mind. Let me explain, what I mean. We have several indications that there is something like a relevant change in the character, or let me say, in the properties of iterants and polarized strings in general. We have in physics, since long, a quite reasonable concept of force. I do not repeat the explanation. We all know about these things. But as soon as we go over to quantum motion, the concept of force takes a new form, a fascinating form. Let us ponder over that issue. If that void I chose to call an unspecified domain can deliver two iterants such as

$$\begin{array}{l} \ldots +1,-1,+1,-1,+1,-1, \ldots \\ \ldots -1,+1,-1,+1,-1,+1, \ldots \end{array}$$

and can be aware of a difference, if it has the potential to disclose quaternary locations, then an operator like σ must be seen as an element of potential energy or force, since σ has the potential to exchange four segments in such a quad location, namely, it brings about an exchange of segments $(1\,2)(3\,4)$. It is not so important how we call these things, segments, characters or quarters, and it does not matter for now that such potential takes the form of a Clifford number. We just realize, it is probable that it requires some energy to take a bit from here to there, from 1 to 2, and likewise another from 3 to 4. But it might be that it does not take energy, as it might be a translocal exchange which does not involve energy. These things have to be clarified. But apart from whether σ, φ, τ represent potentials or do not represent potentials, we can ask in which way they would relate to a process such as the golden sequence. Let a state 0 be identified with a potential to transpose characters in a bit string, let 'zero' be the operator σ. By some evolutionary reason we do not yet know, the σ associates with a quad location and respectively an iterant e so that it turns over into an active bit string $\sigma e = -e\sigma$. In a way, the potential of σ has now really turned into an active, kinetic element that will change the characters of those bit strings with which it interacts. But this will go hand in hand with an act of separating off a bit string $-e$. This would be in perfect harmony with the kind of motion we are expecting since our investigations in the first chapters. The e as

$. . . +1,-1,-1,+1,+1,-1,-1,+1,+1,-1,-1,+1, . . .$ is opposed to
$. . . ,-1,+1,+1,-1,-1,+1,+1,-1,-1,+1,+1,-1 . . .$

But the manner in which this happens is remarkable. The σe turns over into $\sigma e\sigma$ which is equal to $-e\sigma\sigma$, hence $-e$. Anyway, we have the rules 1 and 2 satisfied. This time they take the form:

$$\text{The first rule } 0 \rightarrow 1: \quad \sigma \simeq 0 \rightarrow \sigma e \simeq 1 \tag{444}$$
$$\text{The second rule } 1 \rightarrow 1\,0: \quad \sigma e \simeq 1 \rightarrow \sigma e\sigma \simeq 1\,0$$

This gives the golden iteration a shape like this

1		σe
10		$\sigma e\sigma$
101	$\rightarrow$	$\sigma e\sigma\sigma e$
10110		$\sigma e\sigma\sigma e\sigma e\sigma$
10110101		$\sigma e\sigma\sigma e\sigma e\sigma\sigma e\sigma\sigma e$
1011010110110		$\sigma e\sigma\sigma e\sigma e\sigma\sigma e\sigma\sigma e\sigma e\sigma\sigma e\sigma e\sigma$
101101011011010110101		$\sigma e\sigma\sigma e\sigma e\sigma\sigma e\sigma\sigma e\sigma e\sigma\sigma e\sigma e\sigma\sigma e\sigma\sigma e\sigma e\sigma\sigma e\sigma\sigma e\sigma e\sigma\sigma e\sigma\sigma e$
...		...

Because of this special assignment the zeroes, represented by the sigma, are now allowed to stand next to each other. The golden series loses some of its gold. Now, we would be interested to allow for some freedom of interaction, at least for neighboring iterants. But our decision is not definite, for we can decide to give preference to a neutralisation of potential: $\sigma\sigma = Id$. Or we respect the conjugation of the iterant e by the σ. Keeping in mind that the algebraic product, when carried out, would give $\sigma e\sigma = -e$. This would revert direction. The algebra brings forth the following two iterations of iterants.

σe	σe
$(-e)$	$\sigma e\sigma$
$(-e)\sigma e$	$\sigma e\ Id\ e$
$(-e)(-e)e\sigma$	$\sigma e\ Id\ e\sigma e\sigma$
$(-e)(-e)e\sigma(-e)\sigma e$	$\sigma e\ Id\ e\sigma e\ Id\ e\ Id\ e$
$(-e)(-e)e\sigma(-e)\sigma e(-e)\sigma e(-e)$	$\sigma e\ Id\ e\sigma e\ Id\ e\ Id\ e\sigma e\ Id\ e\sigma e\sigma$
$(-e)(-e)e\sigma(-e)\sigma e(-e)\sigma e(-e)(-e)(-e)e\sigma(-e)\sigma e$	$\sigma e\ Id\ e\sigma e\ Id\ e\ Id\ e\sigma e\ Id\ e\sigma e\ Id\ e\ Id\ e\sigma e\ Id\ e\ Id\ e$
...	...

These pictures have a special meaning. We have here four iterants $Id,\ \sigma, e, \sigma e$ which are in a 1-1 correspondence with the Clifford monomials $Id, -e_{14}, e_1, -e_4$. These generate one of two basic dihedral reorientation groups of the Clifford algebra $Cl_{3,1}$. If we consider the last line, - the 'present population' of iterants in the left image, we have the sequence:

$$(-e_1)(-e_1)(-e_4)(-e_1)(-e_4)(-e_1)(-e_4)(-e_1)(-e_1)(-e_1)(-e_4)(-e_1)(-e_4)$$

If we consider the present population on the right hand side, however, we have the sequence

$$e_4\, Id\, e_1\, e_4\, Id\, e_1 Id\, e_1\, e_4\, Id\, e_1\, e_4\, Id\, e_1\, Id\, e_1\, e_4\, Id\, e_1\, Id\, e_1$$

which boasts a quite definite reversion of direction on the space line element e_1 as well as on the temporal base unit e_4. Therefore, when a step is made in such a process, this step is somewhat tipsily. It is important to know if the directed space-time areas $\sigma \simeq -e_{14} = e_{41}$ are allowed to act on the spatial elements which they encircle, or if they are forced to neutralize their potential by the equation $\sigma\sigma = Id$. Hence we need selection rules or priority rules for the interactions. The two possible developments of the Fibonacci iteration signify a symmetry which is expressed by the positive definite product $(-e_1)(-e_4) = e_{14} = e_1 e_4$ for both series.

Hint 1: each area e_{14}, e_{24}, e_{34} representing a time shift in the Minkowski checkerboard mystifies the orientation of spatial and temporal units e_1, e_2, e_3, e_4, since $(-e_1) \wedge (-e_4) = (+e_1) \wedge (+e_4) = e_{14}$ a. s. o.

Recall, the time shift turned out to be an oriented space-time area of typical signature $+1$ as e. g. the e_{14}. It is the most important actor in the evolution of a path. There may be a fundamental instability, for example, if both the potential σ and the bit strings have a tendency to survive. Then σ would avoid running into itself and thus reduce the chance for $\sigma\sigma = Id$. Hence it would increase the probability for a both-sided conjugation $\sigma e \sigma = -e$, while the iterant e would try to escape this encounter.

The Fibonacci model of particle physics has been called a 'topological quantum field theory' based on a single 'particle' capable of two states, the marked and the unmarked. The particle in the unmarked state is supposed to have no influence in interactions, that is, interacting with any state it yields that state. Resulting there is a tree of sequences with no occurrence of the double ** just as in the Fibonacci word where there is no '00'.

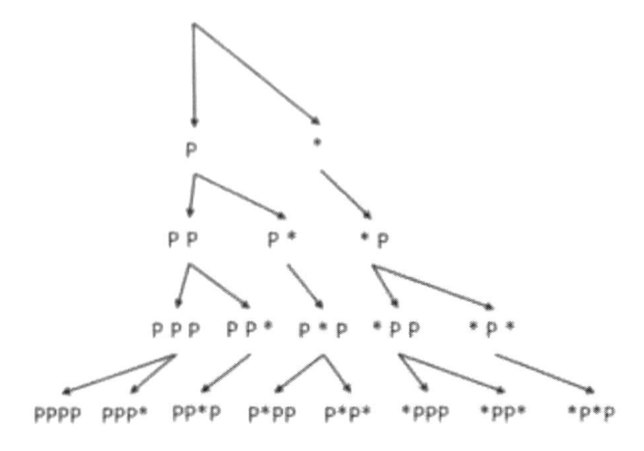

Figure 67. Tree of Fibonacci word and tree of 'anyonic' interaction.

Yet there is a difference between the Fibonacci word and a word of 'anyonic' interaction. The Fibonacci word does not contain the triple '111', whereas an interaction sequence must contain PPP. How can this difference be interpreted? Consider level n. The Fibonacci word of generation n has $F(n)$ letters. For example the word on bottom of figure 67 has $F(6) = 8$ letters. In the right hand figure, on bottom level 5, we also have 8 locations, that is, 8 'anyon' interactions each consisting of a 4-letter word over the alphabet $\{P,*\}$. Hence a single character from the left is carried to 4 characters on the right. Is this posing a risk? Not really, because any such 4 character word represents only one particle somehow interacting with itself. Thereby both the identity of the quantum and the energy seem to be conserved. At least we have done our best, for physics, to have the essential features kept invariant. But are they invariant? The energy is conserved only if the interaction itself does neither consume nor release energy. Otherwise the tree has to be embedded into some environment that compensates the energy imbalance. The reader may ask why the writer has eluded the mathematical issue and quibbled to energy conservation. The reason is that he wants us to think out of the box. Clearly, if the zero and one on the left side would represent particles, there would be no space for energy conservation at all.

How does the arrow of time enter the process? How is time related to generation time? Take a look at the Fibonacci words. As soon as we introduce 'real time' the whole Fibonacci pattern is discredited. For example, we consider level 5 and scan the tree when the first character '1' follows the rule and splits into a pair '1 0'. Thus we obtain a snapshot at time 10 with only six characters. All the other characters (2 to 5) have not yet performed a transition. Notice that the rules are broken, as we have at this intermediate time point two zeroes neighboring each other! However, by some undiscovered reason one time step later, the iteration of level 6 is completed. In reality there may be even some more time steps interposed until the 'final result' of 8 characters in generation 6 is reached.

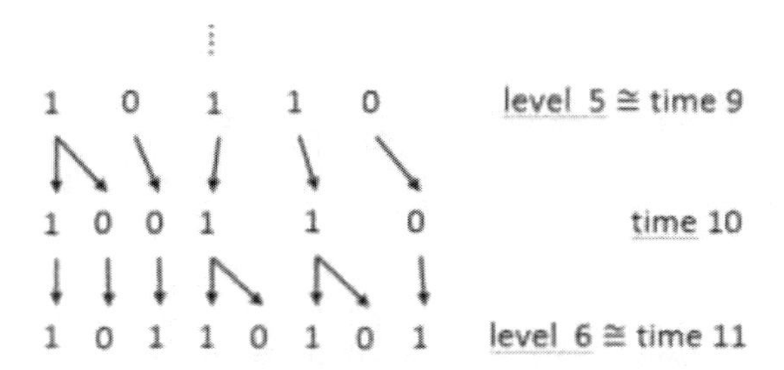

Figure 68. Introducing real time in Fibonacci generation.

We can conclude that generation time leaves time undetermined. There remains an undetermined measure of uncertainty where the appearance of words at a certain time is concerned. Also there is no other way to identify words at a certain time than by using some outer clock that has to be introduced by the observer. I want to remind us that even a cautiously constructed, mereotopological model of space-time, though it may be relational complete, exhaustive and expressive, nonetheless turns out undecidable (Schmeikal 2012, chapter 13). Does the Fibonacci process determine some definite arrow of time? No! The mere growth of numbers does not introduce time, unless time is already there! That is, mentally, we have already introduced the arrow of time, and within this 'cognitive flow of time' we diagnose that something is growing by the time, namely, say, the number of ones in a Fibonacci word. Can we, perhaps, introduce directed time by the entropy? The answer is no. But we shall go into this slowly. Roger Penrose (1989) has speculated about the magnitude of the entropy of the big bang. In his book 'Cycles of time' he gives reasons for the surprisingly low value of the initial entropy of our universe. Feynman (1965, chapter 5) also has thought about the arrow of time. He stressed, like so many of us, that all fundamental laws of physics are reversible, whether classic or quantum. So, where does the irreversibility come from?

He argued it is caused by the transition from order to disorder which begins with the big bang.[4] But is there any Big Bang? Is there any coherent universe, identifiable by us, which begins with minimum entropy? Sean Carroll from Caltech and Jennifer Chen (2004) have also thought about the arrow of time. They argued that the Big Bang is not a unique occurrence at the beginning of time, but rather one of many cosmic events resulting from quantum fluctuations of vacuum energy in a cold space. According to this view the universe is infinitely old. It never reaches thermodynamic equilibrium as entropy increases without limit due to the decreasing matter and energy density. This idea overlooks that the entropy growth is paralleled by a decay of time. So the age will rather turn out as principally undetermined. They assume that the universe is "statistically time-symmetric". That is, it may have equal progressions of time both forward and backward. Let us relate this thought to the Fibonacci word and to the morphogenetic structures of orientation. Especially the latter will give our construction a certain objective validity, as it is not only the human observer who subjectively constructs these morphogenetic frames of thought and matter, but nature herself is also constructing these, and she has been doing so during all the times human cognition can conceive of. In order to be to some degree in agreement with the current belief system, I will quote from the homepage of Eric J. Chaisson, research associate at the Director's office of the Harvard-Smithsonian Center for Astrophysics.[5]

"At the conclusion of the inflationary phase at about 10^{-35} second, the X bosons had disappeared forever, and with them the grand-unified force. In its place were the electroweak and strong nuclear forces that operate around us in the more familiar, lower-temperature Universe of today. Physicists describe such an event as "broken symmetry," with the strong and electroweak forces, previously one and the same force, having then become separate entities. With these new forces in control (along with gravity), the Universe resumed its more leisurely expansion. Later, $\sim 10^{-10}$ second, when the cosmic temperature had decreased to $\sim 10^{15}$ K, a second symmetry breaking occurred, enabling the electroweak force to reveal its more familiar electromagnetic (elm) and weak nature that guides almost everything we currently know about on Earth and in the stars."

Referring to the structural features of the Minkowski space and allowing for the dynamics of bit strings we can most naturally locate the fibers of primitive idempotents representing quarks and leptons and as many fibers of negative anti-idempotent densities for anti-fermions. We find the five main types of Bosons in the even subalgebras. If we refer to about as many supersymmetric particles, we shall have to respect some remarkable number of more than hundred boxes, where nature can allocate the energy. Now suppose we had, say, $n = 100$ boxes representing these types of elementary particles and forces, and let the first

[4] "For some reason, the universe at one time had a very low entropy for its energy content, and since then the energy has increased. [..] (The arrow of time) cannot be completely understood until the mystery of the beginnings of the universe are reduced still further from speculation to scientific understanding." (Feynman Lectures on Physics Volume I, p. 46-49).

[5] https://www.cfa.harvard.edu/~ejchaisson/cosmic_evolution/docs/fr_1/fr_1_part5.html.

particle, having box-number 00 represent the Higgs boson. At the Big Bang, - assuming there was some, - sometime earlier than 10^{-35} seconds after, all the X bosons,- we adopt the notation of Eric Chaisson, as the Higgs might be a little different from the Higgs with no fermions,- all the energy was in box 00. ‘As time proceeded’, it jumped over into other boxes until those times where baryonic mass of protons and neutrons with the u- and d-quarks were as dominant as they seem to be today. After 10^{-35} seconds the X boson had involved all the other particle states and given its energy to them until after some number of steps a state was reached when there was not a single quantum in the X-box. Thus we had passed through the critical phase transition, and from now on the number of quanta in the X-box fluctuated around values close to zero. Considerations like these are at the heart of Ludwig Boltzmann’s thermodynamic description of nature. They relate to the arrow of time and to the second law of thermodynamics. What is entropy? The entropy measure counts indistinguishable arrangements of a system. It measures the number of ways we can rearrange the constituents of a system so that we don’t realize a difference. Macroscopically it looks the same. To know the temperature of the air in a room you don’t need to notice each individual atom. A low entropy configuration is one where there are only a few arrangements that bring forth the same macroscopic image. In a stochastic process of energy exchange the amount of energy that is no longer doing physical work can represent the *statistical entropy* and is given by the minimum length of codes or number of digits needed to describe the state of the system and trace back its history. The second law of thermodynamics tells us that the entropy increases also in any isolated bit of the universe. This is due to the fact that there are many more ways for the system to be high entropy than there are to be low entropy. However, what the law does not explain is why the entropy was ever low in the first place at all. In the beginning, it was said, the universe was very smooth. To find out why, said Carroll “that’s our job as cosmologists”. [6] Connected with that idea, Johannes Koelman constructed a stochastic ‘Fibonacci fleas model’ and a toy model to study a Fibonacci universe. This can be applied to our hundred boxes which represent that isolated bit of the universe where the entropy unremittingly increases. [7]

Suppose we had 120 quanta, all of them X-bosons in the early universe. We assign them 240 digits, that is, two digits for each quantum. Hence the universe would begin with hundred-twenty double zeroes:

000000000000000000000000000000000000000000000000000000000000000000000000
000000000000000000000000000000000000000000000000000000000000000000000000
000000000000000000000000000000000000000000000000000000000000000000000000
000000000000000000000000

The next step would be a word

000000000000000000000000000000000000000000000000000000000000000000000000
000000000000000000000000000000000000000000000000000000000000000000000000
000000000000000000000000000000000000000000000000000000000000000000000000
000000000000000000000001

[6] http://preposterousuniverse.com/eternitytohere/ TEDxCaltech, “Cosmology and the Arrow of time”.

[7] http://www.science20.com/hammock_physicist/fibonacci_chaos_and_times_arrow.

indicating the appearance of the first fermions. We denote this word by

$$W(0) = 1$$

and let it stay a moment

$$W(1) = 1$$

until the inset of a Fibonacci process having form

$$W(t+1) = 3W(t) - W(t-1)(mod\ 10^{240}) \tag{445}$$

This brings forth $W(2) = 2$ in time step 2

000000000000000000000000000000000000000000000000000000000000000000000000
000000000000000000000000000000000000000000000000000000000000000000000000
000000000000000000000000000000000000000000000000000000000000000000000000
000000000000000000000002

Next is $W(3) = 5$ or as a complete 240 digit word

000000000000000000000000000000000000000000000000000000000000000000000000
000000000000000000000000000000000000000000000000000000000000000000000000
000000000000000000000000000000000000000000000000000000000000000000000000
000000000000000000000005

and so on until $W(574) =$

228738023742104873579672300111856420469306743327324692412448017496081811
876828577037986172078704406541673742096986172743151057541530131639856954
502977924301945285982801782362772314189298583923238426430745965802375035
020866499020647087027973

with no more 2 character-location containing a 00. This signifies a time point with maximum number of digits needed to reconstruct the process history. Actually we need two words with 240 digits, hence 480 digits at time step 574. The process begins close to the Big Bang with two times 2 digits and increases almost linearly to its maximum value. Now, what about times *before* the Big Bang? This is the interesting question Kaelman suggests to ask, and he answers: the dynamics is indeed time-symmetric, since equation (445) can be written in time-reversed form

$$W(t-1) = 3W(t) - W(t+1)(mod\ 10^{240}) \tag{446}$$

Therefore, the entropy decreases for times before the Big Bang in just the same way it increases thereafter. "It should now be clear what is causing the entropy increase with time", wrote Kaelman, "entropy increases simply because the system started in a specially prepared low-entropy starting state. There is no paradox related to the fact that the dynamics is reversible." Hence it is proposed to redefine the second law of thermodynamics. In the neighborhood of such a prepared state we should expect the entropy to be represented by a convex function of time.

If it would be a fact that the interactions of polarity strings are themselves stored in bit strings, it could indeed be possible that all or some of the history of events are always present in the universe we observe. Obviously, in this case the entropy seems to increase very rapidly. But what is 'the universe we observe'? Which encounters become actualized in what kind of observations? How can we speak about 'all that is'? How can we talk about lost memories, if nothing is left of them? Is it possible that somehow the events realize and store their own whole history and carry it with them in their eternal presence? – By the way, in Parmenides philosophy such a thought is meaningless, since 'the possible' is what is not; it is a no-thing. But: 'it is not that nothing is.' Therefore: no place for that which is possible, but only for that which is. This seems to be a weakness of the old philosophy. Let us look at that which is. I said to a colleague: "I need a memory which functions without time. *Touch* and *compare*, *shift*, for my part, but without time measure; time too is such a local vanity". He answered: "hmm, the notion of 'memory' has to do with the fact that we have a flow of time. If I put something into the 'memory' I want to ensure that I can reuse/compare it at some later time. Hence: memory needs time. A related concept is that of an 'invariant' (e. g. invariance under the action of a group), that is, some property which does not change under the action, hence can be 're-minded'. Information which is not preserved cannot be stored so easily. If the re-minded is dynamically changing – (in humans it is doing that) – it can no longer reliably be identified as that which we want to remember". Hence my question:

What is (or could be) a 'memory without time'? When we turn over from the differential to the integral (timeless) view, events become curves. And such curves are cuts. Therefore a memory could be an invariant cut, and therefore do without (explicit use of) time. P.s.: in quantum information theory it is common practice to have a 'do nothing operation'. But it is extremely difficult to realize this physically. To do nothing with a q-bit is (almost) impossible and limits the quantum mechanical computing power (decoherence). Therefore one could define quantum decoherence as the average or longest time period of a quantum memory power of a subsystem. Classical information is (in reality) not permanent. PPS.: categorical: if you have a natural transformation from the identity functor to another functor (as in a adjunction unit/counit) then this functor (information) can be used as a memory, since I can find out how that which was originally stored (time=function application) looks like after the application of the functor (and can compare it with the result of the identity functor). So I do not remember, but I know how that which was stored is looking like now".

The present world is a living memory. The invariant strings of space-time algebra embody all features of re-orientation and provide the essential constituents that are easily remembered by the process of nature. This process is not independent of human perception, but it is entangled with our subjective experience. If we perceive things clearly without being disturbed by theories, we can observe in ourselves the constitution of a clear image of the universe. Consider the inner and outer space. We can look outside and perceive what we call objects of our world. And we can also look inside and see what goes on in ourselves. When inner and outer are one, - this unification is usually taking place when thought is silent, - we can see that what we see is moving, but it moves beyond time. The objects partake in our field of perception, and there is no perception of time. Since time is thought. So there is motion without time. As this is happening subjectively, but in a state where inner and outer are one, - it also signifies a modality of the outer being. There is no time, but only motion. Motion is in the perceiving field where observer and observed are one. Once we encounter this event in great clarity, we begin to ask ourselves where the time comes from. How is the mind creating time? What is the foundation of this projection of time onto the outward events? When inner is outer, when time ceases, the outward events are not outward, and the inner experience is not inner. But there is only world in motion. How does time come in? One cannot say that all events within some horizon are integrated and synchronized. They might be not. There may be many more time dimensions running in parallel and in cycles. In quantum motion the cyclic run is even more important than the linear go. But where does time come from? When it is annihilated in a state where inner is outer, then it is obvious that time arises with the separation of inner from outer. Time is resulting from a fragmentation. This process of fragmentation is called cognition. When thought divides subject from object, inner from outer, there emerges the process of time. Time does not exist as independent from this cognitive event of a division between inner and outer. What about the energy stored in such division? Empty space itself has energy. The translocal interactions of polarity strings give rise to permanent radiation. The amount of energy set free in unspecified domains of the void remains the same, even if space gets bigger and bigger. This has crucial implications for the future course of the universe. For it seems to us now, the universe will expand forever. The universe radiates the energy brought to life in quantum fluctuations. There are fluctuations into lower entropy. There are also occasional, large fluctuations. The world may happen from time to time. Our universe may be born like an organism. It comes upon in low entropy configurations, and that would happen more than once. The Bang would reenact, sometimes Big, sometimes Little. Sean Carroll, ending his lecture in February 2011 seemed to favor the low entropy scenario, saying: "An egg is not a closed system. It comes out of a chicken. May be the universe comes out of a universal chicken. May be there is something [..] that gives rise to a universe like ours in low entropy configuration I want to relate this thought to logic quantification, since this would make sense in a multiverse. Consider a Venn diagram

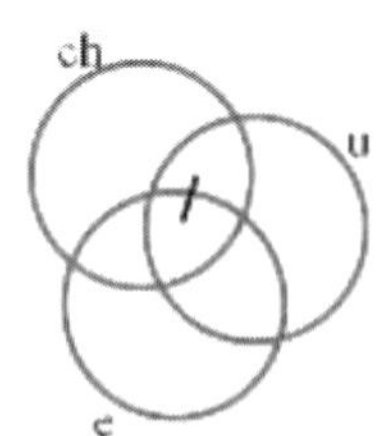

u … set of elements of universes (multiverse)
ch … set of chickens
e … set of elements called eggs

denoting universes, chickens and eggs. Each of these may come out of the void as a random fluctuation. If we recall Carroll's lecture, "Boltzmann said we will only live in a part of the multiverse in this part of a set of infinitely fluctuating particles where life is possible. That is the region where entropy is low. May be our universe is just one of those things that happen. [..] Carl Sagan once said that in order to make an apple pie you must first invent the universe. But he was not right. In Boltzmann's scenario, if you want to make an apple pie, you just wait for the random motion of atoms to make you an apple pie. That will happen much more frequently than the random motions making you an apple orchard and some sugar and an oven and then making you an apple pie. So this scenario makes predictions, and the predictions are that the fluctuation that make us are minimal. [..] The good news is that this scenario does not work. This scenario predicts we should be almost in equilibrium – just a tiny fluctuation in entropy, not the enormous deviation we actually observe". So, our first diagram indicates that some things in any universe are chickens, or equivalently some chickens are born in a universe, - not just random fluctuations in a high entropy void. And a second quantification tells us, some elements in the universe are eggs, and likewise, some eggs come out in a universe. They are not just fluctuations in the late nothingness.

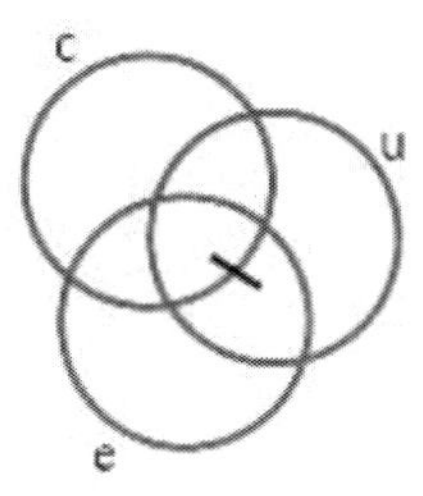

But surprisingly, combining those two, the last diagram indicates something else.

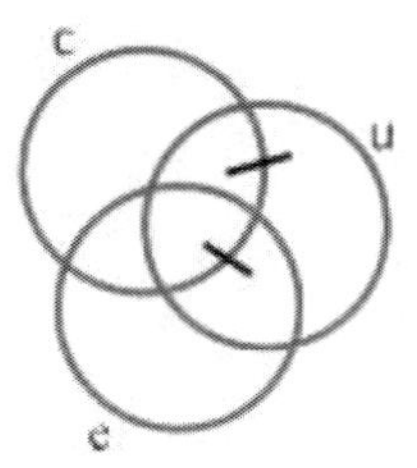

Here we quote Louis Kauffman who investigated syllogism and quantifiers in the mathematics of form, as was initiated by Spencer-Brown (Kauffman 2013, private transactions, section VII). "This diagram indicates the full possible states of affairs, and we see at once that it is quite possible for both propositions to be true without any intersection between a and c (here c and e). For example, consider the following diagram of occupancy where 1 denotes an occupied region, while 0 denotes an empty region".

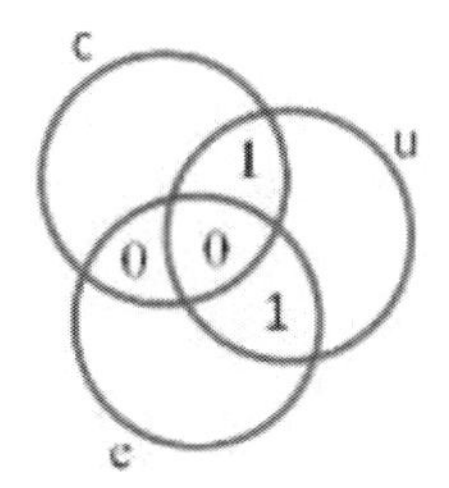

Some chickens are universal
Some universes have eggs
No eggs are chickens

No eggs are chickens, but they can become chickens. They become chickens with much higher probability, if they do not belong to the region of low entropy fluctuations out of the void, - indicated by the left-most zero. They have a better chance to become mature chickens, if they are from eggs laid by chickens born in a real universe (the right-hand zero) which is not just a fluctuation in the void. Kauffman saw that and commented. "Have read your last pages. I agree about time and distinction. And your adoption of the some, some, some syllogism to the chicken and the egg is i-n-t-e-r-e-s-t-i-n-g. It seems to me that the chicken/egg usually assumes the notion of their mutual causation and this puts them outside the usual Venn diagrams". This sounds all very convincing, but we are yet unable to ascribe to the void the intelligence which it actually has. Therefore, in this book, I have tried to recover some of that intelligence by investigating the dynamics of polarized strings. The iterants, as we called them, have a considerable amount of intelligence. Spencer Brown, Louis Kauffman and the present author have tried to disclose some of the fundamental phenomenological events which are responsible for such primordial intelligence.

How is the mind annihilating time? We are projecting a cognitive property of the mind onto the outward events. However, when inner is outer, when time ceases, the outward events are not outward, and the inner experience is not inner. But there is only world in motion. This is not only a subjective happening, and it is not only an objective physical event, but it is both subjective and objective, and beyond time. It is the one event which makes subject equal to object. Man in that encounter becomes universal matter and universal mind. Why and how is this possible for a single individual? How can this unification of inner and outer touch the whole outer? It seems as if it is a peculiar step of separation. Even some sort of isolation. Like the egg comes out of a chicken, and a new chicken will later come out of that egg, so a mature mind comes out of the universe. In this moment outer and inner universe are one. So we should confess that psychology and religion are not really separate from physics. Even mathematics must realize the foundations of science as rooted in awareness and in attentive procession of self-referent elements. Kauffman has shown how such new mathematics can begin with the work of Spencer-Brown, the 'laws of form' and the calculus of indications. This is the reason why I began to affiliate geometric algebra with Kauffman's algebra of iterants. Once we understand the life of the void and the steps we can make on this basis, we can understand better the other things we are struggling with, the foundations of the path integral in the space-time, the origins of the standard model of both physics and cosmology, the origin of measurable time and space. Those are not separate from human cognition, since we are one with nature, even if sociologists tell us that we are alienated from nature. We are not. If we were, it would be all over.

We shall now go into the final question of how the world can be its own memory. We realize that matter is the same as the organization of space-time. Recall the Peano-fractal parquetting of the plane as shown in figures 36, 37, the *zero isospin network*. Such a net can be extended into three or more dimensions. In the plane the energy can be imagined to circulate freely. Every point in this finite arrangement may be thought to represent a location with certain switching properties. Actually we can translate the Peano network into a free computer with many inverters. The advantage of using such an arrangement of inverters connected in the fashion of a Peano curve is that they all can collapse back to void because of the zero-sum spin connections. A veil is invisible and collects zero energy. In that peculiar case it is transparent for one polarity in horizontal direction and a second polarity in vertical direction of the working plane. The basic logic element of the veil is a logic inverter, a peculiar cosmo-dynamic whirl-switch in primordial space:

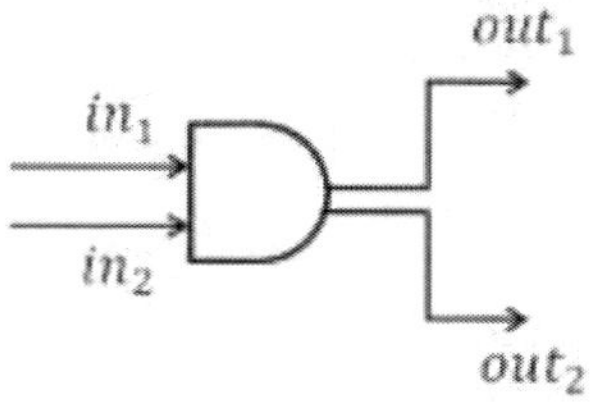

It has two inputs and two outputs. Any single inverter of the form

turns 0 into 1 and 1 to 0; or likewise -1 into $+1$ and reverse. Those are the *characters* in our iterants. Now we construct *a paradox* by feeding the 1 back to the input which is in state 0. Generally such a feedback refers to a relation of self-reference. Namely, let $\langle x \rangle$ denote the result of inverting x so that we have the figure

This is in perfect agreement with the laws of form. But as the output $\langle x \rangle$ is turned back to the input x we obtain a paradox loop asserting that $x = \langle x \rangle$:

As we know from electronics, such a circle will oscillate the faster, the 'shorter' the feedback (small resistance), and we get an iterated sequence 0,1,0,1,0,1, ... hence a figure that can represent a Schrödinger iterant. More than thirty years ago, in a beautiful paper that was, then, not accepted for publication, Kauffman proposed the following scenario. "Long ago and far away there was a primordial soup of curled up single self-referential inverters, each oscillating away, ignoring all the others. (Perhaps they were a byproduct of the big bang). These were very primitive times indeed and in fact the only 'things' around that could be used as signals were the little curls themselves! And there weren't any 1's. But that was allright, since an inverter would just merrily produce a zero

$$\langle 0 \langle \bigcirc \rangle \rangle$$

if nothing came to its input and it would stop, if something came along:

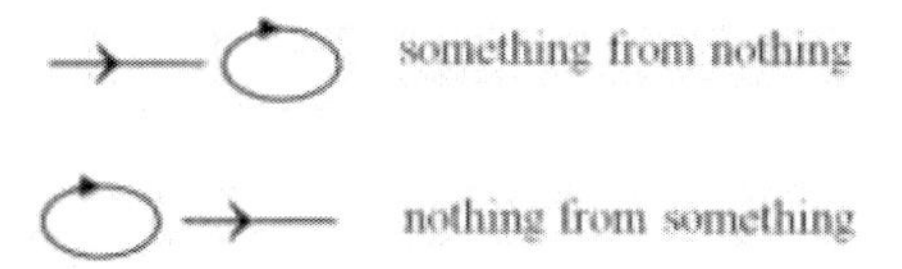

If it weren't for that sense of direction given to each inverter, this might seem a bit confusing. But that was the legacy of the big bang. Oh! I didn't tell you how this works! It went this way. In order for →— to act on and cancel it, the inverter would curl around its victim and engulf it right properly. Then it would move right down and superimpose itself on the unsuspecting curl

And due to a very ancient universal law, to the effect that you can't have two different things in the same place (c.f. W. Pauli) the two curls would just vanish!" ("paper computers", L.H. Kauffman 1981, p. 4ff.)[8] In this way the laws of form were satisfied by realizing the law of crossing. Then, by *idemposition*, a special form of the law of calling, the memory flip-flop was then represented in the familiar way

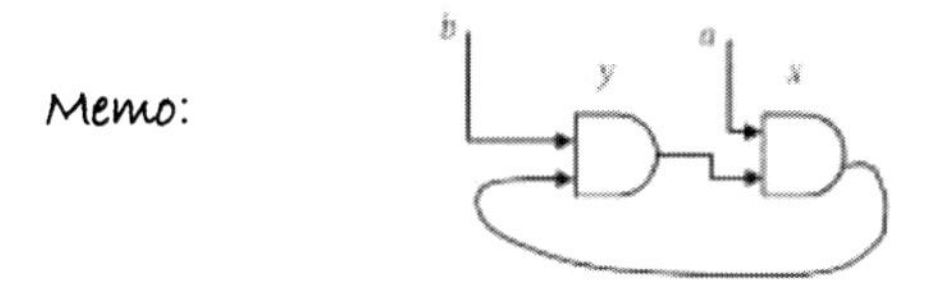

All this may sound like a beautiful, somewhat strange fairytale. But it is much more than that. It has a real meaning, even a meaning beyond words. So, for this elementary memory Memo we have the equations

$$x = \langle ay \rangle$$
$$y = \langle bx \rangle$$

where a, b, x, y can take the values 0 and 1. For simplicity we consider first the reduced circuit with only one input at each inverter

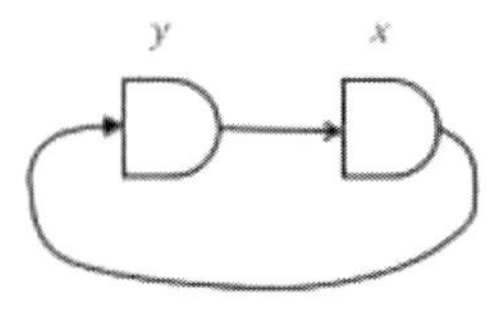

This has the equations $x = \langle y \rangle$ and $y = \langle x \rangle$. It has two stable states given by pairs $(x, y) = (1,0)$ and $(x, y) = (0,1)$. The states $(0,0), (1,1)$ are unstable. Providing a stable input 1 at y must effect a transition of the inverter x to 0, and $y = 0$ enforces $x = 1$. If we

[8] As for the original handwritten 'Version1' of "Paper Computers" we refer to http://dl.Dropboxusercontent.com/u/11067256/PaperComputers.pdf.

add the inputs a, b at inverters x, y we can switch the circuit between the two stable states and obtain a most simple memory circuit. If $a = b = 1$ we obtain for *Memo* the stable state $(x, y) = (0,1)$. Now let b turn to 0. Then the circuit is out of balance at y. Thus y changes to 1 and then x becomes 0, hence giving the new stable state $(x, y) = (1,0)$. In this state the circuit is insensitive to changes at the b-input. We say it is 'opaque' at b. If we want to effectuate a repeated change to the other stable state we have to change the input a. This represents an essential feature of *Memo* that we have two inputs and two stable states and in each stable state exactly one input is sensitive to effect changes (Kauffman 1994). Flipping states implies flipping input sensitivity. *Memo* is a flip-flop and divides by two. It is used in frequency counters. Hence the input frequency of a rectangular signal appears with half that frequency at the outputs.

It is perhaps not immediately clear why any location in 'space-time' should have the properties of something like *Memo,* and it might at first seem totally mad to consider Kauffman's little whirls, these logic inverters with the additional Spencer Brown's form-qualities, as relevant for physics. But the idea is not incongruous. We know the circular coupled fluxes from the Maxwell theory of the electromagnetic fields. In the first volume on primordial space (2010) in sections 48, 49 dealing with the topological evolution of particles and *strong force topological torsion* it became obvious that we can even speak of an *electromagnetic format of all the forces* of nature. Vacuum fluctuations are *elm* phenomena which can form logic lattices thereby creating periodic polarized patterns that can trap energy similar like in an optical lattice which traps fermions (Brennen et al. 1998). Consistent with this *elm* format are phenomena of turbulent flow and some nonlinear effects which we have used in aerospace science, as for instance the Koanda effect and logic behavior in microfluid dynamics.[9] But the Koanda effect is resulting from the more basic dynamic properties of space. The circular structure of space-time, as it locally appears to us, can be visualized by a spatial lattice of planar zero isospin veils (figure 3 in Schmeikal 2011, p. 91). It was in 2010 that I saw the energy transport system in the 'veil' can be comprehended as an unstable logic network. This contains *inhibitory control edges* and specific time-gates (ibidem, p. 104f.) which regulate the circular flows like in a computer. Actually it seems that space-time is an intelligent logic network with an admirable memory. It needs two orthogonal planar veils. Take such a little element as the inverter

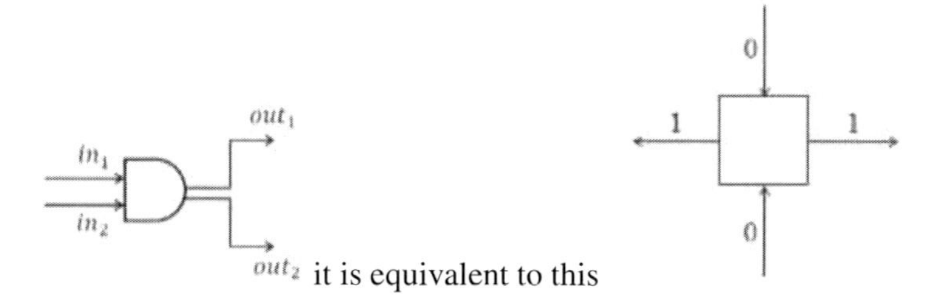

[9] There is a branch of research and development concerned with the engineering of amplifiers and logic circuits based on fluid dynamics rather than electric current. The word 'Fluidics' can be seen as a contraction of the words 'fluid' and 'logics' (Edwin Oosterbroek 1999, p. 1).

We are holding the in-and outputs at some definite values that will make the network stable. Consider a veil in the operating plane. As a coupled ensemble the veil in ground state looks like below (figure 69).

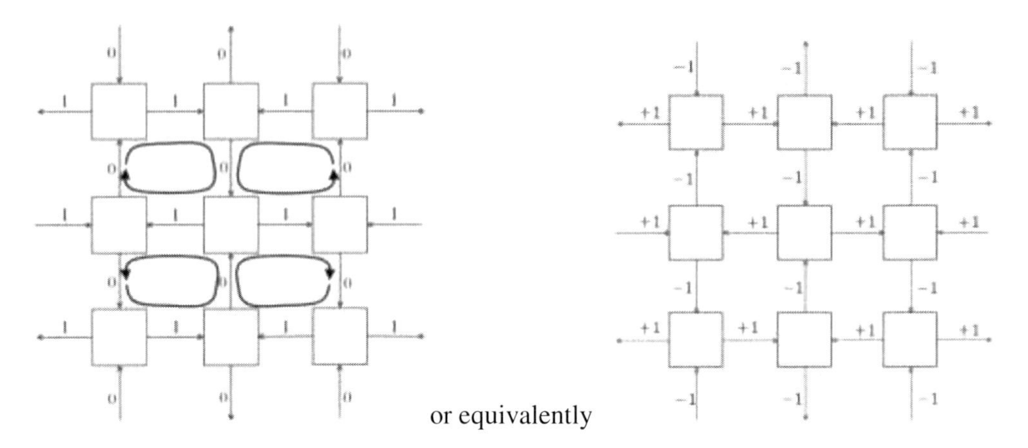

Figure 69. Planar spin veil as logic network.

You can see the 'convection cells' indicated at the center of the veil. In physical reality these things are energy turbulences in subnuclear domains. We use to denote these as vacuum fluctuations. But these fluctuations are well organized and rather intelligent. Let us investigate a few properties of the balanced network. In stable ground state configuration it is so to say invisible and free of active energy. All its energy is at rest, is potential energy. All spin is disappearing itself. It represents perfectly the unmeasured void. Notice, the arrangement has an inner and an outer. The inner consists of one inverter with double output and two inputs. We say it is an *inverter with 4 pins* and coupled to 8 surrounding inverters of the same type. The outer collects altogether 12 pins, 6 of them inputs and 6 outputs. Now let us make an experiment. Let us take out the inverter from the center by making its inputs and outputs equal zero. We annihilate the center of the flow pattern. But nothing comes out into the open. The 12 states of the outer pins are preserved. Only 4 output states connecting some outer inverters are changed. So, by setting the 4 innermost channels zero, - blackouting the center, - we force 4 outputs into a transition from 0 to 1 and sensitize the 4 corner inverters. These are now sensitive for inputs unequal zero. In the right hand side figure you can see the incoming energy at the lower left corner is reflected and returned by the bottom middle inverter. The whole arrangement now tends to return energy coming in from the outside. The arrangement becomes so to say visible by blacking out its center. But, in a way, it mystifies its interior.

What am I supposed to do at the outer pins in order to alter a maximum number of states? By changing the four inputs at the outer corners, indicated by the bold face **1**s, I change altogether 12+4=16 states of the network. Only the two inputs at the outer horizontal line through the center stay opaque; It is irrelevant if we change them or not change them.

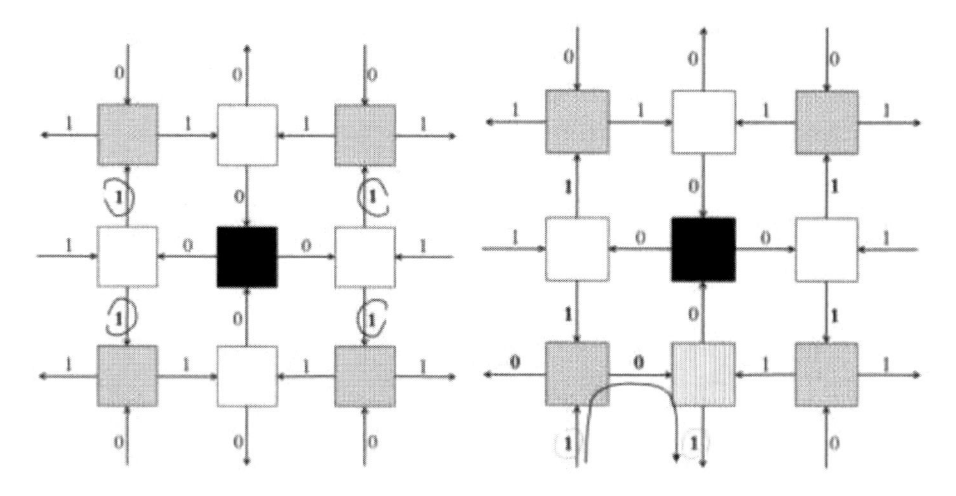

Figure 70. Blackout and sensitization of outer corner inverters.

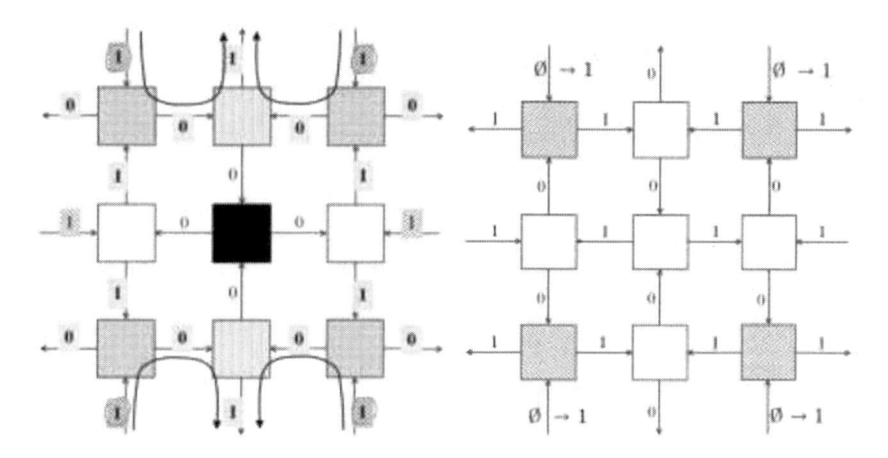

Figure 71. Blackout versus intact center.

However, if the center is intact and in stable state as on the right hand side of figure 71, we can attack the circuit; changing the 4 input signals at the corners would not alter the state of any other pin in that circuit. But given we changed those 4 inputs by switching from zero to one, the two inputs on the middle horizontal line would become sensitive. If we changed those too, all the states would flip. In a black out of the center those two inputs would just be opaque. Now let us go further and take two. There arises a peculiar instability as soon as we want to build spatially arranged memories equivalent to three-space.

It would be interesting to understand elementary quantum motion of reorientation in terms of logic circuits. There comes up a flow paradox in such an arrangement. There appear opposing flows. We shall denote these as *floppositions*. Follow the direction of the arrows in figure 73. Let us imagine three unit vectors, namely e_1 pointing from the left to the right, e_2 indicating the direction forward, and e_3 bottom up. The vertical directed area at the front-side can be symbolized by a bivector e_{13} rotating clockwise. The array in figure 72 contains 4 such areas, two of them rotating clockwise, two anti-clockwise. This is compatible with the idea to decompose motion according to the dynamic architecture of the basic Peano curve. Now consider the other vertical area directed parallel to e_{23}. See its counter-clockwise rotation in the rightmost figure of 73. It is compatible with the flow direction of area e_{13}. The two flows are in accord, amplifying each other. So, consider finally the horizontal flow on top area e_{12}. How can we fix the direction of its rotation? We cannot fix it because it is indeterminate. If we let it rotate in clockwise direction, its motion supports the flow in accord with e_{23}. But then it opposes the flow in the other vertical plane, e_{13}. On the other hand, if we let it rotate counter-clockwise, so that it it supports that flow parallel to area e_{13}, it will block the flow in e_{23}. Whatever direction we chose, it blocks some other direction. This situation is similar to an infinite negation which we symbolize by $\neg\neg\neg\neg\neg\ldots$ or by the infinite *CI* form $\langle\langle\langle\langle\langle\ldots\rangle\rangle\rangle\rangle\rangle$.

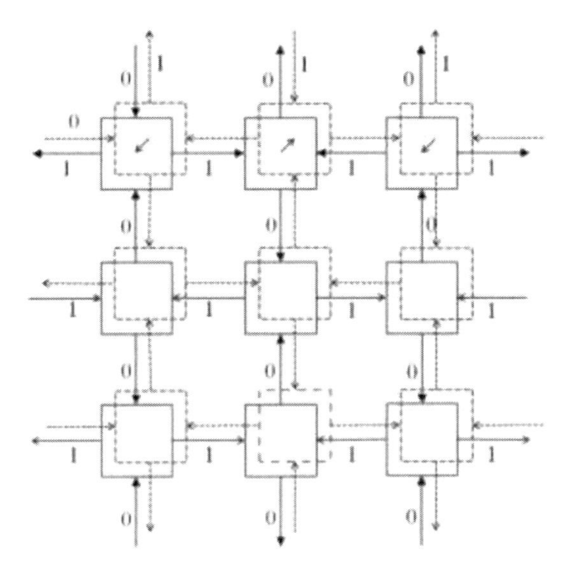

Figure 72. Double layer of dual spin veils as spatial logic network.

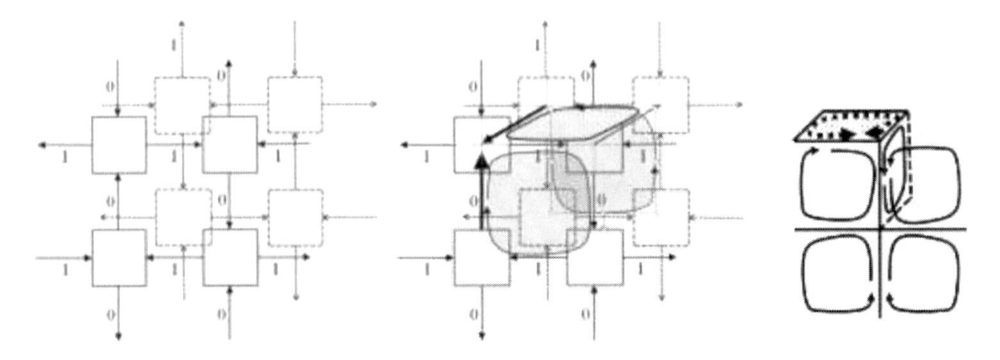

Figure 73. Flow Paradox in double layer of dual spin veils.

This paradox situation can however be overcome by letting the sense of rotation oscillate. This oscillation is the basic driver of turbulent flow in space-time. We can formulate it as a hypothesis:

Hypothesis: Flopposition is the origin of turbulence is subnuclear domains of space-time.

Finally we should be interested to construct a memory element with a consistent three-dimensional 'Euclidean' design compatible with the idea of iterants satisfying the SU(3) motion group. Is it possible? I do not go into the theory, but just show you quite plainly what it is, namely the Euclidean root inverter.

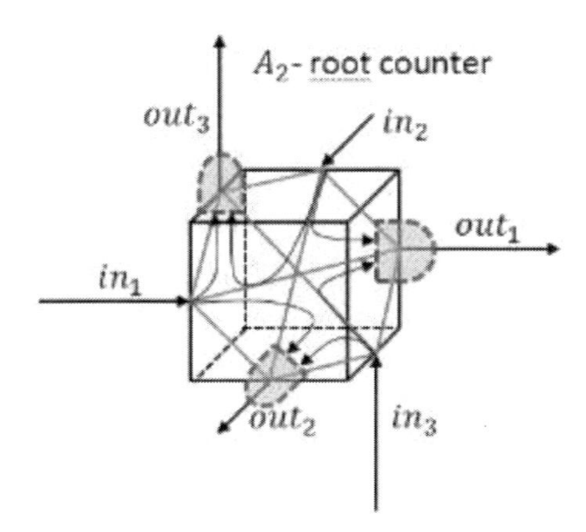

Figure 74. Space-time root inverter.

It has the following switching behavior:

Table 3. Input output relation of root inverter

in_1	in_2	in_3	out_1	out_2	out_3
1	1	1	0	0	0
1	1	0	1	1	0
1	0	1	1	0	1
1	0	0	1	1	1
0	1	1	0	1	1
0	1	0	1	1	1
0	0	1	1	1	1
0	0	0	1	1	1

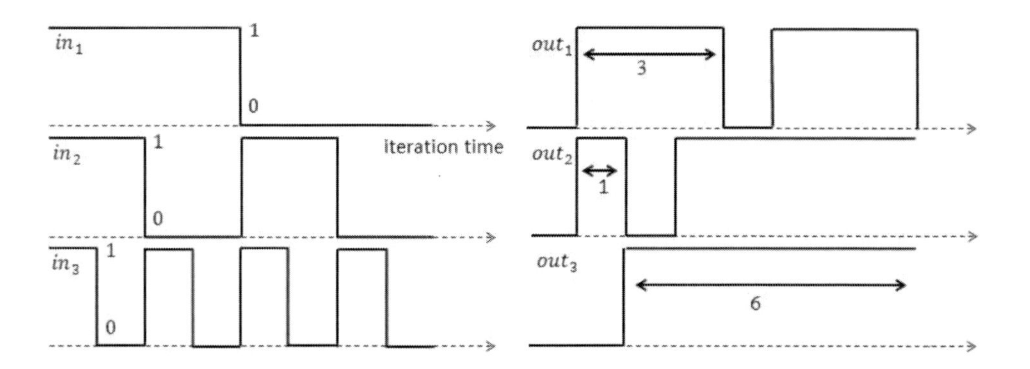

Thus the root inverter is transparent for a single zero occurring at anyone input. Every input vector that has a single zero entry and two one's is preserved and carried to the output. A total zero input vector $\langle 0,0,0\rangle$ is flipped to $\langle 1,1,1\rangle$ and reverse. We are concentrating on the following subtable (table 4).

Table 4. Reduced input output relation of root inverter

in_1	in_2	in_3	out_1	out_2	out_3
1	1	1	0	0	0
1	1	0	1	1	0
1	0	1	1	0	1
0	1	1	0	1	1
0	0	0	1	1	1

Having this in mind we investigate a feedback that generalizes the feedback of a planar inverter.

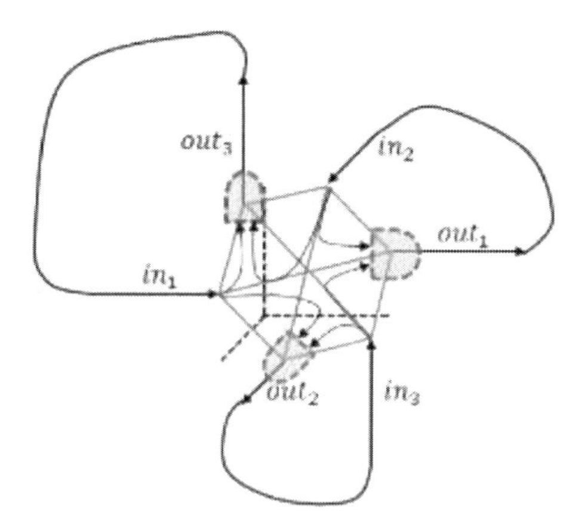

Figure 75. Recurrent root inverter.

The recurrent root inverter causes an oscillation similar like the single feedback inverter,

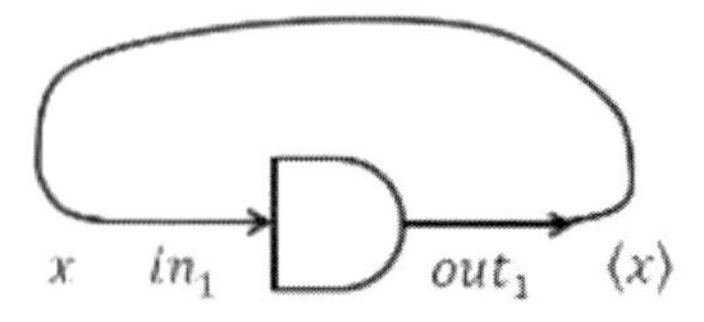

but it gives it orientation and rotation. Consider the root inverter 'at rest', void of input energy. That corresponds to the last line in table 3. Now consider a short rectangular impulse at input 1 so that the input vector is switched to $\langle 1,0,0\rangle$. According to the forth line in table 3 this should shift the output vector with some inner delay time τ to a synchronous state $\langle 1,1,1\rangle$ which with a brief delay would be carried to the inputs. But before that output state is realized the input 1 has fallen back to zero, so that we have only a state $\langle 0,1,1\rangle$ at the inputs. This is transformed to ⟨0,1,1⟩ at the outputs in accord with line four in table 4. With a delay smaller than τ this is carried to the inputs, but now permuted by the feed back lines ⟨1, 0,1⟩. This input vector with a delay τ is transferred to the outputs without a change. The feedback loops carry the output ⟨1, 0,1⟩ with a small delay permuted to the inputs, hence bringing forth a vector ⟨1,1,0⟩ to which the inverter is transparent. Alltogether we observe both an oscillation and a rotation of a signal at the outputs.

It is arduous to figure out the new image of time-space emerging within that landscape spanned by a duality between realism and constructivism. The imagination of real objects with their metaphysics suggesting an independence of the real from the observers seems to be as dissatisfying as the idea of radical constructivism where the real is born within the interaction of observer and observed. After all, what is this interaction, and where is it actually taking place? Is it independent of the actualization of awareness, of perception and meaning? If we say, in accordance with metaphysical objectivism, reality is everywhere, objects are universal appearances independent of the features of the observers, we are totalizing reality. If we insist construction is taking place everywhere, anytime and essentially all we can see is constructed, we also construct a totalitarian view of the real. But once we realize the real is perceived and perception depends on the actualization of attention by awareness, on energy and contact, we also realize that reality is alive. It is a living thing depending on energy and actual links of interaction. Now, in order that the present is able to see, or, let me say, to figure out its own past in the present, it has to see itself in the mirror. When a polarity string realizes its identity in polarity it has to face its dual iterant view to realize both unity and void. The same holds for more complex structures. The present reality has to have access to a past mirror image in order to touch itself and claim: "this is me as I was in the past". Actually this image is in the present, but it is segregated from the present in such a way that 'it' can be realized by itself as 'it'. As this process of realizing and mirroring of the present in the present is a dynamic process, it cannot be totalized in such a way that the object is invariant under all circumstances of interaction. Which kind of duality are we facing when totalizing the idea of a living reality? Do we create a dual pair of terms such as object/subject? Yes, we do, indeed. We are then making a difference between life and death, between a process alive, and a process of dying or even a process of death – the dead that is in motion either. That interval between the living and the dying has a definite meaning. In a living reality, whether we call it subject/object- or observer/observed- or any reality, there is a certain amount of interaction links and preserved structures which provide some (certain and uncertain) amount of meaning and the experience of meaningful events. In a dying reality interaction links and structures are dying and so we perceive the meaning connected with them as dying either. And then there is the interval between the living and the dead forms of knowable reality. While the living interactive system provides and sustains experience of living reality, in the corresponding dead system the life and the meaning have gone, and that system can no longer realize itself as that which it was when it was still alive. The dead cannot mirror itself in a corresponding structure of the past. Or it does not mirror itself, because it is dead. But that which is real and alive can mirror itself in that which is regarded dead, as certain forms are still there. They have been preserved during the act of dying, and some are preserved over such large periods of time, provided we have a measure for that time, that we can say "this puny rest of a bygone reality has, about 20000 years ago, been a woman playing a flute". So the living researcher mirrors himself in a skeleton and a flute in an artifact, and she visualizes a relationship, a "touch" between a living human and a then living instrument that has long passed through the doors of death. So let me compile these few statements:

- There is reality and it is alive.
- What is real and obliging depends on the relation between living and dying.
- In order that an event or object can realize its own past, there must be a living link to that substructure of the past within the present such that there is perceived a preserved meaning of that reality. In that case the present mirrors itself in the past which is part of the present.
- There exists for every reality a mirror image which reconstitutes both unity and void as perceived.
- The past in the present is based on memory– and transparency structures.
- The entanglement feeds back onto the memories, the contents of the 'past'. It is the entanglement that rearranges what to us seems 'past'. The memory is its content.

In order that any living structure of reality obtains access to 'its' past, there must be both memory and transparency structures in the perceived present. We can call these the 'local structures of meaning'. The memories are well known from informatics. The lucency of the present is better known from Richard Feynman's quantum electrodynamics. A 'real' time-space dynamic transparency structure is given by such elementary assemblies like the recurrent root inverter. This can be actualized perhaps over large regions of yet unspecified domains. Translocally acting, such phenomena as the transparent root inversion may neglect the human order of space-time and entangle galaxy clusters like neurons in a brain, albeit by disregarding many of our familiar concepts. After all, the past is just in the present. However, we do not yet know entirely which way it is in the present.

TECHNICAL APPENDIX: ITERATION OF ALGEBRA

TESSARINES, NECTARINES AND OTHER QUATERNIONS

We would like to show the proof of a conjecture by Gerhard Opfer (Janovská & Opfer 2013). Hamilton's Quaternions are not the only ones. There are, within this special context of $'i, j, k'$ some more real algebras like for instance the Tessarines, also known as bicomplex numbers, or the commutative triples as were proposed by Segre. Recently Gerhard Opfer had pointed out that apart from Quaternions, Coquaternions, Tessarines and Cotessarines there are four more 'fourfolds' (Ff, german 'Vierheit') that he denoted by 'New Algebras 1 to 4'. He presumed that the four new ones could be derived from those we already know. Actually we can show that all the eight types of Quaternions can be derived from a commutative mother algebra, namely the Cotessarines, in a natural way. Opfer gave us a list of the eight types sorted according to their algebraic properties:

Number	i^2	j^2	k^2	Name
1	-1	-1	-1	Quaternions
2	-1	$+1$	$+1$	Coquaternions
3	-1	$+1$	-1	Tessarines
4	$+1$	$+1$	$+1$	Cotessarines
5	$+1$	-1	$+1$	new Algebra 1
6	$+1$	$+1$	-1	new Algebra 2
7	$+1$	-1	-1	new Algebra 3
8	-1	-1	$+1$	new Algebra 4

Usually we are familiar with Hamilton's quaternions denoted by $\mathbb{H}$, but are less acquainted with the other seven. Hamilton's $\mathbb{H}$ satisfy the following relations:

1. Quaternions $\mathbb{H}$

$$i^2 = j^2 = k^2 = ijk = -Id \tag{447}$$

whereby in the last equation with the triple we forgot the brackets. Thereby we tacitly suggest associativity.[1] With these equations there correlate some non-vanishing commutators

[1] This peculiar enunciation I have borrowed from my former mentor Mentor Pertti Lounesto.

$$i\,j = k = -j\,i,\, j\,k = i = -k\,j,\, k\,i = j = -i\,k \tag{448}$$

for we have to get $i\,j - j\,i = k + k = 2k$ as well as $[j, k] = 2i$ und $[k, i] = 2j$. All commutators besides the one with the scalar unit Id, all are unequal zero. But the anti-commutators are equal zero.

$$\{i, j\} \stackrel{\text{def}}{=} i\,j + j\,i = 0, \{j, k\} = 0, \{k, i\} = 0. \tag{449}$$

It is well known that the first type can be represented by bivectors in the Pauli algebra or, as we would propose, in the Clifford algebra of the Minkowski space. These are in the Clifford Algebra $Cl_{3,1}$ the three magnitudes

$$i := e_{12},\ j := e_{13},\ k := e_{23} \tag{450}$$

which squared give $-Id$. Those are non commuting, hence antisymmetric bilinear forms as for example

$$i := \varphi e,\ j := g\tau,\ k := f\sigma$$

In accord with the table 2

$$\begin{aligned} \varphi e &= -e\varphi \\ \tau g &= -g\tau \\ \sigma f &= -f\sigma \end{aligned} \tag{451}$$

we verify $ii = -Id,\ jj = -Id,\ kk = -Id,\ ijk = -Id$ and the commutation realtions (xx). The other types can all be derived from the triple $\{e, f, g\}$ by the action of $\{\sigma, \varphi, \tau\}$. We just have to be aware of the table 2 and the commutations relations respectively

$$\begin{aligned} &e\sigma - \sigma e = 0,\ f\sigma + \sigma f = 0,\ g\sigma + \sigma g = 0 \\ &e\varphi + \varphi e = 0,\ f\varphi + \varphi f = 0,\ g\varphi - \varphi g = 0 \\ &e\tau + \tau e = 0,\ f\tau - \tau f = 0,\ g\tau + \tau g = 0 \end{aligned} \tag{452}$$

2. Coquaternions

We consider three unitary forms

$$i := \tau e,\ j := g,\ k := f\tau \tag{453}$$

In the Clifford Algebra $Cl_{3,1}$ these can be the Grassmann monomials

$$i := e_{1234},\ j := e_{124},\ k := e_{3} \tag{454}$$

So we get the multiplication table

Table 5. Multiplication table of Coquaternions

Id	i	j	k
i	$-Id$	k	$-j$
j	$-k$	Id	$-i$
k	j	i	Id

3. Tessarines

$$i := \tau e,\ j := \sigma,\ k := \varphi e \tag{455}$$

isomorph mit Grassmann Monomials

$$i := e_{1234},\ j := e_{34},\ k := e_{12} \tag{456}$$

Table 6. Multiplication table of Tessarines

Id	i	j	k
i	$-Id$	k	$-j$
j	k	Id	i
k	$-j$	i	$-Id$

4. Cotessarines – Motheralgebra[2]

With Cotessarines we have positive definite signature. All three base units squared give $+Id$.

$$i \stackrel{\text{def}}{=} e,\ j \stackrel{\text{def}}{=} f,\ k \stackrel{\text{def}}{=} g \tag{70}$$

The multiplication table is given by the Klein 4 group.

$$\begin{aligned} e\,f &= [+1,+1,-1,-1][+1,-1,-1,+1] = [+1,-1,+1,-1] = g \\ e\,g &= [+1,+1,-1,-1][+1,-1,+1,-1] = [+1,-1,-1,+1] = f \\ fg &= [+1,-1,-1,+1][+1,-1,+1,-1] = [+1,+1,-1,-1] = e \\ e^2 &= [+1,+1,-1,-1][+1,+1,-1,-1] = [+1,+1,+1,+1] = Id \end{aligned} \tag{457}$$

Also $f^2 = g^2 = Id$, hence what seems to us to be the most simple table at all.

[2] This name has been coined by Wolfgang Sprössig. - There was an exchange of e-mails with Sprössig and Opfer on the whole question on 10th January 2014 and the following weeks.

Table 7. Multiplication table of Cotessarines

Id	i	j	k
i	Id	k	j
j	k	Id	i
k	j	i	Id

In der Standard representation of Clifford Algebra $Cl_{3,1}$ for Monomials of grade 0,1,2,3

Table 8. Multiplication table of Cartan subalgebra in $Cl_{3,1}$

Id	e_1	e_{24}	e_{124}
e_1	Id	e_{124}	e_{24}
e_{24}	e_{124}	Id	e_1
e_{124}	e_{24}	e_1	Id

5. Nectarines

Define magnitudes

$$i \stackrel{\text{def}}{=} \varphi,\ j \stackrel{\text{def}}{=} \varphi f,\ k = f \tag{458}$$

squared they give

$$i^2 = +Id,\ j^2 = -Id,\ k^2 = f^2 = +Id \tag{459}$$

We have indeed

$$i\,j = \varphi\varphi f = Id\,f = f = k \qquad \text{since}$$
$$j\,i = \varphi f \varphi = -f\varphi\varphi = -f = -k \qquad \text{hence } i\,j = k = -j\,i$$

with table 2 we calculate

Table 9. Multiplication table of Nectarines

Id	i	j	k
i	Id	k	j
j	$-k$	$-Id$	i
k	$-j$	$-i$	Id

6. Conectarines

$$i \stackrel{\text{def}}{=} f,\ j \stackrel{\text{def}}{=} \varphi,\ k = f\varphi \tag{460}$$

verify that

$$i^2 = f\,f = +Id,\ j^2 = \varphi\,\varphi = +Id,\ k^2 = f\varphi f\varphi = -\varphi f f\varphi = -\varphi Id\varphi = -Id \quad (461)$$

therefore signature $\{+\ +\ -\}$ for triple $\{i, j, k\}$. This is indeed the ‚new algebra 2' found by Gerhard Opfer.

Table 10. Multiplication table of Conectarines – 'New algebra 2'

Id	i	j	k
i	Id	k	j
j	$-k$	Id	$-i$
k	$-j$	i	$-Id$

The triple e_2, e_{24}, e_4 can be used for representation:

$$i \overset{\text{def}}{=} e_2,\ j \overset{\text{def}}{=} e_{24},\ k = e_4 \quad (462)$$

7. Tangerines

$$i \overset{\text{def}}{=} \sigma,\ j \overset{\text{def}}{=} \varphi e, k := \tau e \quad (463)$$

We realize that pairs φ, e and τ, e anticommute, so

$$\varphi e = -e\varphi \quad (464)$$
$$\tau e = -e\tau$$

We calculate squares

$$i^2 = +Id,\ j^2 = -Id,\ k^2 = -Id \quad (465)$$

Further we get the triple product

$$i\,j\,k = \sigma\varphi e\ \tau e = \tau e\ \tau e = -e\ \tau\tau\ e = -e\ Id\ e = -Id \quad (466)$$

Table 11. Multiplication table of Tangerines – 'New algebra 3'

Id	i	j	k
i	Id	k	j
j	k	$-Id$	$-i$
k	j	$-i$	$-Id$

8. Cotangerines

$$i \overset{\text{def}}{=} \varphi e,\ j \overset{\text{def}}{=} f\varphi, k := g$$

We proof as before

$$k = \varphi e\, f\varphi = \varphi g\varphi = \varphi\varphi g = Id\ g = g \text{ and} \tag{467}$$
$$i^2 = \varphi e\ \varphi e = -e\ \varphi\varphi\ e = -Id$$
$$j^2 = f\varphi\ f\varphi = -\varphi\ ff\ \varphi = -Id$$
$$k^2 = g\ g = +Id \quad \text{bzw.} \quad i\,j\,k = +Id$$

further $i\,j = \varphi\ e\ \varphi\ f = \varphi\ g\ \varphi = g\ \varphi\varphi = g := k$
$j\,i = f\ \varphi\varphi\ e = f\ e = g := k$, as well as $i\,k = \varphi\ e\ g = \varphi\ f = -f\varphi := -j$; so

Table 12. Multiplication table of Cotangerines – 'New algebra 4'

Id	i	j	k
i	$-Id$	k	$-j$
j	k	$-Id$	$-i$
k	$-j$	$-i$	Id

In Clifford Algebra $Cl_{3,1}$ for example: $i := e_{13},\ j := e_{1234},\ k := e_{24}$ (468)

ITERANTS OF CLIFFORD ALGEBRA

We assume that the basis cl of every associative Clifford algebra can be decomposed into a product of some set of symmetries S and some set $\mathcal{J}$ of binary sequences. Presumably, if it is true, the following diagram 1 holds for a general category of objects. Let us look at Minkowski algebra since it provides a fortunate example. First we reconsider $cl_{(3,1)}$ as a set of unit monomials, in our special case 16, and the diagonal functor $\Delta\colon cl \longrightarrow cl \times cl$ together with projection morphisms $f_1\colon cl \to S$ from the Clifford basis to a set of permutations on the left hand, and $f_2\colon cl \to \mathcal{J}$ and to a set of binary sequences $\mathcal{J}$ on the right. The diagonal functor also assigns to the universal morphism F the pair (F, F). The iterants $It \stackrel{\text{def}}{=} S \times \mathcal{J}$ are determined by this natural morphism from the functor Δ to the pair $(S, \mathcal{J})$. Any function $F\colon cl \to S \times \mathcal{J}$ is uniquely determined by composites $f_1 F$ and $f_2 F$. Given cl together with two functions f_1, f_2 there is a unique function F which makes the diagram below commute. We can write for this $F\ e_\alpha = (f_1 e_\alpha, f_2 e_\alpha)$ where e_α is any base unit monomial.

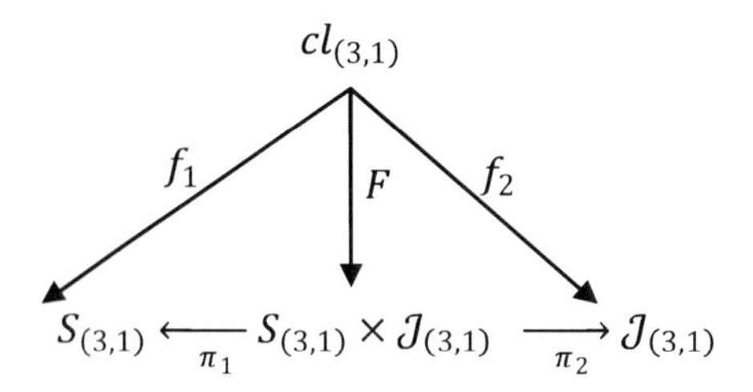

Diagram 1.

A diagram like this signifies the introduction of the concept of a natural functorial relation in category theory. In the case of Clifford algebra of Minkowski space we have to put for two distinct Klein 4 groups $'K4'$ two generating sets that can be paired to form base units of a graded multivector group, namely

$$S := S_{(4)} = \{Id, \varphi, \sigma, \tau\} \simeq K4 \qquad (469)$$
$$\mathcal{J} := \mathcal{J}_{(4)} = \{Id, e, f, g\} \simeq K4$$

Hence, given double transpositions S and periodic binary sequences $\mathcal{J}$, the pair (π_1, π_2) can be considered universal among pairs of functions from Clifford algebra to set S of symmetries and set $\mathcal{J}$ of binary sequences because any pair (f_1, f_2) factors uniquely through F. Up to a bijection, this feature describes the Cartesian product $S \times \mathcal{J}$ in a unique way. In the category of groups, this feature delineates up to bijection a semidirect product of groups. As can be learned from the example, in the category of Clifford algebras with the arrangement of products a further degree of freedom is introduced. Namely, there is a sign ambiguity brought in by the signature group homomorphism of the symmetric group S_n, that is, $sg\colon S_n \to \{1, -1\}$ characteristic for double-groups and quantum groups. The alternating group A_n is the commutator subgroup of the symmetric group S_n, a subgroup with index 2. So A_n represents the kernel of the signature homomorphism. In our example which deals with the Minkowski algebra only, the Klein 4 group comes in as commutator subgroup of A_4 consisting strictly of the double transpositions φ, σ, τ. The $Cl_{3,1}$ over some field is derived from the Cartesian product $S \times \mathcal{J}$ by respecting the commutator subgroup of the quaternion group which is equal to $\{+1, -1\}$. So we get an array of 16 pairs times plus or minus one (theorem 7, equation 72):

$$Cl_{3,1} := span_{\ldots}\{IdId, Id\,e, -\varphi Id, \tau f, -\varphi f, \varphi e, -\tau g, \varphi g, -\sigma f, Idf, \sigma Id, -\sigma g, Idg, \sigma e, -\tau Id, \tau e\}$$

With two distinct quaternion subgroups generated by base units

$\{e_{12}, e_{13}, e_{23}\} \simeq \{\varphi e, -\tau g, -\sigma f\}$ subgroup of bivectors
$\{e_4, e_{123}, e_{1234}\} \simeq \{-\varphi f, -\sigma g, \tau e\}$ subgroup of 'timespace'.

Arranging the Cartesian product $S \times \mathcal{J}$ with correct signatures in accord with the Clifford basis of $Cl_{3,1}$

Id	e	f	g
$-\varphi$	φe	$-\varphi f$	φg
σ	σe	$-\sigma f$	$-\sigma g$
$-\tau$	τe	τf	$-\tau g$

we can see the exact sequence of the involved group homomorphism. One is tempted to suspect that a corresponding construction can be carried out for all quadratic Clifford algebras. But it turns out the 'universal' functorial relation depicted in diagram 1 is a solitary and distinguished property of the Clifford algebra $Cl_{3,1} \simeq Cl_{2,2}$.

Theorem 8:

(i) Every 'logic' Clifford algebra in neutral signature can be represented as an iterant algebra.

(ii) Diagram 1 is featuring a rare functorial relationship between iterants of grade 4 and Minkowski algebra $Cl_{3,1} \simeq Cl_{2,2}$ which cannot be generalized to algebras of arbitrary grade.

(iii) There is an exact similarity of real iterant representations for 'logic' $Cl_{m,m}$ Clifford algebras with neutral signature.

Proof: The construction $S \times \mathcal{J}$ comes in as right adjoint of the Clifford algebra structure. With $Cl_{3,1}$ both components S and $\mathcal{J}$ are abelian. Hence we need to investigate the general case. Let us look upon table 9 as both example and template for the construction of the iterant space $\mathcal{J}$. In fourfold iterants we obtain a total of $2^4 = 16$ polarity strings (last column in table 9). These 16 can also be conceived as all possible sign combinations of the 4 primitive idempotents (see column 4 in table 9) of Clifford algebra $Cl_{3,1}$ having dimension 4; we write abstractly $cl_{(3,1)}$ for the set of Clifford base units. The polarity strings can be represented as vectors in 4-dimensional linear space, the color space, having basis $\{Id, e_1, e_{24}, e_{124}\}$. Taking notice of the operations in a Clifford algebra, we realize that two unit monomials, namely e_1, e_{24} are enough to generate this space. With these observations we can extract the construction plan for iterants in iterant spaces $\mathcal{J}$ of arbitrary dimension.

To match to every $cl_{(p,q)}$ an iterant space $\mathcal{J}_{(p,q)}$ we first verify that in the standard basis of $Cl_{p,q}$ there are exactly $k = q - r_{q-p}$ non-scalar elements $e_{\alpha_i}, e_{\alpha_i}^2 = Id$ which commute (just as $e_1, e_{24} \in Cl_{3,1}$), that is, $e_{\alpha_i}e_{\alpha_j} = e_{\alpha_j}e_{\alpha_i}$.[3] The primitive idempotents are given by products of k mutually non-annihilating idempotents that correspond with the monomials e_{α_i} having form

$$f_{(\dots)} := \tfrac{1}{2}(Id \pm e_{\alpha_1})\tfrac{1}{2}(Id \pm e_{\alpha_2}) \ \dots \ \tfrac{1}{2}(Id \pm e_{\alpha_k}) \tag{470}$$

Hence the Clifford representatives of polarity strings in $\mathcal{J}_{(p,q)}$ are generated by k commuting base unit monomials, and the dimension of the iterant space is 2^k. For practical purposes we need the Radon-Hurwitz number r_i for $i \in \mathbb{Z}$ with the small table

i	0	1	2	3	4	5	6	7
r_i	0	1	2	2	3	3	3	3

And equations for negative values of i

$$r_{-1} = -1 \quad \text{and} \quad r_{-i} = 1 - i + r_{i-2} \tag{471}$$

The Clifford algebra decomposes into an iterant component $\mathcal{J}_{(p,q)}$ and some symmetry component $S_{(p,q)}$. From the diagram 1 we conclude that $S_{(p,q)}$ must have the order 2^{n-k}

[3] The reader is advised to go to see section 17.5 in Pertti Lounesto's "Clifford Algebras and Spinors", p. 226.

since the iterant space $\mathcal{J}_{(p,q)}$ has dimension 2^k and the Clifford algebra has $dim(Cl_{p,q}) = 2^n$. With the above Radon-Hurwitz numbers we can calculate a table of dimensions

Table 13. Dimensions of symmetry component $S_{(p,q)}$ and iterant space $\mathcal{J}_{(p,q)}$ in Clifford algebra $Cl_{p,q}$

$S_{(p,q)}, \mathcal{J}_{(p,q)}$	$q = 0$	1	2	3	4	5	6	7
$p = 0$	1,1	$2^1, 1$	$2^2, 1$	$2^2, 2^1$	$2^3, 2^1$	$2^3, 2^2$	$2^3, 2^3$	$2^3, 2^4$
1	$1, 2^1$	$2^1, 2^1$	$2^2, 2^1$	$2^3, 2^1$	$2^3, 2^2$	$2^4, 2^2$	$2^4, 2^3$	$2^4, 2^4$
2	$2^1, 2^1$	$2^1, 2^2$	$2^2, 2^2$	$2^3, 2^2$	$2^4, 2^2$	$2^4, 2^3$	$2^5, 2^3$	$2^5, 2^4$
3	$2^2, 2^1$	$2^2, 2^2$	$2^2, 2^3$	$2^3, 2^3$	$2^4, 2^3$	$2^5, 2^3$	$2^5, 2^4$	$2^6, 2^4$
4	$2^3, 2^1$	$2^3, 2^2$	$2^3, 2^3$	$2^3, 2^4$	$2^4, 2^4$	$2^5, 2^4$	$2^6, 2^4$	$2^6, 2^5$
5	$2^3, 2^2$	$2^4, 2^2$	$2^4, 2^3$	$2^4, 2^4$	$2^4, 2^5$	$2^5, 2^5$	$2^6, 2^5$	$2^7, 2^5$
6	$2^4, 2^2$	$2^4, 2^3$	$2^5, 2^3$	$2^5, 2^4$	$2^5, 2^5$	$2^5, 2^6$	$2^6, 2^6$	$2^7, 2^6$
7	$2^4, 2^3$	$2^5, 2^3$	$2^5, 2^4$	$2^6, 2^4$	$2^6, 2^5$	$2^6, 2^6$	$2^6, 2^7$	$2^7, 2^7$

Table 14. Clifford Algebras $Cl_{p,q}, p + q < 8$ and dimension of symmetry-component $S_{(p,q)}$ and iterant component $\mathcal{J}_{(p,q)}$ by exponent of 2

$p - q$ / $p + q$	−7	−6	−5	−4	−3	−2	−1	0	1	2	3	4	5	6	7
0								$\mathbb{R}$ $0\mid 0$							
1							$\mathbb{C}$ $1\mid 0$		$^2\mathbb{R}$ $0\mid 1$						
2						$\mathbb{H}$ $2\mid 0$		$\mathbb{R}(2)$ $1\mid 1$		$\mathbb{R}(2)$ $1\mid 1$					
3					$^2\mathbb{H}$ $2\mid 1$		$\mathbb{C}(2)$ $2\mid 1$		$^2\mathbb{R}(2)$ $1\mid 2$		$\mathbb{C}(2)$ $2\mid 1$				
4				$\mathbb{H}(2)$ $3\mid 1$		$\mathbb{H}(2)$ $3\mid 1$		$\mathbb{R}(4)$ $2\mid 2$		$\mathbb{R}(4)$ $2\mid 2$		$\mathbb{H}(2)$ $3\mid 1$			
5			$\mathbb{C}(4)$ $3\mid 2$		$^2\mathbb{H}(2)$ $3\mid 2$		$\mathbb{C}(4)$ $3\mid 2$		$^2\mathbb{R}(4)$ $2\mid 3$		$\mathbb{C}(4)$ $3\mid 2$		$^2\mathbb{H}(2)$ $3\mid 2$		
6		$\mathbb{R}(8)$ $3\mid 3$		$\mathbb{H}(4)$ $4\mid 2$		$\mathbb{H}(4)$ $4\mid 2$		$\mathbb{R}(8)$ $3\mid 3$		$\mathbb{R}(8)$ $3\mid 3$		$\mathbb{H}(4)$ $4\mid 2$		$\mathbb{H}(4)$ $4\mid 2$	
7	$^2\mathbb{R}(8)$ $3\mid 4$		$\mathbb{C}(8)$ $4\mid 3$		$^2\mathbb{H}(4)$ $4\mid 3$		$\mathbb{C}(8)$ $4\mid 3$		$^2\mathbb{R}(8)$ $3\mid 4$		$\mathbb{C}(8)$ $4\mid 3$		$^2\mathbb{H}(4)$ $4\mid 3$		$\mathbb{C}(8)$ $4\mid 3$

The space of iterants $\mathcal{J}_{(p,q)}$ is a monoid of self-inverse elements, that is, and abelian group. The order of the symmetry component $S_{(p,q)}$ is a power of 2. So this could firstly be some abelian group C_2^{n-k} or secondly what we call a 'generalized quaternion group' which would be non abelian, or thirdly some group 'in between', some dihedral group. Consider the second possibility Let ≥ 2 , then $S_{(p,q)}$ is represented by the dicyclic group of order $4m$, hence $4m = 2^{n-k}$. Consider the cases where $m = 2$ and the order of the group is 8, therefore 2^{n-k}, hence $n - k = 3$. There are many positions in the table where the order of group $S_{(p,q)}$ is equal to 2^3. These correlate with the algebras $Cl_{0,4}, Cl_{0,5}, Cl_{0,6}, Cl_{0,7}, Cl_{1,3}, Cl_{1,4}, Cl_{2,3}, Cl_{3,3}, Cl_{4,0}, Cl_{4,1}, Cl_{4,2}, Cl_{4,3}, Cl_{5,0}$. Each of these geometric algebras has a symmetric left component possibly equal to the group of quaternions, and each has an iterant given by a matrix template the dimension of which is given by the integer on the right hand side, that is, we get $Mat(2, \ldots)$, $Mat(2, \ldots)$, $Mat(2, \ldots)$, $Mat(4, \ldots)$, $Mat(4, \ldots)$, $Mat(4, \ldots)$, $Mat(4, \ldots)$, $Mat(4, \ldots)$, $Mat(8, \ldots)$, $Mat(8, \ldots)$, $Mat(8, \ldots)$, $Mat(16, \ldots)$, $Mat(16, \ldots)$. We have to find out about the structure of these symmetry components.

$\mathbb{F}(d)$ denotes the real algebra of $d \times d$-matrices $Mat(d, \mathbb{F})$ with entries in the ring $\mathbb{F} = \mathbb{R}, \mathbb{C}, \mathbb{H}, {}^2\mathbb{R}, {}^2\mathbb{H}$.

The First Record, Clifford Algebra $Cl_{0,0}$

Let us explain the elements of table 14 one after the other. When the second number in an entry "… | …" is equal to zero so as in $0|0$, $1|0$, $2|0$, this zero signifies $k = 0$, the absence of an idempotent. Then $\mathcal{J}_{(p,q)}$ is framed by a $2^0 \times 2^0 = 1 \times 1$- matrix, that is, it contains one number only, one universal iterant identity, represented by the repetitive code $[Id, Id, Id, \ldots]$. The ring $\mathbb{R}$ combined with the $0|0$ means repeating one real number.

The Second Record, Clifford Algebra $Cl_{0,1}$

Likewise for $Cl_{0,1}$ the $\mathbb{C}$ with $1|0$ below denotes a repeated complex number. This has $2^1 = 2$ generators on the left, scalar and imaginary unity. It could be argued that the complex numbers are the same as twofold iterants with the 'temporal shift' η, since we can establish the correspondence

$$It_{(2)} \stackrel{\text{def}}{=} [a, b] + [c, d]\eta \leftrightarrow \begin{pmatrix} a & c \\ d & b \end{pmatrix} \tag{472}$$

But the multiplicative unity in $Mat(2, \mathbb{R})$ is $Id_2 = \begin{pmatrix} +1 & \\ & +1 \end{pmatrix}$ and the unit imaginary $I := \begin{pmatrix} & -1 \\ +1 & \end{pmatrix}$. Every complex number can thus be represented by a certain real 2×2-matrix:

$$\mathbb{C} \leftrightarrow Mat(2, \mathbb{R}) \text{ by } a + ib \leftrightarrow \begin{pmatrix} a & -b \\ b & a \end{pmatrix} \tag{473}$$

Obviously, there are more objects of the form $It_{(2)}$ than there are complex numbers (ordered pairs). Therefore the iterant algebra of 2-fold iterants with time-shift includes the complex numbers, but the two algebras are not isomorphic.

The Third Record, Clifford Algebra $Cl_{1,0}$

Next consider the algebra where $p + q = 1 = p - q$. This is $Cl_{1,0}$ with the table entry ${}^2\mathbb{R}$ '$0|1$', the Clifford algebra generated by $\mathbb{R}^{1,0}$. It has $k = 1$, $2^k = 2$ idempotents $(Id \pm e)/2$

defined by its unit vector e. So it provides one alternating code or iterant $[+1,-1,+1,-1]$. $Cl_{1,0}$ is the Clifford algebra of the Euclidean line spanned by $\{Id, e\}$ where $e^2 = Id$. This is isomorphic, as an associative algebra, to the double-field of 'split complex numbers'. In $Cl_{1,0}$ we define the identity by the pair $Id = (1,1)$ and the unit vector by $e = (1,-1)$. The iterant space $\mathcal{J}_{(1,0)}$ corresponding with the $Cl_{1,0}$ provides iterant sequences $[a,b] = a,b,a,b,a,b,\dots$ that can be obtained from the unit iterant $[+1,+1]$, the repetitive, and $[+1,-1]$ the alternating code. Namely we have $[a,b] = x[+1,+1] + y[+1,-1]$ with $x = (a+b)/2$ and $y = (a-b)/2$. The Clifford algebra of the real line gives us the first and smallest iterant object, but yet without any shift, swap or permutation component, since the $S_{(1,0)}$ has dimension 1 due of the '$0|1$', hence equal to $\{Id\}$.

The Forth Record, Clifford Algebra $Cl_{0,2}$

The forth record $\mathbb{H}$ with $2|0$, means repetition of a quaternion number, because the 2 in $2|0$ denotes $2^2 = 4$ quaternion generators Id, I, J, K for $S_{(0,2)}$ repeated with no alternation: the $\dots|0$.

The Fifth Record, Clifford Algebra $Cl_{1,1}$

As we come to $Cl_{1,1}$ we encounter the entry $\mathbb{R}(2)$ with '$1|1$' telling us that $\mathcal{J}_{(1,1)}$ is determined by $k = 1$, hence $2^k = 2$ idempotents having form $f_{(\pm)} := \frac{1}{2}(Id \pm e_1)$. The Clifford algebra $Cl_{1,1}$ is generated by the hyperbolic plane $\mathbb{R}^{1,1}$ endowed with a quadratic form $(u,v) \to uv$. By a change of variables $u = x + t$, $v = x - t$ we get $(x,t) \to x^2 - t^2$, which denotes the indefinite, neutral signature, the Lorentz signature of $\mathbb{R}^{1,1}$. $Cl_{1,1}$ is isomorphic, as an associative algebra, to the matrix algebra $Mat(2,\mathbb{R})$ by correspondences

$$Id_{(2)} = \begin{pmatrix} 1 & \\ & 1 \end{pmatrix} e_1 = \begin{pmatrix} 1 & \\ & -1 \end{pmatrix} \quad e_2 = \begin{pmatrix} & 1 \\ -1 & \end{pmatrix} e_{12} = \begin{pmatrix} & 1 \\ 1 & \end{pmatrix} \tag{474}$$

Much the same as in table 9 we form the terms $f_+ + f_-$, $f_+ - f_-$, $-f_+ + f_-$ and $-f_+ - f_-$. These correspond with the 2-iterants $[+1,+1]$, $[+1,-1]$, $[-1,+1]$, $[-1,-1]$. So we have the two familiar generators $[+1,-1]$, and its shifted $[-1,+1]$. But now the non-trivial $S_{(1,1)}$ having order 2. It consists of a swop η that carries $[+1,-1]$ to $[-1,+1]$. It does so by commuting itself via $\eta[a,b] = [b,a]\eta$ and $\eta^2 = Id = [+1,+1]$. So $Cl_{1,1}$ is isomorphic with the very first 'iterant algebra' that had been introduced and investigated by Kauffman. It can be used to formulate discrete versions of the Schrödinger equation and the Dirac equation in the hyperbolic plane.

The Sixth Record, Clifford Algebra $Cl_{2,0}$

The next table entry denotes $Cl_{2,0}$ with the same representation in $Mat(2,\mathbb{R})$ and '1|1'. So we have the same iterant algebra with the generating 'Schrödinger iterant' $[+1,-1]$ and a swap. This equivalence goes back to the isomorphism of $Cl_{1,1} \simeq Cl_{2,0}$. So, let us anticipate that this pattern of isomorphism has to repeat for $Cl_{2,2} \simeq Cl_{3,1}$, $Cl_{3,3} \simeq Cl_{4,2}$ as we have $Cl_{p,q} \simeq Cl_{q+1,p-1}$. For $Cl_{2,2}$; $p+q=4, p-q=0$, we find the record $\mathbb{R}(4)$ with '2|2'. This is the same record as for $Cl_{3,1}$; $p+q=4, p-q=2$. Here we have 4-fold real iterants of self inverses forming a group of of order 4 (Klein 4 group) and $2^2=4$ permutations in $S_{(2,2)}$, namely the transpositions $\{Id, \varphi, \sigma, \tau\}$ isomorphic to the Klein 4 group too. Both can be represented in $Mat(4,\mathbb{R})$. The algebras $Cl_{3,3}$ and $Cl_{4,2}$ are both isomorphic to a product of 8-fold self-inverse iterants with a permutation subgroup having order $|S_{(3,3)}| = 2^3 = 8$. This is the dicyclic group of order $4m=8$, for $m=2$ the group of quaternions. So, briefly, The Clifford algebras $Cl_{3,3}$ and $Cl_{4,2}$ can be conceived as products of real valued self-inverse 8-iterants with quaternions. Both $S_{(3,3)}$ and $\mathcal{J}_{(3,3)}$can be represented in $Mat(8,\mathbb{R})$.

The Seventh Record, Clifford Algebra $Cl_{0,3}$

The case $p+q=3$, $p-q=-3$ denotes the algebra $Cl_{0,3}$ with table record $^2\mathbb{H}$ and '2|1'. Due to '...|1' we have two orthogonal primitive idempotents, hence a set with 2-iterants. The iterants are defined over quaternion numbers. The group of iterants $\mathcal{J}_{(0,3)}$ has two elements only. These are the well known $[+1,-1]$, the 'Schrödinger iterant' and the identity $[+1,+1]$. According to the classification record '2|...'the permutation component has $2^2=4$ elements only, how come? The reason is simply that the third quaternion unit is a product of two others. Each quaternion can be expressed in the following form $X := x_0 + x_1e_1 + x_2e_2 + x_3e_{12}$ where e_1, e_2, e_{12} are base units of $Cl_{0,3}$ equal to unit quaternions i, j, k. These satisfy the multiplication rules $i^2 = j^2 = k^2 = Id$, $ij = k = -ji$, $jk = i = -kj$, $ki = j = -ik$, and $ijk = -1$. The 8-dimensional $Cl_{0,3}$ is isomorphic, as an associative algebra, to the direct sum $\mathbb{H}\oplus\mathbb{H}$ and can be obtained from the basic iterants by the correspondences

Table 15. Quaternion iterants for $Cl_{0,3}$

$Cl_{0,3}$	$\mathbb{H}\oplus\mathbb{H}$
Id	$[+Id,+Id]$
e_1, e_2, e_3	$[i,-i], [j,-j], [k,-k]$
e_{23}, e_{31}, e_{12}	$[i,i], [j,j], [k,k]$
e_{123}	$[-Id,+Id]$

We need not complete the proof here, for it is very long. But we jump forward to the records of Clifford algebras generated by a real vector space having dimension 6. So these algebras have 64 unit monomials in the basis. We investigate

The 22nd, 25th and 26th Record, Clifford Algebras $Cl_{0,6}$, $Cl_{3,3}$, and $Cl_{4,2}$

I would like to demonstrate the general method for the construction of basic iterants in $\mathcal{J}_{(p,q)}$ having dimension 2^k where k is the number of terms that bring forth the 2^k primitive idempotents in $Cl_{p,q}$. Consider $Cl_{3,3}$ with $p+q=6$, $p-q=0$. We find the record $\mathbb{R}(8)$ with the pair $'3|3'$. We assume that it has an abelian group of 8-fold iterants since $2^k=8$ and a permutation subgroup $S_{(3,3)}$ acting from the left having order $2^{n-k}=2^3=8$ too. We already know this is the dicyclic group of order $4m=8$, for $m=2$ hence the group of quaternions. In other words, the functorial decomposition of the Clifford algebra $Cl_{3,3}$ immediately suggests that this Clifford algebra can be comprehended as being the product of the action of a quaternion group on the characters of 8-fold self-inverse binary iterants. How do we find the forms of these iterants? Their forms are derived from the diagonal representations of the primitive idempotents. In $Cl_{3,3}$ the 8 primitive, mutually annihilating idempotents are formed by 3-fold products in standard notation

$$f_{...} := \frac{1}{2}(Id \pm e_{14})\frac{1}{2}(Id \pm e_{25})\frac{1}{2}(Id \pm e_{36}) \qquad (475)$$

where the base units e_1, e_2, e_3 are space-like and contribute to positive signature whereas the e_4, e_5, e_6 are time-like and provide negative signature. Hence the bivectors e_{14}, e_{25}, e_{36} represent space-time areas with positive signature, that is, they satisfy $e_{14}^2 = Id$, $e_{25}^2 = Id$, $e_{36}^2 = Id$

Contrasting the hypercomplex units as for example

$$e_{12}^2 = -Id,\ e_{13}^2 = -Id,\ e_{23}^2 = -Id \quad \text{and so on.} \qquad (476)$$

We give the explicit list in standard representation of idempotents primitive in $Cl_{3,3}$

$$\begin{aligned}
f_1 &= Id - e_{123456} + e_{14} + e_{25} + e_{36} - e_{1245} - e_{1346} - e_{2356} \qquad (477)\\
f_2 &= Id + e_{123456} + e_{14} + e_{25} - e_{36} - e_{1245} + e_{1346} + e_{2356}\\
f_3 &= Id + e_{123456} + e_{14} - e_{25} + e_{36} + e_{1245} - e_{1346} + e_{2356}\\
f_4 &= Id - e_{123456} + e_{14} - e_{25} - e_{36} + e_{1245} + e_{1346} - e_{2356}\\
f_5 &= Id + e_{123456} - e_{14} + e_{25} + e_{36} + e_{1245} + e_{1346} - e_{2356}\\
f_6 &= Id - e_{123456} - e_{14} + e_{25} - e_{36} + e_{1245} - e_{1346} + e_{2356}\\
f_7 &= Id - e_{123456} - e_{14} - e_{25} + e_{36} - e_{1245} + e_{1346} + e_{2356}\\
f_8 &= Id + e_{123456} - e_{14} - e_{25} - e_{36} - e_{1245} - e_{1346} - e_{2356}
\end{aligned}$$

contrived by Maple Clifford

This provides a sign-table for the unit monomials. We can think of the primitive idempotent f_i as a diagonal in a matrix representation $Mat(8,\mathbb{R})$ with only one entry, namely $'i,i'$ occupied by unity. If we read the sign-table column-wise, we can calculate the iterants as they are represented by the elements of the Cartan subalgebra $\{Id, e_{123456}, e_{14}, e_{25}, e_{36}, e_{1245}, e_{1346}, e_{2356}\} \subset Cl_{3,3}$

Table 16. Signature table for idempotents primitive in $Cl_{3,3}$

	Id	g	a	b	c	d	e	f
f_1	+	−	+	+	+	−	−	−
f_2	+	+	+	+	−	−	+	+
f_3	+	+	+	−	+	+	−	+
f_4	+	−	+	−	−	+	+	−
f_5	+	+	−	+	+	+	+	−
f_6	+	−	−	+	−	+	−	+
f_7	+	−	−	−	+	−	+	+
f_8	+	+	−	−	−	−	−	−

We calculate for instance:

$$g = -f_1 + f_2 + f_3 - f_4 + f_5 - f_6 - f_7 + f_8 = e_{123456} \quad (478)$$

In this way we obtain all the eight iterants which generate a commutative vector space. The eight Clifford monomials generate a multivector group isomorphic to $\mathbb{Z}_2^4$.

$$\begin{aligned}
Id &= [+1,+1,+1,+1,+1,+1,+1,+1] \simeq Id_{[8]} \\
g &= [-1,+1,+1,-1,+1,-1,-1,+1] \simeq e_{123456} \\
a &= [+1,+1,+1,+1,-1,-1,-1,-1] \simeq e_{14} \\
b &= [+1,+1,-1,-1,+1,+1,-1,-1] \simeq e_{25} \\
c &= [+1,-1,+1,-1,+1,-1,+1,-1] \simeq e_{36} \\
d &= [-1,-1,+1,+1,+1,+1,-1,-1] \simeq e_{1245} \\
e &= [-1,+1,-1,+1,+1,-1,+1,-1] \simeq e_{1346} \\
f &= [-1,+1,+1,-1,-1,+1,+1,-1] \simeq e_{2356}
\end{aligned} \quad (479)$$

Let us carry out a simple test. Let us Clifford multiply e_{14} with e_{25}. This would give e_{1425} with two indices in reverse order. By commuting these we get $e_{1245} = -e_{1425}$. If we 'logically' multiply the corresponding binary iterants by logic equivalence, we obtain

$$\begin{aligned}
a &= [+1,+1,+1,+1,-1,-1,-1,-1] \simeq e_{14} \text{ times} \\
b &= [+1,+1,-1,-1,+1,+1,-1,-1] \simeq e_{25} \text{ is equal to} \\
ab &= [+1,+1,-1,-1,-1,-1,+1,+1] \simeq e_{1425} \text{ reverting signs gives indeed} \\
&\quad [-1,-1,+1,+1,+1,+1,-1,-1] \simeq e_{1245}
\end{aligned} \quad (480)$$

In $Mat(8,\mathbb{R})$ these 8-fold real vectors represent the diagonal representations of these unit monomials of commuting elements from the Cartan subalgebra of the Clifford algebra. They are indeed Cartan subalgebras in a corresponding Lie group and can be considered as constitutive for physical invariants in models based on this algebra.

Now we should have eight permutations acting on these 8 commuting iterants such that we obtain the 64 non commuting base units of the Clifford algebra $Cl_{3,3}$. How can we find these quaternion-permutations? We first recall Cayley's theorem in the following form:

Cayley: Every finite group can be represented as a permutation group.

Lemma 3: The symmetry component $S_{(3,3)}$ which allows for a decomposition of Clifford algebras $Cl_{0,6}$, $Cl_{3,3}$ and $Cl_{4,2}$ in accordance with the equation $cl_{(3,3)} = S_{(3,3)} \times \mathcal{J}_{(3,3)}$ is given by $C_2 \times C_2 \times C_2$.

This is an interesting lemma which holds in adopted form for neutral signature in general. It says that the non-commutativity and the enrolled temporal dynamics of such Clifford algebras is resulting from the relation of two monoids of self-inverse elements, that is, two abelian groups. The left-hand component is an abelian permutation group acting on the characters of the 8-fold iterants. The second group is the group of binary sequences. Both groups are isomorphic abelian groups, for $Cl_{3,3}$ isomorphic to C_2^3.

Proof: *We give a proof with two runs, the first excludes indirectly, the second confirms directly.* Suitable groups of order 8 could just as well be isomorphic with the quaternion group $Q(8)$ or the dihedral group D_4. We first prove that these possibilities are to be excluded, since they contradict the multiplication rules of Clifford algebra. Let us first consider $Q(8)$.

How can we find this representation by permutations? Just pin down the multiplication table of the quaternion group and that in each row there has to appear the top line in some permuted order. This is in accordance with Cayleys theorem. In the below table we have just found out how the quaternion element J permutes the order of the group elements. Namely J acts like a product of two 4-cycles

$$J \simeq (1\ 5\ 2\ 6)(3\ 8\ 4\ 7) \tag{481}$$

Make sure, this is the same as the permutation in original notation $J \simeq \begin{pmatrix} 1 & 2 & 3 & 4 & 5 & 6 & 7 & 8 \\ 5 & 6 & 8 & 7 & 2 & 1 & 3 & 4 \end{pmatrix}$

Table 17. How to calculate permutations for quaternions

	Id	$-Id$	I	$-I$	J	$-J$	K	$-K$
Id								
$-Id$								
I								
$-I$								
J	J	$-J$	$-K$	K	$-Id$	Id	I	$-I$
$-J$								
K								
$-K$								

We prefer to denote this permutation quaternion by ψ. So we have

$$\psi := (1\ 5\ 2\ 6)(3\ 8\ 4\ 7) \tag{482}$$

The whole group can be generated by a minimal set of two 4-cycles which represent quaternion permutations, namely ψ and

$$\phi := (1\ 3\ 2\ 4)(5\ 7\ 6\ 8) \text{ a 'period 4'-element} \tag{483}$$

which is in correspondence with the quaternion I. In this group there is also an element that reminds us of the $\sigma = (1\ 2)(3\ 4)$ in the Klein 4 group for the Minkowski algebra, namely

$$\Sigma := (1\ 2)(3\ 4)(5\ 6)(7\ 8) \tag{484}$$

Thus Identifying the set of symmetries for the generation of the Clifford algebra $Cl_{3,3}$ by $S_{(3,3)} = Q(8)$ we could believe to obtain the 64 base unit monomials of $Cl_{3,3}$ by the product $Q(8) \times \mathcal{J}_{(3,3)}$ with the iterant component

$$\mathcal{J}_{(3,3)} = \{Id, a, b,\ ...\ g\}. \tag{485}$$

Both quadratic form and (anti)commutation relations of $Cl_{3,3}$ can be tested quite mechanically. We obtain in the first steps

$$\begin{aligned}
\phi a &= (1\ 3\ 2\ 4)(5\ 7\ 6\ 8)[+1,+1,+1,+1,-1,-1,-1,-1] = \\
&= [+1,+1,+1,+1,-1,-1,-1,-1]\phi = a\phi \\
\psi a &= (1\ 5\ 2\ 6)(3\ 8\ 4\ 7)[+1,+1,+1,+1,-1,-1,-1,-1] = \\
&= [-1,-1,-1,-1,+1,+1,+1,+1]\psi = -a\psi
\end{aligned} \tag{486}$$

As we know already every element of $Q(8)$ and every iterant of $\mathcal{J}_{(3,3)}$ are corresponding with base unit monomials of the Clifford algebra. Therefore the equations

$$\begin{aligned}
\phi a &= a\phi \\
\psi a &= -a\psi
\end{aligned} \tag{487}$$

could very well correspond with definite commutation and anticommutation-relations in the Clifford algebra. But consider also the product ϕe. We have

$$\begin{aligned}
\phi e &= (1\ 3\ 2\ 4)(5\ 7\ 6\ 8)[-1,+1,-1,+1,+1,-1,+1,-1] = \\
&= [-1,+1,+1,-1,+1,-1,-1,+1]\phi = g\phi \\
\phi e &= g\phi \text{ means the same as } \phi\, e_{1346} = e_{123456}\, \phi
\end{aligned} \tag{488}$$

Since ϕ is supposed to be one of the base unit monomials of $Cl_{3,3}$ this equation cannot be satisfied and thus leads to a contradiction. The algebra generated by the basis $Q(8) \times \mathcal{J}_{(3,3)}$ is not the quadratic Clifford algebra $Cl_{3,3}$. Hence we have to ask, if the $S_{(3,3)}$ could perhaps be given by the dihedral permutation group $S_{(3,3)} = D_4$. But again this leads to a contradiction because of the action of the period-4 permutations. So we prove directly: $S_{(3,3)}$ is equal to the commutative group $C_2 \times C_2 \times C_2$. Let

$$\begin{aligned} \alpha &\overset{\text{def}}{=} \sigma\,\sigma' \text{ with } \sigma = (1\ 2)(3\ 4),\ \sigma' = (5\ 6)(7\ 8) \\ \beta &\overset{\text{def}}{=} \varphi\,\varphi' \text{ with } \varphi = (1\ 3)(2\ 4),\ \varphi' = (5\ 7)(6\ 8) \\ \gamma &\overset{\text{def}}{=} \omega\,\omega' \text{ with } \omega = (1\ 5)(2\ 6),\ \omega' = (3\ 7)(4\ 8) \end{aligned} \tag{489}$$

The 8 group elements of $C_2 \times C_2 \times C_2$ are $\{Id, \alpha, \beta, \alpha\beta, \gamma, \alpha\gamma, \beta\gamma, \alpha\beta\gamma\}$. With these we can construct the basis of $Cl_{3,3}$ and verify all commutation relations and signature of the 64-dimensional Clifford space. We obtain

$$\begin{aligned} &\alpha\,a = (1\ 2)(3\ 4)(5\ 6)(7\ 8)[+1,+1,+1,+1,-1,-1,-1,-1] = \\ &[+1,+1,+1,+1,-1,-1,-1,-1]\,\alpha = a\,\alpha \text{ hence} \\ &\{\alpha, a\} = \alpha\,a - a\,\alpha = 0 \end{aligned} \tag{490}$$

We also have that

$$\beta\,a = (1\ 3)(2\ 4)(5\ 7)(6\ 8)[+1,+1,+1,+1,-1,-1,-1,-1] = \tag{491}$$

$$\begin{aligned} &= [+1,+1,+1,+1,-1,-1,-1,-1]\,\beta = a\,\beta \text{ hence} \\ &\beta\,a - a\,\beta = 0 \end{aligned}$$

Similarly we verify

$$\begin{aligned} &\gamma\,a = (1\ 5)(2\ 6)(3\ 7)(4\ 8)[+1,+1,+1,+1,-1,-1,-1,-1] = \\ &= [-1,-1,-1,-1,+1,+1,+1,+1]\,\gamma = -a\,\gamma \text{ thus} \\ &\gamma\,a + a\,\gamma = 0 \end{aligned} \tag{492}$$

So two commutators vanish, we have $\alpha\,a - a\,\alpha = 0$, $\beta\,a - a\,\beta = 0$, but $\{\gamma, a\} = \gamma\,a + a\,\gamma = 0$, that is the anticommutator vanishes. What about b? We calculate

$$\begin{aligned} &\alpha\,b = (1\ 2)(3\ 4)(5\ 6)(7\ 8)[+1,+1,-1,-1,+1,+1,-1,-1] = \\ &= [+1,+1,-1,-1,+1,+1,-1,-1]\alpha = b\,\alpha \text{ so} \\ &\alpha\,b - b\,\alpha = 0 \text{ further} \end{aligned} \tag{493}$$

$$\begin{aligned} &\beta\,b = (1\ 3)(2\ 4)(5\ 7)(6\ 8)[+1,+1,-1,-1,+1,+1,-1,-1] = \\ &= [-1,-1,-1,-1,+1,+1,+1,+1]\beta = -b\,\beta \text{ or} \\ &\beta\,b + b\,\beta = 0 \text{ but} \end{aligned} \tag{494}$$

$$\begin{aligned} &\gamma\,b = (1\ 5)(2\ 6)(3\ 7)(4\ 8)[+1,+1,-1,-1,+1,+1,-1,-1] = \\ &= [+1,+1,-1,-1,+1,+1,-1,-1]\gamma = b\,\gamma \text{ hence} \\ &\gamma\,b - b\,\gamma = 0 \text{ further} \end{aligned} \tag{495}$$

$$\alpha\,c = (1\ 2)(3\ 4)(5\ 6)(7\ 8)[+1,-1,+1,-1,+1,-1,+1,-1] = \tag{496}$$

$= [-1,+1,-1,+1,-1,+1,-1,+1]\alpha = -c\,\alpha$ or
$\alpha\, c + c\,\alpha = 0$ and also (497)

$\beta\, c = (1\ \ 3)(2\ \ 4)(5\ \ 7)(6\ \ 8)[+1,-1,+1,-1,+1,-1,+1,-1] =$
$= [+1,-1,+1,-1,+1,-1,+1,-1]\beta = c\,\beta$ that is
$\beta\, c - c\,\beta = 0$ and (498)

$\gamma\, c = (1\ \ 5)(2\ \ 6)(3\ \ 7)(4\ \ 8)[+1,-1,+1,-1,+1,-1,+1,-1] =$
$= [+1,-1,+1,-1,+1,-1,+1,-1]\gamma = c\,\gamma$ briefly
$\gamma\, c - c\,\gamma = 0$ (499)

So we obtain a subtable of the whole multiplication table that is similar to table 10.

	α	β	γ
a	αa	βa	$-\gamma a$
b	αb	$-b\beta$	γb
c	$-\alpha c$	βc	γc

Every entry corresponds with either commutation or anticommutation and is thus compatible with the multiplication table of Clifford base units. We continue by defining the other elements of $\mathbb{Z}_2^3$:

$$\begin{aligned} \delta &:= \alpha\,\beta = (1\ \ 4)(2\ \ 3)(5\ \ 8)(6\ \ 7) \\ \varepsilon &:= \alpha\,\gamma = (1\ \ 6)(2\ \ 5)(3\ \ 8)(4\ \ 7) \\ \zeta &:= \beta\,\gamma = (1\ \ 7)(2\ \ 8)(3\ \ 5)(4\ \ 6) \\ \theta &:= \alpha\,\beta\,\gamma = (1\ \ 8)(2\ \ 7)(3\ \ 6)(4\ \ 5) \end{aligned} \quad (500)$$

So we have the symmetry component defined by

$$S_{(3,3)} := \mathbb{Z}_2^3 \overset{\text{def}}{=} \{Id, \alpha, \beta, \gamma, \delta, \varepsilon, \zeta, \theta\} \quad (501)$$

Now we are able to calculate all commutation relations for the multiplication table $S_{(3,3)} \times \mathcal{J}_{(3,3)}$

Table 18. Commutation and anti-commutation relations for iterants in $Cl_{3,3}$

	Id	α	β	γ	δ	ε	ζ	θ
Id	Id	α	β	γ	δ	ε	ζ	θ
a	a	$+\alpha a$	$+\beta a$	$-\gamma a$	$+\delta a$	$-\varepsilon a$	$-\zeta a$	$-\theta a$
b	b	$+\alpha b$	$-b\beta$	$+\gamma b$	$-b\delta$	$+\varepsilon b$	$-\zeta b$	$-\theta b$
c	c	$-\alpha c$	$+\beta c$	$+\gamma c$	$-\delta c$	$-\varepsilon c$	$+\zeta c$	$-\theta c$
d	d	$+\alpha d$	$-\beta d$	$-\gamma d$	$-\delta d$	$-\varepsilon d$	$+\zeta d$	$+\theta d$
e	e	$-\alpha e$	$+\beta e$	$-\gamma e$	$-\delta e$	$+\varepsilon e$	$-\zeta e$	$+\theta e$
f	f	$-\alpha f$	$-\beta f$	$+\gamma f$	$+\delta f$	$-\varepsilon f$	$-\zeta f$	$+\theta f$
g	g	$-\alpha g$	$-\beta g$	$-\gamma g$	$+\delta g$	$+\varepsilon g$	$+\zeta g$	$-\theta g$

The table is to be read as follows. Consider, for example, the entry line for iterant d in the column for symmetry γ that refers to the element $d\,\gamma \in S_{(3,3)} \times \mathcal{J}_{(3,3)}$, it is equal to $-\gamma d$, that is, $d\,\gamma = -\gamma\, d$ hence, we get the anti-commutator $\{d,\gamma\} := d\,\gamma + \gamma\, d = 0$. From this there follows that

$$(d\,\gamma)^2 = d\,\gamma\, d\,\gamma = d\,\gamma(-\gamma\, d) = -d(\gamma\,\gamma)d = -dd\,Id = -Id\,Id = -Id \qquad (502)$$

Therefore the element $d\,\gamma$ squared gives unity and is a candidate for a 'hypercomplex' unit monomial. If we consider, as an opposite example, the element in line c column ζ, we get $c\,\zeta = \zeta\, c$, hence a commuting pair satisfying $c\,\zeta - \zeta\, c = 0$. From this we derive

$$(c\,\zeta)^2 = c\,\zeta\, c\,\zeta = c\,\zeta\,\zeta\, c = c\,Id\,c = cc\,Id = Id\,Id = Id \qquad (503)$$

So the element $c\,\zeta \in S_{(3,3)} \times \mathcal{J}_{(3,3)}$ is a candidate for a unit monomial with positive square. Now we refer to the fact that the dimension of either *positive definite spacelike* or *negative definite timelike* subpaces of Clifford algebras $Cl_{p,q}$ is well known (Schmeikal 2006, p. 128, table 4). Therefore the table 16 for Clifford algebra $Cl_{3,3}$ should have 36 entries with $-$ sign and 28 entries with $+$ sign. This is indeed the case. The 64 terms of the table 16 satisfy all commutation- and form relations of the Clifford algebra $Cl_{3,3}$. They provide a basis $cl_{3,3}$ of $Cl_{3,3}$ in non-standard iterant notation. From the existence of a real standard representation in $Mat(8,\mathbb{R})$ we can infer that each symmetry operation in $S_{(3,3)} := \mathbb{Z}_2^3 \stackrel{\text{def}}{=}$ $\stackrel{\text{def}}{=} \{Id,\alpha,\beta,\gamma,\delta,\varepsilon,\zeta,\theta\}$ and every iterant in $\mathcal{J}_{(3,3)} := \{Id,a,b,c,d,e,f,g\}$ has a real matrix representation as a 8×8-matrix of real numbers. This completes the proof for the 22nd, 25th and 26th record, namely Clifford algebras $Cl_{0,6}$, $Cl_{3,3}$ and $Cl_{4,2}$.

We have found the binary sequences $\mathcal{J}_{(1,1)}$ for the Clifford algebra $Cl_{1,1}$ are the 2-iterants that have been invented by Louis Kauffman to do discrete quantum mechanics. Let us denote 2-iterants by $It_{(2)}$. We found the binary sequences $\mathcal{J}_{(2,2)}$ for the Clifford algebras $Cl_{2,2}$ and $Cl_{3,1}$ are 4-iterants. They have also been introduced and to some extent investigated by Kauffman and me. Let us denote these objects as 4-iterants $It_{(4)}$. Now we have added to these the 8-iterants $It_{(8)}$ that allow for the construction of associative algebras $Cl_{0,6}$, $Cl_{3,3}$ and $Cl_{4,2}$. As we know from the established theory, we can write $Cl_{3,3}$ as a product

$$Cl_{3,3} = Cl_{1,1} \otimes Cl_{2,2} \qquad (504)$$

So it is standing to reason that the iterant algebras are decomposed as $It_{(8)} := It_{(2)} \otimes It_{(4)}$ and it is obvious that the decomposition applies to both the symmetry component $S_{(3,3)}$ and the binary sequences $\mathcal{J}_{(3,3)}$.

Note on the Temporal Organization of Polarity Strings

We are used to make a peculiar difference between an operator and the operated on. Consider a rotator, that is, an operator that can rotate a vector. We write such an 'operation' in terms of, say

$$R\,x = y \tag{505}$$

and the R 'operates' on the left-hand side, but gets lost after the operation at the right-hand side of the equality sign. This is not so with the iterants. Kauffman had the wonderful inspiration to save the life of the operator and let it be commuted through to the resulting side of the carried out equation/operation. He introduced the 'temporal shift operator' η such that

$$[a,b]\eta = \eta[b,a] \quad \text{where } [a,b] \text{ is an iterant in } It_{(2)} \tag{506}$$

"so that concatenated observations can include a time step of one-half period of the process":

$$. . . a\ b\ a\ b\ a\ b\ a\ b\ . . .$$

This makes i both a view of the iterant process and an operator that takes into account a step in time (Kauffman 2011a, p. 18). The imaginary unit is defined by the equation $i = [1,-1]\eta$, and observation of the iterant view $[+1,-1]$ gets the process into motion

$$[+1,-1]\eta \Rightarrow \eta[-1,+1] \quad \text{schematically} \qquad \overleftarrow{. . .}\,\eta \Rightarrow \eta\ . . .$$

In the beginning it was not quite clear which Clifford algebras were exactly described by the iterant algebra $It_{(2)}$. It was said that *this algebra contained more than just the complex numbers* (Kauffman 2011a, p. 38), and on the same page it was confirmed that it was not hard to see that the algebra of elements $[a,b]+[c,d]\eta$ is isomorphic with 2×2-matrix algebra. But which matrix algebra could be meant, $Mat(2,\mathbb{R})$ or $Mat(2,\mathbb{C})$? Studying the classification table for Clifford algebras (second record) we confirmed that obviously, there are more objects of the form $It_{(2)}$ than there are complex numbers (ordered pairs). Therefore the iterant algebra of 2-fold iterants with time-shift includes the complex numbers, but the two algebras are not isomorphic. On the other hand the Clifford algebra isomorphic with 2×2 complex matrices involves at any case two orthogonal primitive idempotents, hence two basic iterants $\{Id, e\}$ with $e = [+1,+1,-1,-1]$ together with a set of four symmetries $S_{(3,0)} = S_{(1,2)} := \{Id, t, \varphi, \tau\}$. These are capable to generate the Clifford algebras $Cl_{1,2}$ and $Cl_{3,0}$ (recall investigation of eighth and tenth record), in other words, the Pauli algebra $Mat(2,\mathbb{C})$. So the iterant algebra that involves all 2×2 complex matrices brings in more than the algebra of 2-iterants. It already engages a fourfold iterant, namely $e = [+1,+1,-1,-1]$ together with four symmetry operations, but it does not engage all 4-iterants.

It is only when we generalize the problem to the commutative space of 4-iterants combined with the Klein 4 group of permutation symmetries σ, φ, τ that we encounter 'more than just the complex numbers'. Here Kauffman discovered the creation of quaternions by permuting binary sequences. But again the isomorphic geometric algebra, the domain of 4-fold real iterants was not exactly identified, namely the Clifford algebra of the Minkowski space which is – to use Kauffman's idiom – more than just the quaternion numbers, since each of the algebras $Cl_{2,2} \simeq Cl_{3,1}$ contains a partition of two strictly separate quaternion rings, namely one of grade 2 and a second one with mixed grades 1,3,4. We clarified this during the investigation of records 13 and 15. The fact that there is always a little more to be found than just these or those numbers can best be understood when we go into the functorial relation between Clifford algebras and iterant algebras, as we have done here.

That there is still something essentially 'more' coming in, can be seen as soon as we go further to the algebras generated by 6-dimensional spaces, namely Clifford algebras $Cl_{0,6}, Cl_{3,3}$ and $Cl_{4,2}$ which imply a transition from iterant algebra $It_{(4)}$ to $It_{(8)}$. Why is there an essential difference? Don't we have just an analogous decomposition $It_{(8)} = S_{(3,3)} \times \mathcal{J}_{(3,3)}$ just as with Minkowski algebra $It_{(4)} = S_{(2,2)} \times \mathcal{J}_{(2,2)}$? Yes, we do. But there is a tiny, but important difference. The three base units together with the scalar identity $\{Id, e_1, e_{24}, e_{124}\}$ form an image of the Klein 4 group. They correspond with the four iterant views of $\mathcal{J}_{(2,2)}$. However, the eight commuting base units of the Cartan subalgebra in $Cl_{3,3}$ which provide us the eight binary sequences of $\mathcal{J}_{(3,3)}$ by correspondence, namely the commuting multivectors $\{Id, e_{123456}, e_{14}, e_{25}, e_{36}, e_{1245}, e_{1346}, e_{2356}\} \subset Cl_{3,3}$ do not form the eight element group of C_2^3 but they generate the 'double group' $C_2^4 := \{\pm Id, \pm e_{123456}, \pm e_{14}, \pm e_{25}, \pm e_{36}, \pm e_{1245}, \pm e_{1346}, \pm e_{2356}\} \subset Cl_{3,3}$. The Klein 4 group provides a peculiar feature that singles out the Minkowski space. We come back to that later. At any case, the algebras $It_{(4)}, It_{(8)}$ bring in a further fact which makes the idea of a 'temporal shift operator' η much more general. Namely, the primary and most important idea created by Kauffman that, so to say, the observation initiates the run in time and shifts through the operator, holds for all the symmetry operators, whether η or σ, φ, τ or $\alpha, \beta, \gamma, \delta, \varepsilon, \zeta, \theta$ and so on, that come in by raising dimensions p, q, n. But what is new, is the peculiar rearrangement, the specific organization, the mutual configuration of binary sequences in the course of a process, as it becomes more and more complex. For example we had in $It_{(8)}$

$$g\,\beta = -\beta g \tag{507}$$

meaning

$$g\,\beta = [-1,+1,+1,-1,+1,-1,-1,+1](1\ 3)(2\ 4)(5\ 7)(6\ 8)$$

And resulting in

$$\beta\,[+1,-1,-1,+1,-1,+1,+1,-1] = -\beta\, g \tag{508}$$

So the symmetry operator β does two things, first either it rearranges the binary sequences by flipping all its polarities, or it preserves them, second it causes one step in time. This elementary fact holds for all symmetries that are used in this context of iterant and Clifford algebra. The temporal shift operators are peculiar polarity changers. Whether they change polarities or preserve them depends on the relation between the shift operator and the binary sequence. It is this flipping of polarity that is responsible for the emergence of noncommutative geometric algebra. Physically these shift operators provide the means for quantum motion, that is, waves, to unfold those spaces we seem to observe in the macroscopic world together with all their mathematical and material symmetries. It is perhaps appropriate, therefore, to give those operators peculiar names, as they are also organizers of space, time and matter. Where mathematics is concerned they are mere 'jumpers' as they jump from the left hand to the right-hand side of an equation, or they connect iterants with equal or different polarity. It is their interplay that makes the game a material game.

The 23rd, 24th, 27th and 28th Record, Clifford Algebras $Cl_{1,5}$, $Cl_{2,4}$, $Cl_{5,1}$, $Cl_{6,0}$

These four Clifford algebras in the sixth line of table 12, namely $Cl_{1,5}$, $Cl_{2,4}$, $Cl_{5,1}$ and $Cl_{6,0}$, are isomorphic with the real matrix algebra of 4×4-matrices with quaternion entries. The four records having $Mat(4, \mathbb{H})$ with a pair $'4|2'$ gives us certain hints as for representations and we shall take the liberty of interpreting the categorical diagram freely and let the multivectors disguise as double-iterants, that is, iterants of iterants. We have already pointed out this was possible with the Clifford algebras $Cl_{0,4}$, $Cl_{1,3}$ and $Cl_{4,0}$ in the smaller representation of $Mat(2, \mathbb{H})$. We can use now double-iterants as 4-fold iterants of 4-fold iterants. We end this monograph by an outlook. There is more than just one linear time, but that time we are convinced to observe, is actually based on a linguistic relation between synchronous form of geometry and diachronous process by iteration.

REFERENCES

Ablamowicz, R. (2005). A Maple 8 Package for Clifford Algebra Computations; version 8. http://math.tntech.edu/rafal/

Ablamowicz, R.; Fauser, B. On the Transposition Anti-Involution in Real Clifford Algebras I: The Transposition Map. May 2010, Tennessee Technological University Cookeville, TN 38505.

Ablamowicz, R.; Fauser, B. On the Transposition Anti-Involution in Real Clifford Algebras II: Stabilizer Groups of Primitive Idempotents, May 2010, Tennessee Technological University Cookeville, TN 38505.

Aharonov, Y.; Cohen, E.; Grossman, D.; Elitzur, A. C. (2012). Can a Future Choice Affect a Past Measurement's Outcome? arXiv:1206.6224v5 [quant-ph].

Aharonov, Y. et al. Time and the Quantum: Erasing the Past and Impacting the Future. *Science* 2005, 11, 875-879.

Aharonov , Y.; J. Anandan, J.; Popescu S.; Vaidman L. Superpositions of Time Evolutions of a Quantum System and a Quantum Time Machine. *Phys. Rev. Lett.* 1990, 64, 2965-2968.

Aharonov, Y.; Vaidman, L. Properties of a Quantum System during the Time Interval Between Two Measurements. *Phys. Rev. A.* 1990, 41, 11-20.

Belger, M.; Ehrenberg, L. *Theorie und Anwendung der Symmetriegruppen*. Teubner Verlag: Leipzig 1981.

Bell, J. L. Set Theory: Boolean-Valued Models and Independence Proofs. Oxford Scholarship Online: September 2007.

Bell, J. S.; Clauser, J. F.; Horne, M. A.; Shimony, A. An exchange on local beables. *Dialectica* 1985, 39, 85-110.

Bell, J. S. *Speakable and Unspeakable in Quantum Mechanics*. Cambridge University Press: Cambridge, 2004.

Berman, M. S. On the Zero-energy Universe. *Int. J. Theor. Phys.* 2009, 48 (11), 3278. arXiv:gr-qc/0605063.

Berman, M. S. *General Relativity and the Pioneers Anomaly*. Nova Science Publishers: New York, 2012.

Berman, M. S. *Realization of Einstein´s Machian Program*. Nova Science Publishers: New York, 2012. (Series: Distinguished Men and Women of Science, Medicine and the Arts).

Böhm, H. R. (2003). A Compact Source for Polarization Entangled Photon Pairs. Wien, Techn. Univ., Diplomarbeit, 2003. http://permalink.obvsg.at/AC03819698.

Bohm, D. *Quantum Theory*. Dover Publ.: New York, 1989.

Bondi, H. *Relativity and Common Sense*. Dover Publ.: New York, 1980.

Buniy, R. V.; Hsu S. D. H. (2012). Everything is entangled. arXiv:1205.1584v2.

Bortoft, H. The Whole: Counterfeit and Authentic. *Systematics* – The Journal of the Institute for the Comparative Study of History, Philosophy and the Sciences 1971, 9(2), 43-73.

Bortoft, H. The Ambiguity of 'One'and 'Two' in the Description of Young's Experiment. *Systematics* – The Journal of the Institute for the Comparative Study of History, Philosophy and the Sciences 1970, 8(3).

Breit, G.; Wheeler J.A. Collision of two light quanta. *Phys. Rev.* 1934, 46, 1087-1091.

Brennen, G. K.; Caves, C. M. Poul, P. S.; Deutsch, I. H. (1998). Quantum Logic Gates in Optical Lattices. arXiv:quant-ph/9806021v4

Brylinski, J. L.; Brylinski, R. Universal quantum gates. In *Mathematics of Quantum Computation*,

Brylinski, R.; Chen, G. (Eds.). Chapman & Hall/CRC Press: Bova Raton, Florida, 2002.

Brylinski, J. L.; Brylinski, R. (2001). Universal quantum gates. arXiv:quant-ph/0108062

De Broglie, L. V. *Licht und Materie*. (7th edition). Claassen & Goverts: Hamburg, 1949.

De Broglie, L. V. On the Theory of Quanta. A translation of: Recheres sur la Théorie des Quanta. *Ann de Phys* 10e série t. III 1925, Kracklauer, A. F. (Ed.), 2004. http://aflb.ensmp.fr/LDB-oeuvres/De_Broglie_Kracklauer.pdf

Burnet, J. *Early Greek Philosophy*. (3rd edition). A & C Black: London, 1920.

Carroll, S. M.; Chen, J. (2004). Spontaneous Inflation and the Origin of the Arrow of Time. arXiv:hep-th/0410270

Cartan, E. J. *Leçons sur la théorie des spineurs*. Hermann: Paris, 1938, vols. I-II. [engl. translation: *The Theory of Spinors*. The MIT Press: Cambridge, Mass, 1966.]

Chevalley, C. *The algebraic theory of spinors*. Columbia University Press: Morningside Heights, NY, 1954.

Chisholm, J.S.R.; Farwell, R.S. Tetrahedral structure of idempotents of the Clifford algebra. In *Clifford Algebras and their Applications in Mathematical Physics*; Micali, A; et al. (Eds.). Kluwer: Dordrecht, 1992, 27–32.

Clauser, J.F.; Horne, M.A. Experimental consequences of objective local theories. *Phys. Rev.* 1974, D 10(2), 526–35.

Crumeyrolle, A. *Orthogonal and Symplectic Clifford Algebras: Spinor Structures*. Mathematics and its Applications 57. Kluwer: Dordrecht, 1990.

Dirac, P. *A theory of electrons and positrons*, Proceedings of the Royal Society 1930, vol.126, 360.

Dirac, P. A. M. *Lectures on Quantum Mechanics*. Dover: New York, 2001.

Dunne, G. V. (2012). *The Heisenberg-Euler Effective Action: 75 years on. Physics Department, University* of Connecticut, Storrs, CT 06269-3046, USA. On this 75th anniversary of the publication of the Heisenberg-Euler paper on the full nonperturbative one-loop effective action for quantum electrodynamics. http://arxiv.org/abs/1202.1557v1

Durdewich, M.; Oziewicz, Z. (1994). Clifford Algebras and Spinors for Arbitrary Braids. arXiv:q-alg/9412002

Durdewich, M. (1995). Clifford Algebras as Braided Quantum Groups. arXiv:q-alg/9507016

Durdewich, M.; Oziewicz, Z. (1996). Spinors in braided geometry. *Banach Center Publications* 1996, 37(1), 315-325.

Durney, B. R. (2012). *The Contribution of Electron-Positron Pair Production to the Vacuum Energy. arXiv*:1208.5388

Etter, T.; Kauffman, L. H. Discrete Ordered Calculus. *ANPA WEST J* 1996, 6(1), 3–5.

Feynman, R. P. Space-time approach to non-relativistic quantum mechanics, *Rev. Mod. Phys.* 1948 , 20.

Feynman, R. P. The theory of positrons. *Phys. Rev.* 1949, 76, 749-759.

Feynman, R. P. Space-time approach to quantum electrodynamics. *Phys. Rev.* 1949, 76, 769-789.

Feynman, R. P. QED: *The Strange Theory of Light and Matter*. Princeton University Press: Princeton, 1985.

Feynman, R. P. *The Character of Physical Law*. MIT Press: Cambridge, 1965.

Feynman, R. P.; Hibbs, A. R. *Quantum Mechanics and Path Integrals*. McGraw-Hill: New York, 1965.

Feynman, R. P.; Leighton, R.; Sands, M. *The Feynman Lectures on Physics*. 3 vols. Addison–Wesley: Boston 1964, 1966.

Filippenko, A. V.; Pasachoff, J. M. A Universe from Nothing. Astronomical Society of the Pacific. (Retrieved 10th March 2010), http://www.leaderu.com/offices/billcraig/docs/ultimatequestion.html#text13

von Foerster, H. Objects: tokens for (eigen-) behaviors, in "Observing Systems," The Systems Inquiry Series, *Intersystems Publications* 1981, pp. 274 - 285.

Fontinha, M. *Desenhos na araei dos Quiocos do Nordeste de Angola*. IICT: Lisboa, 1983.

Frey, G. *Theorie des Bewußtseins*. Alber: München, 1980.

Gaveau, B.; Jacobson, T.; Kac, M.; Schulman, L. S. Relativistic extension of the analogy between quantum mechanics and Brownian motion, *Phys. Rev. Lett.* 1984, 53, 419-422.

Gerasimov, A.; Kharchev, S.; Marshakov, A.; Mironov, A.; Morozov, A.; Olshanetsky, M. (1996). Liouville Type Models in Group Theory Framework. arXiv:hep-th/9601161v1.

Gerdes, P. *Ethnomathematik* – dargestellt am Beispiel der Sona Geometrie. Spektrum Akad. Verl.: Heidelberg 1997.

Gericke, H. *Mathematik in Antike, Orient und Abendland*. Fourier Verlag: Wiesbaden 2004.

Gersch, H. A. Feynman's Relativistic Chessboard as an Ising Model. *Int.J. Theor. Phys.* 1981, 20, 491-501.

Ghilardi, S. Free Heyting algebras as bi-Heyting algebras. *Math. Rep. Acad. Sci.* 1992. Canada XVI, 6, 240–244.

Guth B. A.; Vilenkin, A. Inflationary space-times are not past-complete. *Phys. Rev. Lett.* 2003, 90, 151301, p. 4.

Gürlebeck, K.; Sprössig, W. Quaternionic and Clifford calculus for physicists and engineers, Wiley: Chichester, 1997, 371 pp.

Hamilton, W. R. Theory of Conjugate Functions, or Algebraic Couples; with a Preliminary and Elementary Essay on Algebra as the Science of Pure Time. *Transact. Royal Irish Acad.* 1837, 17, 293-422.

Hawking, S. *A Brief History of Time*. Bantam Books, Random House: New York, 1988.

Hawking, S.; Penrose, R. *The Nature of Space and Time*. Princeton Univ. Press: Princeton, 1996.

Heidegger, M. *Was ist Metaphysik?* Vittorio Klostermann: Frankfurt am Main, 2006.

Heisenberg, W.; Euler, W. Folgerungen aus der Dirac'schen Theorie des Positrons, *Z. Phys.* 1936, 98, 714-732.

Hillmer, R.; Kwiat, P. A Do-It-Yourself Quantum Eraser, *Scientific American*, May 2007. http://www.youtube.com/watch?v=bnxHc6OqB7U.

Isham, Ch. Quantum Theories of the Creation of the Universe. In *Quantum Cosmology and the Laws of Nature;* Russell, J.R.; Murphey, N.; and Isham, C.J. (Eds.). Vatican City: Vatican Observatory, 1993, p. 56.

Janovská, D.; Opfer, G. Linear equations and the Kronecker product in coquaternions. *Mitt. Math. Ges. Hamburg*. BAND XXXIII 2013, 181-196.

Kac, M. A stochastic model related to the telegrapher's equation. *Rocky Mountain J. Math.* 1974, 4, 497-509.

Kanitscheider, B. Does Physical Cosmology Transcend the Limits of Naturalistic Reasoning? In *Studies on Mario Bunge's "Treatise"*; Weingartner, P.; Doen, G. J. W. (Eds.). Rodopi: Amsterdam, 1990.

Kästner, E. *Ölberge, Weinberge* – Ein Griechenland-Buch. Insel: Frankfurt am Main, 1991.

Kauffman, L. H. DeMorgan Algebras - Completeness and Recursion. *Proceedings of the Eighth International Conference on Multiple Valued Logic*; IEEE Computer Society Press, 1978, 82-86.

Kauffman, L. H. (1981).Paper computers. Version 1, http://dl. dropboxusercontent. com/u/11067256/PaperComputers.pdf.

Kauffman, L. H. Transformations in Special Relativity. *Int. J. Theor. Phys.* 1985, 24, 223-236.

Kauffman, L. H. Knot Automata. In *Proceedings of the Twenty-Fourth International Conference on Multiple Valued Logic- Boston (1994);* IEEE Computer Society Press: Dallas, Texas, pp. 328-333. http://homepages.math.uic.edu/~kauffman/KnotAutomata.pdf.

Kauffman, L. H. Space and time in computation, topology and discrete physics. In *Proceedings of the Workshop on Physics and Computation* - PhysComp '94, Nov. 1994; IEEE Computer Society Press: Dallas, TX, 1995, pp. 44-53.

Kauffman, L. H.; Noyes, H. P. Discrete Physics and the Dirac Equation. *Phys. Lett.* 1996, A218, 139-146. arXiv:hep-th/9603202.

Kauffman, L. H. Space and Time in Computation and Discrete Physics, *Int. J. General Systems* 1998, 27, 1-3. http://homepages.math.uic.edu/~kauffman/TimeSpace.pdf.

Kauffman, L. H.; Lomonaco, S. J. Quantum entanglement and topological entanglement. *New J. Phys.* 2002, 4, 73.1-73.18.

Kauffman, L. H. Eigenforms - Objects as Tokens for Eigenbehaviors, *Cybernetics and Human Knowing*, 2003, 10, 3-4, 73-90.

Kauffman, L. H.; Lomonaco, S. J. (2006). Spin Networks and Anyonic Topological Quantum Computing. http://arxiv.org/pdf/0707.3678 ; www.math.uic.edu/~kauffman/Unitary.pdf.

Kauffman, Reflexivity and Eigenform - The Shape of Process. *Constructivist Foundations*, 2009, 4(3), 121-137.

Kauffman, L. H.; Lomonaco, S. J. (2009). Topological Quantum Information Theory. quant-ph/0606114, arXiv:0804.4304; http://homepages.math.uic.edu/~kauffman/Quanta.pdf.

Kauffman, L. H. Paper computers. Version 2 [to appear].

Kauffman, L. H. Eigenforms and Eigenvalues – Cybernetics and Physics. On the occasion of the 100st birthday of Heinz von Foerster, Dec 2011 (handout, private e-communication).

Kauffman, L. H. Eigenforms and Quantum Physics, *Cybernetics and Human Knowing*. 2011, 18, 3-4, 111-121.

Kauffman, L. H. (2011). Eigenforms and Quantumphysics. http://arxiv.org/ftp/arxiv/papers/1109/1109.1892.pdf.

Kauffman, L. H. (2012). Non-Commutative Worlds and Classical Constraints. arXiv:1109.1085 [math-ph], http://arxiv.org/abs/1109.1085.

Kauffman, L. H. (n.d.) Laws of Form. (Retrieved on 12th August 2013). http://www.math.uic.edu/~kauffman/Laws.pdf.

Kauffman, L. H. Laws of Form and Topology – Presentation and Discussion. MoF Transcript.pdf, (7th July 2013, private communication by email).

Kiehn, R. M. *Non-Equilibrium Systems and Irreversible Processes – Adventures in Applied Topology, Falaco Solitons, Cosmology, and the Arrow of Time*; Lulu Enterprises: Morrisville, NC, 2004, vol. 2. [private copy from the author, Oct. 2008].

Kim, Y-Ho.; Yu, R.; Kulik, S. P.; Shih, Y. H.; Scully, M. O. A Delayed Choice Quantum Eraser. *Phys. Rev. Lett.* 2000, 84, 1-5. http://xxx.lanl.gov/pdf/quant-ph/9903047.

Klein, O. Die Reflexion von Elektronen an einem Potentialsprung nach der relativistischen Dynamik von Dirac. *Z. Phys.* 1929. 53, 157-165.

Kocsis, S.; Braverman, B.; Ravets, S.; Stevens, M. S.; Mirin, R. P.; Shalm, L. K.; Steinberg, A. M. Observing the Average Trajectories of Single Photons in a Two-Slit Interferometer. *Science,* 2011, 332 (6034)

Krajewska, K.; Kaminski, J. Z. High Energy Physics – Phenomenology. Breit-Wheeler Process in Intense Short Laser Pulses. *Phys. Re.v 2012*, A 86, 52-104. arXiv:1209.2394 [hep-ph].

Kwiat, P.G.; Waks, E.; White, A.G.; Appelbaum, I.; Eberhard, P.H. Ultrabright source of polarization-entangled photons. *Phys. Rev. A* 1999, 60 (2), R773–6. arXiv:quant-ph/9810003.

Lin, Ch.-L. Generalized Fibonacci Sequences and the Triangular Map. *Chin. J. Phys.* 1994, 32, 5-1, 467-477.

Lounesto, P. *Clifford algebras and Spinors*. Cambridge Univ. Press: Cambridge, 2001.

Ma, X-S.; Kofler, J.; Qarry, A.; Tetik, N.; Scheidl, Th.; Ursin, R.; Ramelow, S.; Herbst, Th.; Ratschbacher, L.; Fedrizzi, A.; Jennewein, Th.; Zeilinger, A. Quantum erasure with causally disconnected choice. *PNAS* 2013, 110(4), 1221-1226. www.pnas.org/cgi/doi/10.1073/pnas.1213201110

Magnea, U. (2002). An Introduction to symmetric spaces. Department of Mathematics, University of Torino, Italy. arXiv:cond-mat/0205288v1

Matzke, D. J. (2002). Quantum Computation Using Geometric Algebra. Dissertation at the University of Texas at Dallas. (Retrieved on 10th Aug 2013). http://www.tauquernions.org/wp-content/uploads/2012/11/QUANTUM-COMPUTATION-USING-GEOMETRIC-ALGEBRA.pdf.

McCabe, G. (2005). The structure and interpretation of cosmology: Part II - The concept of creation in inflation and quantum cosmology. http://arxiv.org/pdf/gr-qc/0503029.pdf.

Mermin, N. D. Hidden Variables and the two theorems of John Bell. *Rev. Mod. Phys.* 1993, 65, 803.

Messiah, A. *Quantum Mechanics*, 2 vols. North Holland: Amsterdam 1961/62.

Narlikar, J. Path Amplitudes for Dirac particles. *J. Indian Math. Soc.* 1972, 36, 9-32.

von Neumann, J. *Mathematical Foundations of Quantum Mechanics*. Princeton Univ. Press: Princeton, 1955.

Nyanatiloka Mahathera. *The noble truth of the extinction of suffering*. Compiled, translated and explained from the Pali Canon. Colombo, 1952.

Oldershaw, R. L. (2012). Towards a Resolution of the Vacuum Energy Density Crisis. arXiv:0901.3381v2

Ord, G. N. A Classical Analog of Quantum Phase, *Int. J. Theor. Phys.* 1992, 31, 1177-1195.

Pearle, P. M. Hidden-Variable Example Based upon Data Rejection, *Phys. Rev. D 2* 1970, 8, 1418–25.

Perlmutter, S. et al. Measurements of Omega and Lambda from 42 High-Redshift Supernovae. *Ap. J.* 1999, 517, 565-586.

Perk, J. H. H.; Au-Yang, H. (2006). Yang–Baxter Equations. arXiv:math-ph/0606053.

Piaget, J. *Jean Piaget über Jean Piaget*. Sein Werk aus seiner Sicht. Kindler: München, 1982.

Planat, M. (2009). Clifford group dipoles and the enactment of Weyl/Coxeter group W(E8) by entangling gates. arXiv:0904.3691v4 [quant-ph].

Rau, A. R. P. (2009). Mapping two-qubit operators onto projective geometries. arXiv:0808.0598v2.

Riess, A.G. et al. Observational Evidence from Supernovae for an Accelerating Universe and Cosmological Constant. *Astron. J.* 1998, 116, 1009-1038.

Rowe, M.A.; Kielpinski, D.; Meyer, V.; Sackett, C.A.; Itano, W.M.; Monroe, C.; Wineland, D.J. Experimental violation of a Bell's inequality with efficient detection. *Nature* 2001, 409 (6822), 791–94.

Sauter, F. Über das Verhalten eines Elektrons im homogenen elektrischen Feld nach der relativistischen Theorie Diracs, (On the behavior of an electron in a homogeneous electrical field according to the relativistic Theory of Dirac) *Z. Phys.* 1931, 69, 742.

Scheidl, T. et al. Violation of local realism with freedom of choice. *Proc. Nat. Acad. Sci. USA* 2010, 107, 19708-13.

Schmeikal, B. Minimal Spin Gauge Theory – Clifford Algebra and Quantumchromodynamics. *Adv. Appl. Cliff. Alg.* 2001, 11 (1), 63.

Schmeikal, B. Transposition in Clifford Algebra. In *Clifford Algebras – Applications to Mathematics Physics and Engineering*; Ablamowicz, R. (Ed.). Birkhäuser: Boston, 2004, pp. 351-372.

Schmeikal, B. Lie Group Guide to the Universe. In *Lie Groups – New Research*; Canterra, A.B. (Ed.); Nova Science Publishers: New York, 2009, pp. 1-59.

Schmeikal, B. Spacetime Fermion Manifolds. In *Horizons in World Physics*; Reimer, A. (Ed.). Nova Science Publishers: New York, 2011, pp. 319-327.

Schmeikal, B. (2011). The Universe of Spacetime Spinors. In *Proceedings of the 9th International Conference on Clifford Algebras and their Applications in Mathematical Physics*. Digital Proceedings ICCA9, Bauhaus University, Weimar.

Schmeikal, B. *Primordial Space – Pointfree Space and Logic Case*. Nova Science Publishers: New York, 2012.

Schmeikal, B. Reconstructing Space-Time. *Clifford Analysis, Clifford Algebras and their Applications*, 1(2), 2012, 117-133.

Schmeikal, B. On Motion. *CACAA* 2013.

Scully, M.O.; Drühl, K. Quantum eraser – A proposed photon correlation experiment concerning observation and 'delayed choice' in quantum mechanics. *Phys. Rev. A.* 1982, 25, 2208-2213

Schrödinger, E. Die gegenwärtige Situation in der Quantenmechanik. *Naturwissenschaften* 1935, vol 23, Issue 49, 823-828.

Schupp, F. *Geschichte der Philosophie im Überblick 1*: Antike. Bd 1. Felix Meiner: Hamburg 2007.

Shaw, R.: Finite geometry, Dirac groups and the table of real Clifford Algebras. In *Clifford Algebras and Spinor Structures.* Ablamowicz, R.; Lounesto, P. (Eds.). Kluwer: Dordrecht, 1995, pp. 59-99.

Smilga, W. (2009). Quantenelektrodynamik: Nah- oder Fernwirkungstheorie. arXiv: 0912.5486v4.

Spencer-Brown, G. *Laws of Form*. George Allen and Unwin Ltd: London, 1969.

von Steuben, H. (Ed.) *Parmenides, Über das Sein*. Reclam: Stuttgart, 1981.

Steinberg, Ae.; Kwiat, P.; Chiao, R. Quantum Optical Tests of the Foundations of Physics. In *Springer Handbook of Atomic, Molecular, and Optical Physics;* Drake, G. (Ed.), (2006), pp. 1185-1213.

Steinberg, Ae. (2003). Speakable and Unspeakable, Past and Future. arXiv:quant-ph/0302003

Tittel, W.; Brendel, J.; Gisin, B.; Herzog, T.; Zbinden, H.; Gisin, N. (1998). Experimental demonstration of quantum-correlations over more than 10 kilometers. *Phys. Rev. A*. 57, 3229, arXiv:quant-ph/9707042.

Valente, M. B. http://ebookbrowse.com/the-feynman-diagrams-and-virtual-quanta-pdf-d104344697. (2011-04-05), http://philsciarchive.pitt.edu/5571/1/THE_FEYNMAN_DIAGRAMS_AND_VIRTUAL_QUANTA.pdf (2013-07-03).

Volkov, D. M. Über eine Klasse von Lösungen der Diracschen Gleichung. *Z. Phys.* 1935, 94, 250. http://ntrs.nasa.gov/archive/nasa/casi.ntrs.nasa.gov/19680024713_1968024713.pdf.

van der Waerden, B. L. *Die Pythagoräer - religiöse Bruderschaft und Schule der Wissenschaft,* Artemis: Zürich, 1979.

Walborn, S. P.; Terra Cunha, M. O.; Pádua S.; Monken, C. H. Double-slit quantum eraser. *Phys. Rev. A.* 2002, 65, 033818, http://grad.physics.sunysb.edu/~amarch/

Walborn, S. P.; Terra Cunha, M. O.; Pádua S.; Monken, C. H. Quantum eraser. *Am. Scientist* 2003, 91(4), 336-344. http://www.americanscientist.org/issues/feature/2003/4/quantum-erasure

Wallace, R. The Pattern of Reality. *Adv. Appl. Cliff. Alg.* 2008, 18(1), 115-133.

Weihs, G.; Jennewein, T.; Simon, C.; Weinfurtner, H.; Zeilinger A. Violation of Bell's inequality under strict Einstein locality conditions. *Phys. Rev. Lett.* 1998, 81, 5039-43.

Wiseman, H. M.; Gambetta J. M. Are dynamical quantum jumps detector-dependent? *Phys. Rev. Lett.* 2012, 108, 220402. arXiv:1110.0069v2 [quant-ph].

Wittmann, B. et al. Nature Communications 2012, 3, 625. arXiv:1111.0760

Wittmann, B.; Ramelow, S.; Steinlechner, F.; K. Langford, N. K.; Brunner, N.; Wiseman, H.; Ursin, R.; Zeilinger, A. Loophole-free Einstein-Podolsky-Rosen experiment via quantum steering. *New J. Phys.* 2012, 14, 053030. arXiv:1111.0760 [quant-ph].

AUTHOR'S CONTACT INFORMATION

Dr. Bernd Schmeikal,
Senior Researcher
Wiener Institute for Social Science
Documentation and Methodology (WISDOM), R&D,
Liechtensteinstrasse 22A/17, 1090 Wien, Austria
schmeika@wu.ac.at

INDEX

A

absorption of a photon, 197
action-at-a-distance, xii
active locations, 82
additive inverse, 26
admissible minimal form, 168
advanced
  path has a retarded double, 98
advanced and retarded, 195, 197, 198
Aharonov, Y., 7
algebra of pure time, 82
alternating, 25, 33, 70, 71, 99, 206, 207, 211
  code, 24, 25, 32, 33, 73, 75, 79, 80, 113
  string, 141
alternation
  translocal undisclosed, 37
ambiguity, 70
  between +1 and -1, 26
amplitude, 53, 123
  Schrödinger wave, 40
angular momentum, 160, 184
annihilation, 17, 138, 148, 195
anti-fermions, 162
anti-fringes, 56
anti-physics, xii, 12, 14, 126
anyon, 201, 216
APD
  junction area, 45
Apollon, vii
approximation theorem, 42
archaic
  form of self-reference, 9
  orientation morpheme, 10
  thought-form, 9, 72
  wall, vii
Arché, 9, 72
arrow of time, 20, 61, 64, 76, 78, 217
Artin braid group, 122, 127, 128
associative algebra, 68, 83
asymmetry, 72, 93
automorphism, 81, 128
  group of a 'Dreibein', 35
  of coordinates, 8
avalanche
  of charge carriers, 45
  photodiode, 45, 48, 60
awareness, 12, 37, 206

B

background field, 18
backshift-operator, 26, 69
baryon iterant, 164
base unit, 9, 25, 35, 84, 86, 95, 107, 112, 117, 172, 181
  of color space, 109
Bat-Ears, 47
BBO, 47
beamsplitter, 48, 49, 51
beginning
  no beginning or end, 1
beginning of time, 218
Bell
  basis, 179
  nonlocality theorem, 61
  -state quantum eraser, 56
bell-shaped distribution, 54, 56
beyond
  metric space, 33
  time, xi, xii, xiv, 9, 13, 63, 72, 192, 221, 224
beyond ..., 52, 63, 64, 87, 92, 121, 189, 196, 211, 226
big bang, xi, 226
big moment, 61

binary composition, 80
bit-strings, 82
bivector, 35, 94
blackouting, 228
blob, 42, 54
bookkeeping, 57, 65
boson, vii, xiv, 140
braid, 127, 128, 130, 132, 133, 141, 174
  algebra, 162
  group, 127, 128, 140, 195, 199
  triangles, 141
braided electron spinor, 192
braiding multivector, 131
breakdown threshold, 45
Breit-Wheeler, 16, 17, 197
Brownian motion, 31, 34, 39, 51, 101
Buddhist Meditator, 63

C

calculus of indications, 208
Calibration, 65
Cartan
  iterant, 145
  subalgebra, 68, 78, 138, 146
  subalgebras
    of Minkowski algebra, 105
causality, 6, 44, 46, 48, 55, 57, 59, 61, 63, 87
Cayley graph, 70
center, 228
Chaironeia, vii
change, 3
  history, 57
chaos, 52, 58, 66, 76, 77, 114, 141, 146, 147, 148, 167, 171, 202
characters, 20
charge operator, 163
Chevalley, C., 174
Chisholm, J. S. R., 193
choice, 60, 94
choice sequence, 99
Chokwe, 97
circle frequency, 195
clear appearance, xiv
Clifford
  algebra
    basic facts, 83
  conjugation, 163
  numbers, 21
  reversion, 99
Clifford number, 214
closure-operator, 11
Coalesence, 22
cognition and extension, 75
coherence, 6, 7, 54
coincidence counting, 53
collapse, 54
collimated light beam, 48
color, 111, 114, 120
  braid, 171
  rotation, 121, 171, 172
  -shift operator, 110
color space, 78, 105, 114, 117, 120, 125, 129, 131, 133, 161
colors pace, 181
coming into being, 3
commutative
  quaternions, 78
  subspace, 125
compact subalgebra, 135
comparison, 32, 76, 206
complementarity, 46, 50, 56
complex conjugation, 81
complex numbers
  in discrete observations, 27
Compresence, 22
computational basis, 181
concatenation, 26, 28, 69, 127
concept
  of space-time, 67
conditional logic operation, 56
confrontation, 16, 25, 145
conjugation, 81, 134
connecting
  beyond space-time, 66
  strand, 127
constant code, 37
constitutive iterant elements, 35
constructivist, xi, 64
contact, 2, 15, 65, 76
contact transformation, 190
cosmological model, 18
counting, 39, 41, 48, 51, 98
  measure, 42, 43
  negative case, 91
coupling lens, 47
Coxeter groups, 178
creation, xi, xii, 17, 20, 52, 148
  ex nihilo, xi
  of an observer, 13
  of electron-positron pair, 15
  of quanta, 34, 48
  of space, 14
  of space-time, 15
  of time, 32
creation operator, 33

creatures in space, 147
curls
  primordial, 225
cybernetics, 14, 21, 22
cycle duration, 40
cyclic
  thought-form, 9
Dreibein, 8
dual spin veils, 230
Durdewich, M., 174
dynamic
  partitions, 58
  polarity strings, 67
  polarization, 195

D

Dark Van der Waals Gas, 44
de Broglie wave, 30
decay, 5, 9
decoherence, 55
degree
  of freedom, 110
degree of freedom, 87, 106
delayed choice, xii, 44, 46, 47, 49, 57
delayed erasure, 51
derivation
  with respect to space and time, 28
detection, 46, 48, 49, 50, 51, 57, 59
  counting rate, 54
detemporalized image, 213
diachronic dimension, 37
diagonal flip, 93
differentiation, 30
dihedral group, 9, 35, 71, 100
dimensional equivalence, 105
Dirac
  bra- and ket-vectors, 160
Dirac equation, 79, 82, 85, 86, 88, 91, 94
  in light cone coordinates, 82
directed
  unit space-time areas, 125
directed area
  with opposite orientation, 93
disappear
  into the void, 206, 208
discrete approach, 20, 25, 34, 35, 64, 77, 82, 88, 98
discrete unitary oscillation, 25
distance between slits, 53
Do It Yourself Quantum Eraser, 55
Dodona, vii
domain, viii, 23, 25, 73
  self referent, xii
double
  cycle duration, 78
  frequency, 46
  ring of reals, 68, 109
  slit, 42, 43, 47, 51, 52, 55, 56
  wavelength, 75
doubling of 'wavelength, 77

E

earth and water,, 2
Egypt, 5
eigenbehavior, 22
eigenform, 14, 21, 22, 26, 28, 30, 68, 77
  for a 1-norm, 108, 125
  for three operations, 28
  Majorana spinor, 165
  of backshift, 27
  of the oscillatory process, 25
eigenstate, 7
eigenvalues, 29
Einstein, A., 52
Eleatic philosophy, xiii, 3, 4
electric charge, 11, 168
electron, 150
  -hole pair, 45
  neutrino, 151
  neutrino iterant, 167
  -photon-nirvana, 153
  -positron encounter, 193
  spinor, 176
  string, 151
electronic output pulses, 51
electron-positron
  pair creation, 16, 18
electroweak forces, 218
elementary
  arrangements of quarters, 35
  assemblies, 234
  images of spin, 89
  iterants, 37
  memory, 226
  particles, 140
  quantum motion, 229
  self-reference, 36
  stationary phase waves, 54
  two strand braid, 128
emergence of undulation, 20
emitting region on BBO, 54
empty
  naught, 2
  space, 3
energy, 28, 30, 150

conservation, 217
density, 43, 111
operator, 82, 84
entangled
pairs, 189
particles, 61
photons, 45, 46, 57, 59, 63, 177
polarity strings, 67
quantum states, 131
states, 131, 133
strands, 127
with path, 56
entanglement, 54, 60, 72, 126, 133, 189, 234
entropy, 217, 220, 221
environment photon, 60
EPR, 59, 61
EPR-quantum steering experiments, 6
equation of motion, 12, 79
erase time, 75
Euclid's Elements, 21
Euler's formula, 30
events, 33, 42, 57
exchange of arguments, 68
exchange of characters, 138
existence, 4, 15
EXOR, 80
experimental
method, 46
setup, 60
experiments
involving large distances, 55
exponential
synchronized via the imaginary unit, 29
exponential map, 123, 134
extension and cognition, vii, 58
exterior product, 25

F

Fano plane, 178
far field condition, 48
fermion, 11, 113, 139
bit strings, 112
environment, 148
phase gate, 133
quantum numbers, 164
fermion Nirvana, 148
Feynman's Checkerboard, 82
fiber loop compensators, 47
Fibonacci
iteration, 204
model, 199, 200, 219
number, 199
word, 204, 214, 217
Fibonacci, L., 176
field, 32, 37, 54, 63
finite
cutoff, 18
difference calculus, 82
multivector group, 101
first diagrams by Feynman, 125
fixed point, 14, 21, 24, 25, 134
flavor rotation, 113, 114, 118, 125
flavor SU2, 122
fleeing cock, 97
flip, 66
-flop, 227
flip, 67
flopposition, 229
focal length of lens, 53
forced pattern, 214
forces, 20
forms disappear, 211
four
-cycle of positive iterants, 75
dynamic elements, 32
locations, 10
pucks in a polarized string, 66
quarters, 37
fourfold
iterants, 72
locations, 38
polarity strings, 32
strings, 112
fragment
Eleatic, 1
fragment of Melissos, 4
frames, 15
free field
Schrödinger, 32, 40, 206
free field observing itself, 32
freedom of motion, 123, 126
freeze pattern, 34, 38
frequency doubling, 45, 78
fringe pattern, 56, 57
fundamental
gates, 178
future choice, 61, 78

G

Galois-field GF2, 68
Gedankenkreis, 9
Geiger mode, 45
Gell-Mann matrices, 136, 137
Gell-Mann-Nishijima relation, 163

generating braids, 127, 128
generation time, 203, 217
generative space, 78
genetic psychology, 72
geometric algebras, 83
geometric realities, 65
geometry, 65, 71
  of detection, 48
  of locations, 39
Glan-Thompson prism, 47
global positioning system, 60
global time, xii
god, 72
Gödel, K., 51
gods of Greece, vii
golden ratio, 201
golden string, 200, 213
grade, 10, 106, 137, 140, 192
graded Lie group, 134
graded swaps, 130

H

Hadamard matrix, 180
Hadamard, J. S., 180
half
  period of the process, 26, 69, 75
Hamiltonian, 20, 190
Heisenberg-Euler
  Effective Action, 17
Herbst, Th., 55, 60
Herzog, T., 59
Heyting, A., 83
Hibbs, A. R., 86
Higgs boson, vii, 218
Hilbert-space, 53, 62
historic construction
  of the square root of minus one, 27
holistic dynamics, 63
horizontal polarizer, 57
Hubble constant, 18
hyperbolic angles, 81
hyperbolic polar form, 81
hypercomplex number, 21, 28, 29, 35

I

ICNAAM12, 5
iconic theorem, 149
idempotence, 9
idempotent, 177
identity *Id*, 68, 81
Idler photons, 48
imaginary square, 94
imaginary unit, 25, 27, 35, 98
Immutable, xiii
implicate order, 23
impossible extraction, 28
impulse, 28
inconsistency $1 = 0$, 190
indefinite signature, 81
indispensable tools, 25
infinity
  of reality, 3
inflationary space-time, xi, 218
inner and outer, 207
innermost quantum layer
  of real motion, 72
integer numbers, 22
interaction, 66, 78
interaction with polarized void, 142
interactive quantum jumps, 61
interface
  geometry/iterants, 162
  matter/mind, 58, 76
interference, 6, 48, 54, 56
  pattern, 56, 60
interferometer, 48, 60
invariance
  objects, 20
invariant
  $\Delta a, b$, 79
  lepton, 113
  norm, 80
inverses, 79
Ionic school, 3
isospin, 108, 137
  and hypercharge, 138
iterant, 17, 20, 26, 37, 40, 78
  binary composition, 80
  color rotation, 120
  flavor rotation, 110
  fourfold, 32
  pattern, 35
  undecided, 32
  views, 26, 208
iterant algebra, 25, 28, 30, 79
iterant views, 32
iteration, 9
  time, 99

J

Jacobi identity, 107
joint detection, 53

juxtaposition, 25, 80

K

Kauffman, L. H., 22, 79, 85, 95, 122, 202
ket-multivector, 161
kinship
  Iterant- and Clifford algebra, 36, 112
Klein 4 group, 126
knot group, 128
Kochen-Specker theorem, 61
Kronecker Delta, 163
Kwiat, P., 55

L

laboratories, 55, 59, 60
Lagrangian, 190
Langford, N. K., 59
laser beam across open air, 55
lattice, 88, 90, 96, 102, 108, 157, 161, 227
  of primitive idempotents, 87
law
  of calling, 208, 209, 211
  of crossing, 208, 209
  of mixture, 28
laws
  of form, 207
  of physics, 218
left and right, 40, 63
Licht und Materie, 30
Lie algebra, 107, 108, 134, 139, 177
Lie subgroup of t-spin, 138
life and death, viii, 233, 234
light, vii, 20, 82, 124
  identical to darkness, 211
lightning, 43
linear superposition, 53
Liouville type models, 136
local
  theories, 58
local features, 32, 43, 56, 59, 104, 205
locality loophole
  in qcd, 62
locations, 20, 32, 33, 41, 42, 58, 68
  have four characters, 138
LOF, 210
logic
  binary connectives, 78
  center of alternating codes, 73
  circuit, 58
  comparison, 25
  conceptional pairs, 72
  identity as multiplication, 77
  identity represented by pol string, 75
  operation, 75
  product, 68
loophole, 58, 62
  free steering, 58
  freedom of choice, 59
  major, 59
Lorentz iterant, 96
Lorentz metric, 105
Lorentz square norm, 81
Lorentz transformation, 80
  gives iterant equations, 80
Lounesto, P., 81, 108
low dark current, 45
lowering shift operator, 160
lusona, 97

M

magic array modulo 8, 97
magic number 7, 176
Majorana fermion, 155, 156, 161, 165
Majorana matrices, 168
Majorana representation, 172
Man
  is part of nature, 64
manifold of braid generators, 174
markedness, 202, 206, 207
  acting on itself, 208
  disappears itself, 211
mass, 83
mathematics of 'pure time', 139
matrix algebra
  of space-time spinors, 160
matrix mechanics
  of the strong force, 119
matter, 65
matter wave, 29
Max Planck Institute, 65
measurable dynamic variables, 30
measurable space, 42, 79
measure, 20, 33, 34, 43
measure space, 42
meditation
  phasing out time, 75
Melissos of Samos, xiii
memory, 13, 76, 208, 212, 214, 221
  circuit, 227
memory an invariant cut, 221
memory without time, 221
metamorphosis, xiv

metaphysics, 4, 9, 63
method, 20
methodology, 20
micro-oscillatory system, 29, 77
mind annihilating time, 224
minimal iterant, 160
minimal left ideal, 159
minimal solution, 160
minimalistic elements of motion, 130
minimum number
  of dimensions, 108
Minkowski algebra, 59, 61, 67, 68, 78, 129, 131
Minkowski space, 35, 36, 59, 61, 67, 78
Minkowski, H., 108
Moivre's formula, 29
momentum, 83
monolinear sona, 97
monomial
  swap gate, 140
monomial sequence, 77
morpheme, 9
morphogenetic structures, 202
  of orientation, viii
motion, 4
  beyond time, 13, 63
  creates the frame, 98
  in iterant space, 139
motion group, 133, 137, 148
multiple eigenform, 29, 30
multiplication table
  of fourfold iterants, 74
multivector
  -group, 36
  in geometric algebra, 66
mystery, 75, 76

N

natural
  motion group, 139
natural numbers, 210
nature, xii, 33
  constructing time, 64
  makes the bookkeeping, 42
negation and shift, 26
negative
  case counting, 82
  free energy, 19
  numbers, 21
  polarity, 73
neighborhood system, 15
neighboring iterants, 204
neither a particle
  nor a wave, 52
nest of spheres, 23
neutral Majorana fermion, 160
neutrino density, 168
nilpotent field, 149
nilpotent string, 151
Nirvana, 150
  annihilating itself, 153
no interval
  between four poles, 66
no measure, 20
no orientation, 76
no topology, 15
non-being is not, 72
noncommutativity, 71, 99
  of dihedral reorientation group, 99
non-compact group $SL(3,R)$, 136
non-entangled state, 185
nonlinear
  Compton scattering, 17
  electron-positron pair creation, 17
non-linear crystal
  beta-barium borate, 45
nonlinear process, 52
nonlinear stochastic motion, 31
nonperturbative regime
  of pair creation, 17
non-space, 17
normal real form, 134
nothing, xi, 1
nothing at all, 208
not-nature-thinkers, xiii
Noyes, H. P., 82
number
  generators, 59
  of digits, 219
  of iterants, 78
  of particles emitted, 43
numeri absurdi, 21
Nyanatiloka Mahâthera, xiii

O

object, 21
observable, 6, 165
observation
  algebraic conception, 32
  brings upon time, 27
observer, 12, 27, 80
  and observed, 32, 60, 64, 206, 233
observing
  an iterant, 27
  element, 64

octahedral symmetries, 9
Oedipus, vii
One, 3, 4, 22
one step, 31, 68
one-norm, 108
operational structure, 9
operator on ordered pairs, 82
opposite metric, 87
optical
  delay, 51
  distance, 51
orange glass filter, 47
orbiting
  in the complex plane, 30
ordered
  algebraic pair, 28
orientation, 33, 34, 37, 53, 58, 63, 66, 68
oriented, 37
  space-time area, 78, 112
  space-time volume, 124, 125
  unit area, 35, 95
orthogonal, 8, 83, 109
oscillatory martingale, 77
over-dot, 129

P

pain, 2, 3
pair
  annihilation, 193
  creation, 17, 197
Paleolithic, 9
parity flip, 39, 40
Parmenides, xiv
Parnassos, vii
parquetting, 105
parsed compass, 66
participation, 14
particle
  history, 7
particle scan of photon, 146
particles, 43, 46, 49, 52
partition functions, 20
partition of isospin, 138
past, 5, 7, 13, 233
  in the present, xiv, 65, 233, 234
path integration, 82
Pauli matrix, 81
Pauli, W., 113
PBS, 60
Peano fractal, 104, 204, 224
perception, xiv, 4, 5, 12, 221, 233
period, 26
permutation, 93, 99, 110, 116
permutations
  of base unit monomials, 130
phase, 33, 52, 54, 56
  flip error, 153
  gate, 126, 129, 140, 141
  shift, 53, 54
phenomenology, 13, 15, 20, 126
photon, vii, 16, 37, 45, 56
  absorbed and re-emitted, 148
  field, 196
  iterants, 140
  polarization/path-entangled, 59
  polarized, 47, 58, 143
  reproducing itself, 148
photon phase template, 148
photons and Nirvanas, 149
photosensitive environment, 45
piezo-nanopositioner, 60
Planck constant, 18
Planck mass, 18
plane wave, 30
polarity, 66, 73, 80, 82
polarity string, 25, 32, 66, 67, 73, 106
  observing its dual, 32
  ping, buzz and slosh around, 79
polarization, 47, 55, 56, 141
  entangled with path, 56
polarized, 5, 55, 146
  braids, 77, 78
  in a directed area, 142
  string, vii, 66
polarized photon
  releases a spinor flip, 153
polarizer, 55
Polybos, 4
population of iterants, 78
position, 30
  and moments, 31
positron, 193
potential, 130, 214
  to transpose polarities, 144
power series, 29
pratityasamutpada, 51
precise timing, 55
presence, xii, xiii, xiv, 46
  moving, xiv
  operating on itself, 13
pre-Socratic thought, 22
pre-spinor, 95, 102, 124
pressure, 43
primary movens, 9, 72
primitive idempotent, 10

primordial, vii, 9, 32, 48, 63, 74
observation, 70
photon, 144
space, 15, 20
string, vii
time, 28
principle of distinction, 208
prism, 48
probability measure, 42
probe, 7
process, 12, 20, 28, 31, 68
of fragmentation, 222
of observation, 31
recursive, 24
process of nature, 12, 41, 43, 54, 63, 72
propagation, 99
iterant, 36
photons, 145
prophetic insight
by Louis de Broglie, 55
protocols, 33
pulsation, 110
pure states, 125
pure unit length quaternion, 173

Q

QCD, 15
QED, xii, 15, 65
QFT, 18, 20
QRNG, 60
quad location, 32, 40, 66, 104
quadrants, 40
quadratic Clifford algebra Cl1,1, 81
quadruple, 71
quanta, 37, 39, 46, 56
quantum, 60
communication, 60
decoherence, 221
fluctuations, 218
information, 63, 127, 129
key, 59
motion, 9
random number generators, 58
reality, 43
steering, 59
steering loophole-free, 59
quantum eraser, 46, 47, 49, 52, 147
Do It Yourself, 55
quantum erasure, 56
quantum motion, 31, 63, 64, 71, 78, 123, 124, 139, 222
quantumelectrodynamics
action-at-a-distance formulation, 44
quark, 64, 76, 167, 168, 182, 194
quark density component, 66
quarks and leptons, 218
quarter-wave plates, 55, 56, 58
quaternion location, 66
quaternions, 32, 66, 102, 116, 172
quibbling primordial creatures, 150

R

radar-coordinates, 79, 80
rank, 107
-2 Lie group, 163
-3 Lie group, 113, 134
real, 3, 64
couple, 28
double ring, 81
forms, 134
reality, xiii, 3, 234
reconstructing orientation, 63
recover electron spinor, 197
recursive equation, 77
recursive process, 22
re-definition
of the derivative, 31
reentering mark, 207
reentry, 24
reentry form, 28
reentry into void, 17
reflexive domain, 14
reflexivity, 13
relativistic locality, 61
relativistic yardsticks, 66
renaming, 8
reorientation, 8, 9, 71, 99, 132, 181
group, 36, 93, 131
reproducing the electron, 196
reproducing unity, 32, 206
rest frame, 15
restore
superposition, 131
reversion, 26, 146, 162
of directed areas, 94
Riess, A. G., 18
rigid body, 58
root inverter, 231, 232, 234
root lattice A2, 120

S

sand drawings, 97

scanning, 48
Schrödinger equation, 20, 30, 64
Schrödinger iterant, 34, 38, 39, 75
science of pure time, 27
seal, 109, 113, 143, 149, 150, 151, 166, 177
  of space-time, 108
  Solomon's, 113
self annihilating field, 148, 149
self-interaction, 52
self-organization of energy, 48
self-reference, 9, 32, 36, 41, 207, 225
self-referent observation, 32, 206
self-referential inverters, 225
separation
  between observer and observed, 32
shift, 16, 26, 73, 216
  operators of the SL(3, R), 136
sigma-algebra, 33, 41
sign and token, 14
signal and idler, 45, 53, 78
signature of iterants, 73
silent elements, 36, 37
simultaneous observation
  of wave and particle, 52
sine iterant, 30
single
  -mode optical fiber, 47, 58
single qubit, 177
singularity theorem, xi
sink and source, 37
skew-symmetric, 136
small table experiments, 55
S-matrix elements, 19
solid body, 66
source, 36, 37, 60
space involution, 35
space of events, 37
space-like separation, 48, 55, 57, 60, 62
space-time
  loops, 101
space-time group L2, 134
spatial differentiation, 30
spatial logic network, 230
spatial shift, 34, 37
speed of light, 80, 83
Spencer-Brown
  view of form, 206
spinor construction, 181
spinor space, 130
spontaneous parametric down conversion, 46
spooky effects, 58, 59
stability, 22, 42
stabilizer group, 134
stable measure, 42
standard representation, 136, 161
Stanford Linear Acceleration Center, 17
steering experiments, 59
step, 75
step in the seal, 172
step in time, 27
stochastic process, 31, 51
strangeness, 164
strangeness iterant, 164
string, 79
  of alternating polarity, 33
  three types of, 125
strong force
  seal, 166
strong measurement, 61
Study numbers, 81
subgroup SU3, 134
substantiation, 22
superposition
  of observations, 29
  of position states, 56
survival rule, 204
swap, 64, 68, 81, 109, 143, 147, 187
  matrix, 129
swap gate, 130
  for quantum motion, 130
swap of locations
  neutrino/electron, 186
syllogism and quantifiers, 223
symmetric
  difference, 33, 39, 42
  group, 32
  group S4, 35, 115
symmetry breakage, 71
  at light source, 54
  brings on orientation, 72
synchronization, 29, 30, 77
synchronous
  and diachronic, 213
  detemporalized image, 66
  dimension, 37
  freeze brackets, 27
  freeze pattern, 31
  image, 31
  image of Schrödinger wave, 63
  wave pattern, 34
system photon, 60

T

tangle-time, 103
teaching aid, 55

temperature, 43
template, 78
temporal, 20
  grid, 30
  permutation operators, 118
  shift, 26, 68, 156
  shift operator, 69, 103
  wave, 31
temporal shift
  as directed ST-area, 95
thermodynamic equilibrium, 218
thermodynamic magnitudes, 119
thermodynamics, 43
  of the universe, 219
thesis of Louis V. de Broglie, 54
third kind of object, 52
Thompson, S. P., 47
Tibetan Buddhist logic, 211
time, 20, 124
  and no-time, 212
  ending, 5
  evolution operator, 19
  is thought, 64
  reversion, 39, 40, 101
  shift, 28, 30, 31, 32, 40, 68
  shift by touch, 74
  shift operators η and t, 102
  step, 26
  -step, 31
  -tagging units, 60
time reversion, 99, 147
titanium-doped sapphire, 45
tokens
  of eigenbehavior, 21
topological entanglement, 128
torsion momentum, 159, 161, 167
total volume
  of space-time, 20
touch, viii, 16, 25, 69, 80, 124, 143
  and compare, 67
  and emotion, 63
trajectory, 43
translocal relations, 15, 16, 37, 123
transparency structures, 234
transposition, 71
  in the graded basis, 131
trefoil knot, 128
trihedral rotations, 77
trihedral series of iterants, 77
t-spin, 135
turbulence, 231
two-qubit, 181, 185
Two-State-Vector Formalism, 61

U

u, 165
ultrafast switching, 55
ultraviolet divergence, 18, 20
uncountable universe
  of idempotents, 11
undecided iterant, 70
undisclosed alternation, 26, 32, 69
unfolding
  from a line, 106
Unformed, xiii
unit, 27
  bivector, 25, 68
  imaginary, 25, 68, 84, 94, 156, 160
  iterant, 40
  space-time volume, 112
  vector, 30
unitary
  reorientation gate, 131
  states of SU3, 122
unitary braid element h, 176
unity
  constant iteration of, 33
  ideal or material, 4
  restored, 41
universal
  chicken, 222
  element, 10, 12
  gate, 180
  inexact predicate, 10
  spinor, 158
universe, 36
  knows orientation, 63
  referring to itself, 11, 36
  time-symmetric, 218
unspecified
  area, 78
  domain, 20
  locations, 32
Ursin, R., 55, 60
u-spin affects characters, 139

V

vacuum, 17
  bubbles, 19
  energy, 19, 20
  fluctuations, 228
  polarization, 16
  -vacuum amplitude, 19
vanishing polarities, 76

veil a logic inverter, 224
Venn diagram, 222
void, xii, 3, 21, 32, 76, 77, 79, 141
  additive, 210
  an unmarked state, 206
  in self-reference, 78
  location, 16, 20
  mathematical, xii
  multiplicative, 210
  polarized by g, 145
  unmeasured, 228
Volkov states, 17
von Foerster, H., 51
von Steuben, H., xiii

W

Walborn group, 55
warm becomes cold, 2
wave
  advanced and retarded, 161
wave, 52
  and bell, 50
  circular polarized, 56
  diagonally polarized, 55
  equation, 28
  form, 34
  functions, 53
  interference versus bell, 52
  -particle duality, 49
weak
  measurement, 7, 8, 46, 61
weak force, 81, 218
Weyl rotation, 184
Weyl/Coxeter group WE8, 178
Weyl-trick, 135, 139
Wheeler, J. A., 16
which is/which is not, xiii
which path information, 46, 48, 49, 50, 51, 54, 55, 56, 60
whirls in Peano curve, 98
whole, 5, 22, 178
Wineland, D. J., 59
world
  is a living memory, 221
  is its own memory, 213

X

XOR, 75

Y

Young-Baxter equation, 128, 179, 180

Z

Z[0], 19
Zeno, xiii
zero baby rabbit, 203
zero energy, xi, xii
  polarized pattern, 142

INDEX - NAMES

A

Ablamowicz, R., 120, 170
Aharonov, Y., 8, 69
Anaxagoras, 4
Appollodoros, 1
Artin, E., 135
Aspect, A., 67

B

Bateson, G., 111
Baxter, R., 141
Belger, M., 103
Bell, J. S., 67, 194
Berman, M., xii
Bohm, D., 59, 61
Böhm, H. R., 53
Bell, J. S., 257
Bohr, N., 56
Boltzmann, L., 233, 237
Bondi, H., 90
Borde, A., xi
Bortoft, H., 25
Breit, G., 19, 211
Brendel, J., 67
Brennen, G. K., 242
Brown, R., 36, 39, 46
Brunner, N., 67
Brylinski, J. L., 192
Brylinski, R., 192
Buniy, R. V., 59
Burnet, J., xvi, 1

C

Carroll, S., 232, 237
Cartan, E. J., 76, 121, 151, 195
Caves, C. M., 242
Cayley, A., 123
Chaisson, E. J., 233
Chen, J., 232
Chevalley, C., 195
Chisholm, J. S. R., 171
Clauser, J. F., 67
Clifford, W. K., 13, 24, 106, 119, 128, 133, 138, 148
Cohen, E., 69
Coxeter, H. S. Mac D., 140, 142
Crumeyrolle, A., 172, 195, 207

D

de Broglie, L. V., 34, 61
de Moivre, A., 33
Deutsch, I. H., 242
Dirac, P., 89, 95, 203
Drühl, K., 53
Durney, B., 21

E

Ehrenberg, L., 103
Einstein, A., 66
Elitzur, A. C., 69
Euclid, 24
Euler, L., 34
Euler, W., 20

F

Farwell, R. S., 171
Fauser, B., 120, 170
Fedrizzi, A., 62, 69

Feynman, R. P., 17, 49, 52, 73, 97, 102, 113, 136, 181, 232, 249
Fibonacci, L., 213, 219, 231
Filippenko, A., xii
Fontinha, M., 108
Frey, G., 11, 12, 81

G

Gaveau, B., 113
Geiger, H., 51
Gell-Mann, M., 149, 177
Gerasimov, A., 150
Gerdes, P., 108
Gericke, H., 24
Gersch, H. A., 113
Ghilardi, S., 93
Gisin, B., 67
Gisin, N., 67
Glan, P., 53
Grossman, D., 69
Guth, A., xi

H

Hamilton, W. R., 29, 31, 170
Hawking, St, xi
Heidegger, M., 11, 81
Heisenberg, W., 19, 20
Hermann, A., xv
Heyting, A., 151
Hibbs, A. R., 52, 102, 113, 136
Hilbert, D., 60
Hillmer, R., 62, 63
Hippokrates, 5
Horne, M. A., 67
Hsu, S. D. H., 59

I

Isham, Ch., xii
Itano, W. M., 67
Ithagenes, 1
Ivan, H., 242

J

Jennewein, Th., 62, 69
Jordan, P., 7

K

Kac, M., 113
Kähler, E., 188
Kanitscheider, B., xii
Karmanov, V. A., 95
Kästner, E., vii, 92, 204
Kauffman, L. H., 14, 17, 24, 27, 29, 39, 92, 121, 141, 187, 213, 221, 238
Kharchev, S., 150
Kiehn, R., 49
Kielpinski, D., 67
Kim, Y. H., 54, 60
Klein, F. Ch., 139
Klein, O., 19
Kochen, S., 70, 193
Koelman, L., 234
Kofler, J., 62, 69
Kronecker, L., 177
Kulik, S. P., 54, 60
Kwiat, P., 62

L

Lagrange, J. L., 204
Leonardo of Pisa, 24
Lie, S. 147
Lieb, E., 213
Lin, Ch. L., 215
Lomonaco, S. J., 135, 141, 187, 213
Lorentz, H. A., 90
Lounesto, P., 93, 135, 195

M

Ma, X-S., 62, 69
Magnea, U., 148
Majorana, E., 169, 174, 179, 186
Marshakov, A., 150
Matzke, D. J., 168, 192
Maxwell, J. C., 62
McCabe, G., xii
Melissos, xvi, 3, 4, 15
Mermin, N. D., 193
Messiah, A., 198
Meyer, V., 67
Minkowski, H., 128, 132
Mironov, A., 150
Monken, C. H., 62, 65
Monroe, C., 67
Morozov, A., 150

N

Narlikar, J., 113
Nemiroff, R., 63
Nishijima, K., 178
Noyes, H. P., 96, 106
Nyanatiloka Mahâthera, xv

O

Oldershaw, R., 20
Olshanetsky, M., 150
Oosterbroek, E., 242
Opfer, G., 251
Ord, G. N., 113
Oziewicz, Zb., 188

P

Pádua, S., 62, 65
Parmenides, xv, 1, 3, 15, 25
Pasachoff, J., xii
Pauli, W., 198
Peano, G. P., 117, 219
Pearle, P. M., 67
Penrose, R., xii
Perikles, 1
Perlmutter, S., 21
Piaget, J., 81
Planat, M., 192, 195
Plutarch, 1
Podolsky, B., 66
Polybos, 5
Poul, P. S., 242

Q

Quarry, A., 62, 69

R

Ramelow, S., 62, 67, 69
Ratschbacher, L., 62, 69
Reidemeister, K., 215
Rhodes, R., 57
Rosen, N., 66
Rowe, M. A., 67

S

Sackett, C. A., 67
Sauter, F., 19
Scheidl, Th., 62, 67, 69
Schmeikal, A. B., 121, 171
Schönfließ, A. M., 104
Schrödinger, E., 23, 34, 45, 220
Schupp, F., xv
Scully, M. O., 53, 54, 60
Shaw, R., 41
Shih, Y. H., 54, 60
Simplicius, 4
Smilga, W., xii
Specker, E., 70, 193
Spencer-Brown, G., 220, 239
Spinoza, B., 65
Sprössig, W., 253
Steinberg, Ae., 8, 10
Steinlechner, F., 67
Stifel, M., 24

T

Temperley, N., 213
Terra Cunha, M. O., 62
Tetik, N., 62, 69
Thirring, W., 205
Tittel, W., 67
Turing, A., 58

U

Ursin, R., 67

V

Vaidman, L., 9
van der Waerden, 25
Vilenkin, A., xi
von Foerster, H., 24, 29, 121
von Neumann, J., 9

W

Walborn, S. P., 62, 65
Wallace, R., 153
Weihs, G., 67
Weyl, H., 149, 153, 199
Wheeler, J. A., 8, 211

Wiseman, H. M., 8, 67
Wittman, B., 67

Y

Young, Ch. N., 141
Yu, R., 54, 60

Z

Zbinden, H., 67
Zeilinger, A., 58, 62, 67, 69
Zeno, 1